Discrete Mathematics
Essentials and Applications

Discrete Mathematics
Essentials and Applications

Ali Grami

ELSEVIER

ACADEMIC PRESS
An imprint of Elsevier

Academic Press is an imprint of Elsevier
125 London Wall, London EC2Y 5AS, United Kingdom
525 B Street, Suite 1650, San Diego, CA 92101, United States
50 Hampshire Street, 5th Floor, Cambridge, MA 02139, United States
The Boulevard, Langford Lane, Kidlington, Oxford OX5 1GB, United Kingdom

Notices
Knowledge and best practice in this field are constantly changing. As new research and experience broaden our understanding, changes in research methods, professional practices, or medical treatment may become necessary.

Practitioners and researchers must always rely on their own experience and knowledge in evaluating and using any information, methods, compounds, or experiments described herein. In using such information or methods they should be mindful of their own safety and the safety of others, including parties for whom they have a professional responsibility.

To the fullest extent of the law, neither nor the Publisher, nor the authors, contributors, or editors, assume any liability for any injury and/or damage to persons or property as a matter of products liability, negligence or otherwise, or from any use or operation of any methods, products, instructions, or ideas contained in the material herein.

Library of Congress Cataloging-in-Publication Data
A catalog record for this book is available from the Library of Congress

British Library Cataloguing-in-Publication Data
A catalogue record for this book is available from the British Library

ISBN: 978-0-12-820656-0

For information on all Academic Press publications visit our website at
https://www.elsevier.com/books-and-journals

Publisher/Acquisitions Editor: Katey Birtcher
Editorial Project Manager: Chris Hockaday
Production Project Manager: Kamatchi Madhavan
Designer: Matthew Limbert

Printed in the United States of America.

Last digit is the print number: 9 8 7 6 5 4 3 2 1

*To the billions of people living in poverty and
the few living to end the injustice of poverty*

and

*In loving memory of my friends
Farzin Sharifi and Razgar Rahimi*

Preface

As any author knows, writing a textbook is a long but rewarding process. I enjoy writing, but even more so, I enjoy having written a book that can help students study, learn, and apply subject matter.

Discrete mathematics is about the processes that consist of a sequence of individual steps. Discrete mathematics, which includes a multitude of diverse yet interrelated topics, is the study of discrete structures and is accounted as an effective approach for developing problem-solving strengths and critical thinking skills. The relevance, importance, and applications of discrete mathematics have significantly increased over the past few decades, mainly due to the development of an array of computers, which all operate in discrete steps; they are ubiquitous and indispensable in all facets of life.

This book presents the essentials and applications of discrete mathematics in a simple and intuitive approach while maintaining a reasonable level of mathematical rigor. With an accessible writing style, the goal is to introduce a mathematical method of thinking to help solve an array of problems in computer science, software engineering, information technology, and engineering design.

As the topics in mathematics are best understood when they are introduced in a variety of contexts and used to solve problems in a broad range of applied situations, the focus of this book is on the concise and lucid introduction of core concepts, followed by illustrative examples and practical applications. With a basic background in algebra as the only prerequisite, the topics presented in a manner that can be understood by both first-year and second-year undergraduate students in engineering and computer science, and can be taught in a one-semester course (36 lecture hours). This book consists of 20 short chapters, each discusses one major topic. A chapter can be covered in one to two lecture hours, depending on the breadth and depth of its topic.

The pedagogy behind this book and its choice of contents evolved over many years. Almost all of the material in the book has been class tested and proven very effective. There are over 400 independent examples to help understand the fundamental concepts and a total of 200 exercises to test understanding of the material. Upon request from the publisher, a Solutions Manual can be obtained only by instructors who use the book for adoption in a course.

No book is flawless, and this book is certainly no different from others. Your comments and suggestions for improvement are always welcome. I would greatly appreciate it if you would please send your feedback to ali.grami@ontariotechu.ca.

Acknowledgments

The last thing an author usually writes in a book is the acknowledgments. Most readers of a book do not read the author's acknowledgments, as they find them a total bore. However, the author truly regards them as a source of tremendous joy and a tiny attempt to pay back a nonrepayable debt of gratitude.

As this is my last book, I would like to take this opportunity to express my heartfelt appreciation to my caring sister Shahnam, who helped me for over 25 years by being the caregiver for our mother and providing her with the utmost love and care. Also, I am deeply grateful to Ken Gordon, who a long time ago did me the greatest favor. I owe him immeasurably for what I treasure most in life.

Writing a textbook is in some sense collaborative, as one is bound to lean on the bits and pieces of materials and the ideas and concepts developed by others. I would therefore like to thank the many authors whose invaluable writings and insights helped me very much, and also all the students I have had over the years. A reflection of what I learned from them is on every page of this book. My special gratitude goes to Dr. Azam Asilian Bidgoli, who carefully reviewed the entire manuscript and helped improve many aspects of the book through her dedicated commitment and broad knowledge.

The financial support of the Natural Sciences and Engineering Research Council (NSERC) of Canada was also crucial to this project. I am very much appreciative of the staff of Elsevier for their support throughout various phases of this project: Steve Merken, Katey Birtcher, Alice Grant, Chris Hockaday, Kamatchi Madhavan, and all members of the production team.

Finally, I am incredibly fortunate to have an extraordinary family with boundless love and support: my exceptional wife Shirin, who makes me the luckiest husband, and my two half universes, Nickrooz and Neeloufar, who are compassionate, smart, hardworking, generous, fun, and on their way to realize their dreams, making me the proudest father.

CONTENTS

CHAPTER 1

Propositional Logic

Contents

The philosopher Aristotle is often called the father of logic. ***Logic*** is the basis upon which correct inferences may be made from facts. Logic deals with formal principles of reasoning, strict criteria of validity, and necessary rules of thought. Logic is extensively used to solve a multitude of problems and make valid arguments in our everyday lives. Although logic is an essential tool in our interactions with other people as well as in the decisions we make every day, it does have limitations, simply because logic cannot help convince someone out of something they were not reasoned into in the first place. The rules of logic provide meaning to mathematical statements, verify the correctness of programs and algorithms in computer science, and help construct some proofs in software systems. Logic can also be employed in the optimum design of engineering systems, where the system is complex and consists of many subsystems with redundancy. Logic is also applied in physical and social sciences to draw conclusions from experiments. This chapter briefly presents the fundamentals of propositional logic.

1.1 Propositions

The basic building blocks of logic are propositions. A ***proposition*** is a declarative statement, which is either true or false but not both; that is, it has a well-defined truth value. In addition, it is sometimes difficult to know if a sentence is a proposition, and if it is a proposition, it may not be known for some reason whether it is true or false. The area of logic that deals with propositions is called ***propositional logic***.

Example 1.1
Consider the following statements, and if a statement is a proposition, identify its truth value.
(a) How are you?
(b) What a kind person!

Discrete Mathematics
ISBN 978-0-12-820656-0, https://doi.org/10.1016/B978-0-12-820656-0.00001-0

(c) $2 + 2 = 3$.

(d) Ice floats in water.

(e) The earth is flat.

(f) Yellow is a primary color.

(g) Blue is the best color.

(h) Close the door.

(i) God exists.

Solution

(a) It is not a proposition, as it is a question.

(b) It is not a proposition, as it is an exclamation.

(c) It is a proposition, and it is false.

(d) It is a proposition, and it is true.

(e) It is a proposition, and it is false.

(f) It is a proposition, and it is true.

(g) It is not a proposition, as it is an opinion.

(h) It is not a proposition, as it is a command.

(i) It is not a proposition, as it is an opinion.

Example 1.2

Consider the following statements, and if a statement is a proposition, identify its truth value.

(a) $x + 1 = 5$.

(b) There is life in outer space.

(c) This sentence is false.

(d) Fermat's last theorem: The equation $x^n + y^n = z^n$, where x, y, and z are integers and $xyz \neq 0$, has no solutions for an integer $n > 2$.

(e) Human beings will never live to be 200 years old.

Solution

(a) It is not a proposition because x is unknown. However, for a value of x, it becomes a proposition.

(b) It is a proposition. Because science has not advanced enough to know with certainty, we cannot show if this proposition is true or false.

(c) If we assume the sentence "This sentence is false" is true, then the sentence says it is false, which contradicts our assumption. If we assume the sentence "This sentence is false" is false, then the sentence says it is true, which again contradicts our assumption. We can thus conclude that "This sentence is false" is a self-contradictory sentence, and it is not a proposition but a paradox.

(d) It is a proposition. However, for over 300 years, we did not know if this proposition was true or false, but in 1994, it was proven to be true.

(e) It is a proposition. Because science has not advanced enough to know with certainty what the future holds, at the present time, we cannot know if this proposition is true or false.

The mathematician Gottfried Leibniz introduced symbolism into logic. We use lower-case letters, such as $p, q, r, s,$ and $t,$ to denote **propositional variables** or **statement variables**. The **truth value** of a proposition is either true or false. If the truth value of a proposition is true, it is denoted by T, and if the truth value of a proposition is false, it is denoted by F.

If a proposition cannot be broken down into simpler propositions, it is then called a **simple proposition**, a **primitive proposition**, or an **atomic proposition**. For instance, the proposition "The earth is flat" is a simple proposition, which is false, and the proposition "The sun is hot" is a simple proposition, which is true.

If a proposition is a composite (i.e., it is composed of more than one proposition), it is then called a **compound proposition**. The truth value of a compound proposition is completely determined by both the truth values of its simple propositions and the logical operators connecting the simple propositions. For instance, the proposition "Water boils at 50 degrees celsius and water freezes at 0 degrees celsius," which is a compound proposition connected by an "and", is false, and the proposition "There are more women than men in the world or men can become pregnant," which is a compound proposition connected by an "or", is true.

1.2 Basic Logical Operators

Logical operators, also known as **logical connectives**, are used to combine two or more simple propositions to form a compound proposition. A **statement form** or a **propositional form** is an expression consisting of propositional variables and logical operators.

The **truth table** for a given propositional form presents the truth values that correspond to all possible combinations of truth values for the propositional variables. Two compound propositions are called **logically equivalent** or simply **equivalent** if they have identical truth tables (i.e., they have the same truth values regardless of the truth values of its propositional variables). The notation "$\equiv$" denotes logical equivalence.

The **negation** of the proposition $p,$ denoted by $\bar{p},$ is the statement "It is not the case that $p.$" The simple proposition $\bar{p},$ which is read as "not $p,$" has the truth value that is the opposite of the truth value of $p.$ Table 1.1 presents the truth table for the negation of a proposition $p,$ where it has two rows corresponding to the two possible truth values of $p.$

TABLE 1.1 Truth table for the negation of a proposition.

p	$\bar{p}$
T	F
F	T

For instance, if p denotes hope is a good thing, then $\bar{p}$ denotes it is false (or not true) that hope is a good thing or hope is not a good thing.

The **conjunction** of the two propositions p and q, denoted by $p \wedge q$ and read as "p and q," is a compound proposition that is true when both p and q are true and is false otherwise. For instance, the compound proposition "The sun is hot and water is a liquid" is true because both its simple propositions are true, and the compound proposition "$2+ 2 = 4$ and the United States of America is a country with a very long history" is false because not both of its simple propositions are true. Note that the word "but" sometimes is used instead of the word "and" to show conjunction. As an example, in the propositional logic, the two statements "The United States of America is the most advanced country in the world but it was built on indigenous land" and "The United States of America is the most advanced country in the world and it was built on indigenous land" are equivalent. Table 1.2 presents the truth table for the conjunction of two propositions.

The **disjunction** of the two propositions p and q, denoted by $p \vee q$ and read as "p or q," is a compound proposition that is false when both p and q are false and is true otherwise. Note that the word "or" in the propositional logic is an **inclusive or**, meaning a disjunction is true when at least one of the two propositions is true. In other words, $p \vee q$ implies "p or q or both"; that is, it is an inclusive disjunction. For instance, the proposition "It is August or it is sunny" is true in the month of August or when it is sunny. It is false if it is not August and also it is not sunny. Table 1.3 presents the truth table for the disjunction of two propositions.

TABLE 1.2 Truth table for the conjunction of two propositions.

p	q	$p \wedge q$
T	T	T
T	F	F
F	T	F
F	F	F

TABLE 1.3 Truth table for the disjunction of two propositions.

p	q	$p \vee q$
T	T	T
T	F	T
F	T	T
F	F	F

Example 1.3

When a brother and a sister stop playing in their backyard, their father says, "At least one of you has a muddy forehead," and then asks the children to answer "yes" or "no" to the question: "Do you know whether you have a muddy forehead?" The father asks this question twice, and the children answer each question simultaneously. What will the children answer each time this question is asked, assuming that a child can see whether his or her sibling has a muddy forehead but cannot see his or her own forehead?

Solution

Let s be the statement that the son has a muddy forehead, and let d be the statement that the daughter has a muddy forehead. When the father says that at least one of the two children has a muddy forehead, he is saying that $s \vee d$ is true. Both children will answer "no" the first time the question is asked because each sees mud on the other child's forehead. That is, the son knows that d is true but does not know whether s is true, and the daughter knows that s is true but does not know whether d is true. After the son has answered "no" to the first question, the daughter can determine that d must be true. This follows because when the first question is asked, the son knows that $s \vee d$ is true but cannot determine whether s is true. Using this information, the daughter can conclude that d must be true, for if d were false, the son could have reasoned that because $s \vee d$ is true, then s must be true, and he would have answered "yes" to the first question. The son can reason in a similar way to determine that s must be true. It follows that both children answer "yes" the second time the question is asked.

The ***exclusive or*** of the two propositions p and q, denoted by $p \oplus q$ and read as "exclusive or of p and q," is a compound proposition that is true when exactly either p or q (i.e., only one of the two) is true and is false otherwise. In other words, when p and q are both true or when they are both false, the exclusive or of p and q is false. For instance, the exclusive or is employed when "You must take one of the two courses, as a required course." Table 1.4 presents the truth table for the exclusive or of two propositions.

Each of Tables 1.2, 1.3, and 1.4 has four rows corresponding to the $4\ (= 2^2)$ possible combinations of truth values of p and q, as there are two variables, each with two truth values. Table 1.5 shows the total number of nonequivalent compound propositions, each consisting of two propositions, p and q, is $16\ \left(= 2^{(2^2)}\right)$.

TABLE 1.4 Truth table for the exclusive or of two propositions.

p	q	$p \oplus q$
T	T	F
T	F	T
F	T	T
F	F	F

TABLE 1.5 Truth tables for all nonequivalent logical statements with two propositions.

p	q	1	2	3	4	5	6	7	8	9	10	11	12	13	14	15	16
T	T	T	T	T	T	T	T	T	T	F	F	F	F	F	F	F	F
T	F	T	T	T	T	F	F	F	F	T	T	T	T	F	F	F	F
F	T	T	T	F	F	T	T	F	F	T	T	F	F	T	T	F	F
F	F	T	F	T	F	T	F	T	F	T	F	T	F	T	F	T	F

Note that the number of rows in a truth table is 2^m when m in the exponent represents the number of propositional variables, and 2 in the base indicates the two possible truth values of each variable. In addition, the total number of different ways to combine m simple propositions to make a compound proposition is $2^{(2^m)}$.

Example 1.4

Determine the truth table for the propositional logic $(\bar{p} \oplus q) \wedge (p \vee \bar{q})$.

Solution

As shown in Table 1.6, p and q each can have two different truth values. The truth table thus has four rows.

It is also important to note that the **dual** of a compound proposition, with the logical operators negation, conjunction, and disjunction, is the compound proposition obtained by replacing each $\vee$ by $\wedge$, each $\wedge$ by $\vee$, each T by F, and each F by T. Moreover, the duals of two equivalent compound propositions are equivalent, provided that they contain only the logical operators of negation, conjunction, and disjunction. The dual of the dual proposition of a proposition is logically equivalent to the original proposition. A proposition and its dual are equivalent if and only if the proposition is simply one propositional variable. It may be a difficult task to determine if a compound proposition consisting of only one propositional variable and its dual are logically equivalent.

Example 1.5

Determine the dual of each of the following propositions.
 (a) $p \wedge \bar{q} \wedge \bar{r}$.
 (b) $(p \wedge q \wedge r) \vee s$.
 (c) $(p \vee F) \wedge (q \vee T)$.
 (d) p.
 (e) $q \wedge T$.

TABLE 1.6 Truth table for Example 1.4.

p	q	$\bar{p}$	$\bar{q}$	$\bar{p} \oplus q$	$p \vee \bar{q}$	$(\bar{p} \oplus q) \wedge (p \vee \bar{q})$
T	T	F	F	T	T	T
T	F	F	T	F	T	F
F	T	T	F	F	F	F
F	F	T	T	T	T	T

Solution

(a) $p \vee \bar{q} \vee \bar{r}$.

(b) $(p \vee q \vee r) \wedge s$.

(c) $(p \wedge \text{T}) \vee (q \wedge \text{F})$.

(d) p (equivalent to the original proposition).

(e) $q \vee \text{F}$ (equivalent to the original proposition).

1.3 Conditional Statements

Let p and q be propositions. The ***conditional statement*** $p \rightarrow q$, read as "if p, then q" or "p implies q," is a compound proposition that is false when p is true and q is false and is true otherwise. Table 1.7 presents the truth table for the conditional statement. There are also other ways to express this conditional statement, such as "p is sufficient for q," "a sufficient condition for q is p," "q is necessary for p," or "a necessary condition for p is q." In the ***implication*** $p \rightarrow q$, p is called the ***hypothesis***, the ***premise***, or the ***antecedent***, and q is called the ***conclusion*** or the ***consequence***. In an implication, the hypothesis and its conclusion are not required to have related subject matters.

If the implication is true, we do not automatically know that either the hypothesis or the conclusion is true. For instance, consider the conditional statement "If you obey the law, you never go to prison." In this implication, if you obey the law, then you do not expect to go to prison. If you do not obey the law, you may or may not go to prison depending on other factors. However, if you do obey the law but you go to prison, you feel outraged. This last scenario corresponds to the case when p is true, but q is false, and thus the truth value of the conditional statement $p \rightarrow q$ is false.

From an implication $p \rightarrow q$, the following well-known conditional statements, whose truth tables are presented in Table 1.8, can be made:

- The **converse** of $p \rightarrow q$ is $q \rightarrow p$.
- The **inverse** of $p \rightarrow q$ is $\bar{p} \rightarrow \bar{q}$.
- The **contrapositive** of $p \rightarrow q$ is $\bar{q} \rightarrow \bar{p}$.

Noting that logically equivalent propositions have the same truth values regardless of the truth values of its propositional variables, the implication (original conditional

TABLE 1.7 Truth table for the conditional statement.

p	q	$p \rightarrow q$
T	T	T
T	F	F
F	T	T
F	F	T

TABLE 1.8 Truth tables for the contrapositive, converse, and inverse of the conditional statement.

p	q	$\bar{p}$	$\bar{q}$	$p \rightarrow q$	$\bar{q} \rightarrow \bar{p}$	$q \rightarrow p$	$\bar{p} \rightarrow \bar{q}$
T	T	F	F	T	T	T	T
T	F	F	T	F	F	T	T
F	T	T	F	T	T	F	F
F	F	T	T	T	T	T	T

statement) and its contrapositive are equivalent, and the converse and the inverse of a conditional statement are also equivalent. Some people mistakenly think that an implication and its converse mean the same thing as they usually say one to mean another. In fact, their truth tables are not identical.

Example 1.6

Determine the contrapositive, converse, and inverse of the conditional statement "If you start using drugs, then you are a moron."

Solution

Noting p represents the statement "You start using drugs," and q represents the statement "You are a moron," that is, for $p \rightarrow q$, we have the following statements:

- The contrapositive statement $(\bar{q} \rightarrow \bar{p})$ is "If you are not a moron, then you do not start using drugs."
- The converse statement $(q \rightarrow p)$ is "If you are a moron, then you start using drugs."
- The inverse statement $(\bar{p} \rightarrow \bar{q})$ is "If you do not start using drugs, then you are not a moron."

Example 1.7

Determine the contrapositive, converse, and inverse of the conditional statement "If it is a Sunday, then I rest all day."

Solution

Noting p represents the statement "It is a Sunday," and q represents the statement "I rest all day," that is, for $p \rightarrow q$, we have the following statements:

- The contrapositive statement $(\bar{q} \rightarrow \bar{p})$ is "If I do not rest all day, then it is not a Sunday."
- The converse statement $(q \rightarrow p)$ is "If I rest all day, then it is a Sunday."
- The inverse statement $(\bar{p} \rightarrow \bar{q})$ is "If it is not a Sunday, then I do not rest all day."

Example 1.8

Determine the contrapositive, converse, and inverse of the conditional statement "If I am a man, then I am not a mother."

Solution

Noting p represents the statement "I am a man," and q represents the statement "I am not a mother," that is, for $p \rightarrow q$, we have the following statements:

- The contrapositive statement $(\bar{q} \rightarrow \bar{p})$ is "If I am a mother, then I am not a man."
- The converse statement $(q \rightarrow p)$ is "If I am not a mother, then I am a man."
- The inverse statement $(\bar{p} \rightarrow \bar{q})$ is "If I am not a man, then I am a mother."

Note that the original implication and the contrapositive are both true, while the converse and inverse need not be true, as a woman may not be a mother.

Let p and q be propositions. The ***biconditional statement*** $p \leftrightarrow q$, read as "p if and only if q," "p iff q," "if p, then q, and conversely," or "p is necessary and sufficient for q," is a compound proposition that is true when p and q have the same truth values and is false otherwise. Biconditional statements are also called ***bi-implications***. Table 1.9 presents the truth table for the biconditional statement.

To construct compound propositions, parentheses are generally used to specify the order in which operators are to be applied. Parenthesized expressions are evaluated starting with the innermost pair of parentheses outward, analogous to the evaluation of an arithmetic expression. Parenthesized subexpressions are always evaluated first, and with two operators of equal precedence, the corresponding expression is evaluated from the left. The ***precedence rules***, which can reduce the number of parentheses required, for logical statements must be performed as follows: the negation operator is applied before all other logical operators, and the conjunction operator takes precedence over the disjunction operator. It is advised to use parentheses for the exclusive or operator. A conditional operator takes precedence over a biconditional one. In addition, the conditional and biconditional operators have lower precedence than the conjunction and disjunction operators. Table 1.10 highlights the precedence of logical operators.

TABLE 1.9 Truth table for the biconditional statement.

p	q	$p \leftrightarrow q$
T	T	T
T	F	F
F	T	F
F	F	T

TABLE 1.10 Precedence of logical operators (connectives).

Operator	Symbol	Precedence
Negation	$-$	1
Conjunction	$\wedge$	2
Disjunction	$\vee$	3
Conditional	$\rightarrow$	4
Biconditional	$\leftrightarrow$	5

Example 1.9

Determine how each of the following statements must be evaluated.

(a) $(p \rightarrow q) \wedge \overline{q} \rightarrow \overline{p}$.

(b) $p \rightarrow q \leftrightarrow \overline{q} \rightarrow \overline{p}$.

Solution

Using the precedence rule, we have the following statements:

(a) $(p \rightarrow q) \wedge \overline{q} \rightarrow \overline{p} \equiv ((p \rightarrow q) \wedge \overline{q}) \rightarrow \overline{p}$.

(b) $p \rightarrow q \leftrightarrow \overline{q} \rightarrow \overline{p} \equiv (p \rightarrow q) \leftrightarrow (\overline{q} \rightarrow \overline{p})$.

It is sometimes necessary to translate English sentences into expressions involving propositional variables and logical connectives so as to analyze them using rules of inference.

Example 1.10

Let p and q be the following propositions.

   p: You drive impaired.

   q: You die in a car accident.

Write the following propositions using p and q and logical connectivities including negations.

(a) You do not drive impaired.

(b) You drive impaired, but you do not die in a car accident.

(c) You will die in a car accident if you drive impaired.

(d) If you do not drive impaired, then you will not die in a car accident.

(e) Driving impaired is sufficient for dying in a car accident.

(f) You die in a car accident, but you do not drive impaired.

(g) Whenever you die in a car accident, you are driving impaired.

Solution

(a) $\overline{p}$.

(b) $p \wedge \overline{q}$.

(c) $p \rightarrow q$.

(d) $\overline{p} \rightarrow \overline{q}$.

(e) $p \rightarrow q$.

(f) $q \wedge \overline{p}$.

(g) $q \rightarrow p$.

Example 1.11

Noting that the number of different nonequivalent logical statements with two propositions is 16 $\left(= 2^{(2^2)}\right)$, provide examples of compound propositions for all nonequivalent logical statements.

Solution

Table 1.11 presents all 16 different combinations, where compound propositions are shown at the top of the columns.

1.4 Propositional Equivalences

It is sometimes important to replace a logical statement with an equivalent statement in a mathematical argument. One method to determine whether two compound propositions are equivalent is to use well-known logical identities to establish new logical identities. This method is quite effective, especially when there are a large number of propositional variables involved. Table 1.12 presents some important logical equivalences involving the negation, conjunction, and disjunction operators. Of all logical equivalences, De Morgan's laws are of great importance, as they have wide applications in logic. **De Morgan's laws** state that (1) the negation of an "and" statement is logically equivalent to the "or" statement in which each component is negated, and (2) the negation of an "or" statement is logically equivalent to the "and" statement in which each component is negated.

Table 1.13 presents some important logical equivalences involving conditional and biconditional statements.

Example 1.12

Using logical identities, verify the logical equivalence $\overline{(\overline{p} \wedge q)} \wedge (p \vee q) \equiv p$.

Solution

Using Table 1.12, we can have

$$
\begin{aligned}
\overline{(\overline{p} \wedge q)} \wedge (p \vee q) &\equiv \left(\overline{\overline{p}} \vee \overline{q}\right) \wedge (p \vee q) && \text{by De Morgan's laws} \\
&\equiv (p \vee \overline{q}) \wedge (p \vee q) && \text{by the double negation law} \\
&\equiv p \vee (\overline{q} \wedge q) && \text{by the distributive law} \\
&\equiv p \vee (q \wedge \overline{q}) && \text{by the commutative law} \\
&\equiv p \vee (\mathbf{F}) && \text{by the negation law} \\
&\equiv p && \text{by the identity law}
\end{aligned}
$$

TABLE 1.11 Truth tables for Example 1.11.

p	q	1	2	3	4	5	6	7	8	9	10	11	12	13	14	15	16
		T	$p \vee q$	$q \to p$	p	$p \to q$	q	$p \leftrightarrow q$	$p \wedge q$	$\overline{p} \vee \overline{q}$	$p \oplus q$	$\overline{q}$	$p \wedge \overline{q}$	$\overline{p}$	$\overline{p} \wedge q$	$\overline{p} \wedge \overline{q}$	F
T	T	T	T	T	T	T	T	T	T	F	F	F	F	F	F	F	F
T	F	T	T	T	T	F	F	F	F	T	T	T	T	F	F	F	F
F	T	T	T	F	F	T	T	F	F	T	T	F	F	T	T	F	F
F	F	T	F	T	F	T	F	T	F	T	F	T	F	T	F	T	F

TABLE 1.12 Logical equivalences.

Equivalence	Name
$p \wedge \mathbf{T} \equiv p$	Identity
$p \vee \mathbf{F} \equiv p$	laws
$p \vee \mathbf{T} \equiv \mathbf{T}$	Domination
$p \wedge \mathbf{F} \equiv \mathbf{F}$	laws
$p \vee p \equiv p$	Idempotent
$p \wedge p \equiv p$	laws
$\overline{\overline{p}} \equiv p$	Double negation law
$p \vee q \equiv q \vee p$	Commutative laws
$p \wedge q \equiv q \wedge p$	
$(p \vee q) \vee r \equiv p \vee (q \vee r)$	Associative laws
$(p \wedge q) \wedge r \equiv p \wedge (q \wedge r)$	
$p \vee (q \wedge r) \equiv (p \vee q) \wedge (p \vee r)$	Distributive laws
$p \wedge (q \vee r) \equiv (p \wedge q) \vee (p \wedge r)$	
$\overline{p \wedge q} \equiv \overline{p} \vee \overline{q}$	De Morgan's laws
$\overline{p \vee q} \equiv \overline{p} \wedge \overline{q}$	
$p \vee (p \wedge q) \equiv p$	Absorption laws
$p \wedge (p \vee q) \equiv p$	
$p \vee \overline{p} \equiv \mathbf{T}$	Negation laws
$p \wedge \overline{p} \equiv \mathbf{F}$	

$\mathbf{T}$ denotes the compound proposition that is always true.
$\mathbf{F}$ denotes the compound proposition that is always false.

TABLE 1.13 Logical equivalences involving conditional and biconditional statements.

$$p \rightarrow q \equiv \overline{p} \vee q$$
$$p \rightarrow q \equiv \overline{q} \rightarrow \overline{p}$$
$$p \vee q \equiv \overline{p} \rightarrow q$$
$$p \wedge q \equiv \overline{(p \rightarrow \overline{q})}$$
$$\overline{(p \rightarrow q)} \equiv p \wedge \overline{q}$$
$$(p \rightarrow q) \wedge (p \rightarrow r) \equiv p \rightarrow (q \wedge r)$$
$$(p \rightarrow r) \wedge (q \rightarrow r) \equiv (p \vee q) \rightarrow r$$
$$(p \rightarrow q) \vee (p \rightarrow r) \equiv p \rightarrow (q \vee r)$$
$$(p \rightarrow r) \vee (q \rightarrow r) \equiv (p \wedge q) \rightarrow r$$
$$p \leftrightarrow q \equiv (p \rightarrow q) \wedge (q \rightarrow p)$$
$$p \leftrightarrow q \equiv \overline{p} \leftrightarrow \overline{q}$$
$$p \leftrightarrow q \equiv (p \wedge q) \vee (\overline{p} \wedge \overline{q})$$
$$\overline{(p \leftrightarrow q)} \equiv p \leftrightarrow \overline{q}$$

To verify logical equivalences or simplify logical statements, truth tables can also be used, where in each row of the truth table, the truth value of one statement is the same as the truth value of the other statement.

Example 1.13

Using a truth table, verify $\overline{p \oplus q} \equiv (p \leftrightarrow q)$.

Solution

The last two columns of Table 1.14, representing the two sides of the statement, are identical for every single row.

Example 1.14

(a) Prove that the negation of the conditional statement "If p, then q" is logically equivalent to "p and not q."

(b) Write the negation of the conditional statement "If I sleep late at night, then I cannot get up early in the morning."

Solution

(a) From Table 1.13, we have $p \rightarrow q \equiv \overline{p} \vee q$; therefore its negation using De Morgan's laws and the double negative law is as follows: $\overline{(p \rightarrow q)} \equiv \overline{\overline{p} \vee q} \equiv \overline{\overline{p}} \wedge \overline{q} \equiv p \wedge \overline{q}$, which means p and not q.

(b) Let p be the proposition "I sleep late at night" and q be the proposition "I can get up early in the morning." Based on $\overline{(p \rightarrow \overline{q})} \equiv p \wedge q$, the negation is "I sleep late at night, and I can get up early in the morning." Note that the negation of an implication (if-then statement) does not start with the word "if".

Note that De Morgan's laws can be extended to more than two variables; that is, we have

$$
\begin{cases}
\overline{(p_1 \vee p_2 \vee \ldots \vee p_n)} \equiv (\overline{p}_1 \wedge \overline{p}_2 \wedge \ldots \wedge \overline{p}_n) \\[2mm]
\overline{(p_1 \wedge p_2 \wedge \ldots \wedge p_n)} \equiv (\overline{p}_1 \vee \overline{p}_2 \vee \ldots \vee \overline{p}_n)
\end{cases}
$$

where $p_1, p_2, \ldots, p_n$ are n propositions.

TABLE 1.14 Truth table for Example 1.13.

p	q	$p \oplus q$	$\overline{p \oplus q}$	$p \leftrightarrow q$
T	T	F	T	T
T	F	T	F	F
F	T	T	F	F
F	F	F	T	T

Example 1.15

Apply De Morgan's laws to write the negation for each of the following statements.

(a) My friend is 2 meters tall and he weighs at least 100 kilograms.

(b) The flight was delayed or the airport's clock was slow.

Solution

Note that to avoid any potential confusion in the English language, De Morgan's laws can be applied if there exist complete statements on either side of each "and" and on either side of each "or."

(a) My friend is not 2 meters tall or he weighs less than 100 kilograms.

(b) The flight was not delayed and the airport's clock was not slow.

A compound proposition that is always true, regardless of the truth values of the propositional variables (i.e., the compound proposition contains only T in the last column of its truth table), is called a **tautology**. In other words, a tautology is an always-true proposition regardless of the truth values of the propositional variables. A statement whose form is a tautology is a **tautological statement**. Note that the compound propositions p and q are logically equivalent if $p \leftrightarrow q$ is a tautology. Some simple examples of tautology in English are "Parents are older than their children," "You don't give what you don't have," and "Dead people do not breathe." A simple example of tautology in logic is $p \vee \overline{p}$.

A compound proposition that is always false, regardless of the truth values of the propositional variables (i.e., the compound proposition contains only F in the last column of its truth table), is called a *contradiction*. In other words, a contradiction is an always-false proposition regardless of the truth values of the propositional variables. A statement whose form is a contradiction is a *contradictory statement*. Note that the negation of a tautology is a contradiction, and the negation of a contradiction is a tautology. Some simple examples of contradiction in English are "Some are more equal than others," "Rich people need a tax cut because they do not have enough money," and "Texting while driving reduces chances of having a car accident." A simple example of contradiction in logic is $p \wedge \overline{p}$.

Note that a compound proposition that is neither a tautology nor a contradiction is called a *contingency*. In most practical applications and statements in logic, the proposition happens to be contingency. A statement whose form is a contingency is a *contingent statement*. Some simple examples of contingency in English are "Politicians are dishonest" and "People in this country are not racist." A simple example of contingency in logic is $p \rightarrow \overline{p}$.

Example 1.16

Show that the logical expressions $p \leftrightarrow q$ and $\overline{p \oplus q}$ are equivalent.

Solution

Table 1.15 presents the truth table for $(p \leftrightarrow q) \leftrightarrow \overline{(p \oplus q)}$. Because the truth values in the last column are all true, the logical expressions are equivalent.

Example 1.17

Show that $(p \wedge q) \wedge (p \oplus q)$ is a contradiction.

Solution

As shown in Table 1.16, $(p \wedge q) \wedge (p \oplus q)$ is a contradiction (i.e., the truth values in the last column are all false).

Example 1.18

Show that $p \vee (q \wedge r)$ is a contingency.

Solution

As shown in Table 1.17, $p \vee (q \wedge r)$ is a contingency (i.e., the truth values in the last column consist of both true and false values).

A compound proposition is **satisfiable** if there is at least one assignment of truth values to its variables for which it is then true. Such an assignment is called a **solution** of the satisfiability problem. When there exists no such an assignment (i.e., the compound proposition is false for all assignments of truth values to its variables), the compound proposition

TABLE 1.15 Truth table for Example 1.16.

p	q	$p \leftrightarrow q$	$p \oplus q$	$\overline{p \oplus q}$	$(p \leftrightarrow q) \leftrightarrow \overline{(p \oplus q)}$
T	T	T	F	T	T
T	F	F	T	F	T
F	T	F	T	F	T
F	F	T	F	T	T

TABLE 1.16 Truth table for Example 1.17.

p	q	$p \wedge q$	$p \oplus q$	$(p \wedge q) \wedge (p \oplus q)$
T	T	T	F	F
T	F	F	T	F
F	T	F	T	F
F	F	F	F	F

TABLE 1.17 Truth table for Example 1.18.

p	q	r	$q \wedge r$	$p \vee (q \wedge r)$
T	T	T	T	T
T	T	F	F	T
T	F	T	F	T
T	F	F	F	T
F	T	T	T	T
F	T	F	F	F
F	F	T	F	F
F	F	F	F	F

is **unsatisfiable**. In other words, a compound proposition is unsatisfiable if and only if its negation is a tautology. There are many applications of satisfiability in a multitude of disciplines in engineering. However, solving the satisfiability problem for a compound proposition with a very significant number of propositional variables is a time-consuming process.

Example 1.19
Determine if the compound proposition $(p \vee q \vee r \vee s \vee t) \wedge (\overline{p} \vee \overline{q} \vee \overline{r} \vee \overline{s} \vee \overline{t})$ is satisfiable.

Solution
This compound proposition is true when these five variables do not all have the same truth value (i.e., at least one of the propositional variables p, q, r, s, and t is true, and at least one is false). Hence, this compound proposition is satisfiable because there is at least one assignment of truth values for these variables that makes it true.

1.5 Logic Puzzles

Logic puzzles require solutions that are based on logical reasoning. A *logic puzzle* is a problem that can be solved through deductive reasoning. The fact that if an assumption leads to a contradiction and that assumption must be false forms the basis for solving many logic puzzles by eliminating contradictory answers.

Example 1.20
Knights and Knaves is a type of logic puzzle with two types of people, where knights can only answer questions truthfully (always tell the truth) and knaves can only answer questions falsely (always lie). On the island of knights and knaves,

you come to a fork in the road with one individual standing before each path. You know that one of them is a knight, and the other is a knave. You also know that one path leads to freedom, and the other path leads to certain death. You can ask one of the individuals one yes–no question. What do you ask to determine the path to freedom?

Solution

It is important to note there is no need to figure out which person is a knight and which one is a knave to figure out which path leads to freedom. There are several ways to find out which way leads to freedom. One possible question is "Would the other individual tell me that your path leads to freedom?" With this question, the knight will tell the truth about a lie, while the knave will tell a lie about the truth. Therefore the given answer will always be the opposite of the correct one.

Example 1.21

Suppose there are signs on the doors to two rooms. The sign on the first door reads "In this room, there are $1 billion to have, and in the other room, there is a deadly virus to catch," and the sign on the second door reads "In one of these rooms, there is a deadly virus to catch, and in the other one, there are $1 billion to have." Suppose that you know that one of these signs is true and the other is false. The question is, behind which door are there $1 billion?

Solution

If the first sign is true, then the second sign would also be true. In that case, we could not have one true sign and one false sign. Rather, if the second sign is true and the first is false, there are $1 billion in the second room and a virus in the first room.

Exercises

(1.1)

Which of the following sentences are propositions? What are the truth values of those that are propositions?

(a) Mount Everest is the highest mountain on the earth.
(b) $x + 7 = 11$.
(c) Answer all the questions in this chapter.
(d) The sun is cold at night.

(1.2)

We have the following propositions.

   p: He is smart.
   q: He is hardworking.

Write the following propositions using p and q and logical connectives and negations.

(a) He is smart and hardworking.
(b) He is smart but not hardworking.
(c) He is not smart, and he is not hardworking.
(d) He is smart or hardworking or both.
(e) If he is smart, then he is hardworking.
(f) Either he is smart or he is hardworking, but he is not hardworking if he is smart.

(1.3)

We have the following propositions.

p: He studied hard for the final exam.

q: He got an A^+ in the course.

Express each of the following propositions as an English sentence.

(a) $p \wedge q$.
(b) $\bar{p} \vee q$.
(c) $p \rightarrow \bar{q}$.
(d) $\bar{q} \rightarrow p$.
(e) $\bar{p} \rightarrow \bar{q}$.
(f) $p \leftrightarrow \bar{q}$.
(g) $\bar{p} \wedge (p \vee \bar{q})$.

(1.4)

Translate each of the following statements into a logical statement.

(a) It is not cold, but it is cloudy, where p: it is cold, and q: it is cloudy.
(b) It is neither sunny nor rainy, where r: it is sunny, and s: it is rainy.
(c) $0 < x \leq 5$, where t: $\{x > 0\}$, u: $\{x < 5\}$, and w: $\{x = 5\}$.

(1.5)

Prove the following De Morgan's laws.

(a) $\overline{(p \wedge q)} \equiv \bar{p} \vee \bar{q}$.
(b) $\overline{(p \vee q)} \equiv \bar{p} \wedge \bar{q}$.

(1.6)

Construct a truth table for each of the following propositions.

(a) $(p \vee q) \rightarrow (p \oplus q)$.
(b) $(p \leftrightarrow q) \oplus (\bar{p} \leftrightarrow q)$.

(1.7)

Determine the converse, inverse, and contrapositive of each of the following conditional statements.

(a) If you do not work hard in life, then you fail in life.
(b) If you have no empathy, then you are nobody.

(1.8)

Determine a compound proposition involving three different propositional variables that is true when exactly two of them are true and is false otherwise.

(1.9)

Construct the truth table for each of the following compound propositions, and then determine if the compound proposition is a tautology, a contradiction, or a contingency.

(a) $\bar{p} \rightarrow ((p \wedge q) \oplus (p \vee q))$.

(b) $\overline{(p \rightarrow q)} \rightarrow p$.

(1.10)

Evaluate each of the following statements using the order of precedence.

(a) $p \vee q \wedge r$.

(b) $p \leftrightarrow q \rightarrow r$.

(c) $\bar{p} \vee \bar{q} \rightarrow r \wedge s$.

CHAPTER 2

Predicate Logic

Contents

"Everyone who is born can learn to love." "My wife has given birth to our daughter." "Our daughter can learn to love." This simple analysis is intuitively perceived as correct, yet its validity cannot be derived using propositional logic. Propositional logic sometimes cannot adequately provide the appropriate meaning of a statement, as some mathematical statements and everyday situations may involve the notion of quantification. To this end, another type of logic, known as predicate logic, is needed to allow us to reason. The symbolic analysis of predicates and quantifiers is called *predicate logic* or *predicate calculus*. Predicate logic uses quantified variables and allows the use of expressions that contain variables. This chapter briefly presents basic aspects of predicate logic.

2.1 Predicates

To understand predicate logic, we first need to understand the concept of a predicate. A *predicate* refers to the part of a sentence that attributes a property to the subject. For instance, in the sentence "The United States of America is a powerful country," "The United States of America" is the subject, and the part of the sentence from which the subject has been removed (i.e., "is a powerful country") is the predicate. Another example is the sentence "x represents the world population", in which the variable "x" is the subject and "represents the world population" is the predicate.

A predicate contains a finite number of variables and becomes a propositional statement when specific values are substituted for the variables. The *domain*, also known as the *universe of discourse* or the *domain of discourse*, is the set of all values of a variable that can replace it.

A predicate that involves just one variable may be denoted by $P(x)$. The statement $P(x)$ is said to be the value of the *propositional function* P at x. A propositional function P, by itself, is neither true nor false. However, once a value from the domain has been assigned to the variable x, $P(x)$ becomes a propositional statement and thus has a truth value.

Discrete Mathematics
ISBN 978-0-12-820656-0, https://doi.org/10.1016/B978-0-12-820656-0.00002-2
21

A predicate that involves n variables is called an **n-ary predicate** and denoted by $P(x_1, x_2, ..., x_n)$. Once a set of values has been assigned to the variables $x_1, x_2, ..., x_n$, $P(x_1, x_2, ..., x_n)$ has a truth value, as it is then a propositional statement. Note that a predicate with two variables is called a **binary predicate**.

Example 2.1

Discuss the following statements in the context of predicate logic.
(a) The statement $x > 2$, for $x = 0$ and $x = 7.2$, where the domain for the variable consists of all real numbers.
(b) The statement $x^2 + y^2 = z^2$, for $x = 3$, $y = 4$, and $z = 5$ and for $x = 5$, $y = 6$, and $z = 7$, where the domain for each of the three variables consists of all positive integers.

Solution

(a) The statement $x > 2$ has two parts. The first part is the variable x, which is the subject of the statement, and the second part is the predicate P, which denotes "is greater than 2." The propositional function $P(x)$ denotes the statement $x > 2$. Therefore $P(0)$ and $P(7.2)$ are both propositional statements, where $P(0)$, indicating $0 > 2$, is false, and $P(7.2)$, indicating $7.2 > 2$, is true.
(b) The statement $x^2 + y^2 = z^2$ has two parts. The first part consists of the variables x, y, and z, and the second part is the predicate Q. The propositional function $Q(x, y, z)$ denotes the statement $x^2 + y^2 = z^2$. Therefore $Q(3, 4, 5)$ and $Q(5, 6, 7)$ are both propositional statements, where $Q(3, 4, 5)$, indicating $9 + 16 = 25$, is true, and $Q(5, 6, 7)$, indicating $25 + 36 = 49$, is false.

2.2 Quantifiers

As stated earlier, by assigning a value to the variable x, the propositional function $P(x)$ becomes a propositional statement with a truth value. Another way to obtain a proposition from a propositional function is to add quantifiers. For instance, the propositions "Few people are very compassionate," "Some people are racist," "All people are mortal," "None of them are good," "One even prime number exists," and "Every day the sun rises" each contains a word indicating a quantity, such as "few," "some," "all," "none," "one," and "every." These words are called **quantifiers**, as each word reveals for how many elements a given predicate is true. In other words, **quantification** is a way to express the extent to which a predicate is true over a range of elements. There are two widely known quantifications in predicate logic, namely, universal quantification and existential quantification.

Universal quantification indicates that a predicate is true for every element under consideration. In other words, **universal quantification** asserts that a predicate is true for all values of a variable in a given domain. Because the domain specifies the possible values of a variable, by changing the domain, the meaning of the universal quantification of a predicate may change. For instance, if the domain consists of all real numbers greater than 1, then the assertion that every number, say 2, is greater than its inverse (i.e., $\frac{1}{2}$) is true, as we have $\frac{1}{2} < 2$. However, if the domain changes and includes all positive real numbers, then the assertion that every number, say $\frac{1}{3}$, is greater than its inverse (i.e., 3) is false, as we have $\frac{1}{3} < 3$.

The universal quantification of $P(x)$, which is the statement $P(x)$ for all values of x in the domain, is denoted by $\forall x P(x)$. The symbol $\forall$ is called the **universal quantifier** and read as "for all" or "for every." Note that if a domain is not specified when a universal quantifier is used, then the universal quantification of a statement is not defined. The statement $\forall x P(x)$ is defined to be true if and only if $P(x)$ is true for every x in the domain, and it is defined to be false for at least one x in the domain. A value of x for which $P(x)$ is false is called a **counterexample** to the universal statement $\forall x P(x)$. Moreover, if the domain is empty, then $\forall x P(x)$ is true for any $P(x)$, as there exists no element x in the domain for which $P(x)$ is false.

Example 2.2

Determine the truth values of the universal statement $\forall x (x \geq \sqrt{x})$ for the following domains.

(a) All positive integers less than or equal to 3.
(b) All positive real numbers less than or equal to 3.

Solution

(a) The statement is true for each element in the domain, that is, for $x \in \{1,\ 2,\ 3\}$, as we have $1 = 1$, $2 > \sqrt{2}$ and $3 > \sqrt{3}$. Hence, $\forall x (x \geq \sqrt{x})$ is true.
(b) The statement is false for at least one element in the domain $x \in (0,\ 3]$. As a counterexample, if $x = 0.49$, we then have $0.49 \geq \sqrt{0.49} = 0.7$, which in turn means $\forall x (x \geq \sqrt{x})$ is false. In fact, the statement is false for $x \in (0,\ 1)$. Note that to prove a universal statement is false, a single example is sufficient.

Existential quantification indicates that a predicate is true for at least one element under consideration. In other words, **existential quantification** asserts that a predicate is true for at least one value of a variable in a given domain. Because the domain specifies the possible values of a variable, by changing the domain, the meaning of the existential quantification of a predicate may change. For instance, if the domain consists of all females, then the assertion that there is at least one person who is pregnant is true. However, if the domain changes and includes all males, then the assertion that there is at least one person who is pregnant is false.

The existential quantification of $P(x)$, which is the statement $P(x)$ for at least one value of x in the domain, is denoted by $\exists x P(x)$. The symbol $\exists$ is called the **existential quantifier** and read as "for some," "there exists a," or "for at least one." Note that if a domain is not specified when an existential quantifier is used, then the existential quantification of a statement is not defined. The statement $\exists x P(x)$ is defined to be true if and only if $P(x)$ is true for at least one x in the domain, and it is defined to be false for every x in the domain. Moreover, if the domain is empty, then $\exists x P(x)$ is false for every $P(x)$, as there exists no element x in the domain for which $P(x)$ is true.

Example 2.3

Determine the truth values of the existential statement $\exists x\,(x < \sqrt{x})$ for the following domains.

(a) All positive integers less than or equal to 3.

(b) All positive real numbers less than or equal to 3.

Solution

(a) The statement is false for each element in the domain, that is, for $x \in \{1, 2, 3\}$, as we have $1 = 1$, $2 > \sqrt{2}$ and $3 > \sqrt{3}$. Hence, $\exists x P(x)$ is false.

(b) The statement is true for an element in the domain $x \in (0, 3]$, as for $x = 0.49$, we have $0.49 < \sqrt{0.49} = 0.7$. Hence, $\exists x P(x)$ is true. In fact, the statement is true for $x \in (0, 1)$. Note that to prove an existential statement is true, a single example is sufficient.

Of all quantifiers, the **uniqueness quantifier**, denoted by $\exists!$ or $\exists_1$, is perhaps most often used. The notation $\exists! x P(x)$ or $\exists_1 x P(x)$ states that "There exists a unique x such that $P(x)$ is true," "There is exactly one x such that $P(x)$ is true," or "There is one and only one x such that $P(x)$ is true." For instance, $\exists! x (x^2 = 1)$, where the domain is the set of positive integers, states that there is only one positive integer x such that $x^2 = 1$, that is $x = 1$.

It is important to note that universal statements are generalizations of "and" statements, and existential statements are generalizations of "or" statements. Therefore when the domain of a quantifier is finite, quantified statements can be expressed using propositional logic. In particular, when the elements of the domain are $x_1, x_2, \ldots, x_n$, where n is a positive integer (i.e., all elements in the domain can be listed), the universal quantification $\forall x P(x)$ is the same as the conjunction

$P(x_1) \wedge P(x_2) \wedge \ldots \wedge P(x_n)$, because this conjunction is true if and only if the propositional functions $P(x_1)$, $P(x_2)$, ..., and $P(x_n)$ are all true, and the existential quantification $\exists x P(x)$ is the same as the disjunction $P(x_1) \vee P(x_2) \vee \ldots \vee P(x_n)$, because this disjunction is true if and only if at least one of the propositional functions $P(x_1)$, $P(x_2)$, ..., and $P(x_n)$ is true.

When a quantifier is used on the variable x, it is then called a **bound variable**, as it is bound by the quantifier. When a variable is not bound by a quantifier or is not equal to a particular value, it is called a **free variable**, as it can roam over the domain. A statement with free variables is not a proposition. A propositional function can be turned into a proposition by quantifiers and/or value assignments. The part of a logical expression to which a quantifier is applied is called the **scope** of the quantifier.

Example 2.4
Identify the bound and free variables and the scopes of the quantifiers in the following statement.

$$(\exists x(x - y + z = 2)) \vee (\forall y(x - y - z = 1)).$$

Solution
The variables x and y are bound by the two quantifiers, and the variable z is free, as it is not bound by a quantifier, nor does it have an assigned value. The scope of the first quantifier (i.e., $\exists x$) is $x - y + z = 2$, and the scope of the second quantifier (i.e., $\forall y$) is $x - y - z = 1$, and the scopes of the two quantifiers do not overlap.

In the context of **logical equivalence for quantified statements**, the statements are **logically equivalent** if and only if they have identical truth values regardless of what predicates are substituted into these statements and what domains are used for the variables in these propositional functions. If the two statements P and Q are logically equivalent, we then indicate their equivalence by $P \equiv Q$. For instance, the logical equivalence of a unique existential quantifier is as follows:

$$\exists! x P(x) \equiv \exists x(P(x) \wedge \forall y(P(y) \rightarrow (x = y))).$$

Example 2.5
Show that $\exists x(P(x) \vee Q(x))$ and $\exists x P(x) \vee \exists x Q(x)$ are logically equivalent.

Solution
To prove these two quantified statements are logically equivalent, we need to prove first if $\exists x(P(x) \vee Q(x))$ is true, then $\exists x P(x) \vee \exists x Q(x)$ is true, and second if $\exists x P(x) \vee \exists x Q(x)$ is true, then $\exists x(P(x) \vee Q(x))$ is true.

First, suppose that $\exists x(P(x) \lor Q(x))$ is true. This means that there exists a c, a particular value of x, for which $P(c) \lor Q(c)$ is true. Hence, either $P(c)$ or $Q(c)$ or both are true. Therefore either $\exists x P(x)$ or $\exists x Q(x)$ or both are true. This means that $\exists x P(x) \lor \exists x Q(x)$ is true.

Second, suppose that $\exists x P(x) \lor \exists x Q(x)$ is true. This means that either $\exists x P(x)$ or $\exists x Q(x)$ or both are true. Hence, there exists a c, a particular value of x, for which either $P(c)$ or $Q(c)$ or both are true. Therefore $P(c) \lor Q(c)$ is true. This means that $\exists x(P(x) \lor Q(x))$ is true.

Example 2.6

(a) Show that $\forall x P(x) \lor \forall x Q(x)$ and $\forall x(P(x) \lor Q(x))$ are not logically equivalent.

(b) Show that $\exists x P(x) \land \exists x Q(x)$ and $\exists x(P(x) \land Q(x))$ are not logically equivalent.

Solution

To show two statements are not logically equivalent, only an example is required to show one of the two statements is true, and the other is false. Suppose $P(x)$ is the statement that x is odd and $Q(x)$ is the statement that x is even. Let the domain of discourse be the positive integers.

(a) $\forall x P(x)$ represents all odd numbers, $\forall x Q(x)$ represents all even numbers, and $\forall x P(x) \lor \forall x Q(x)$ represents positive integers are all odd or all even, which is false. However, $\forall x(P(x) \lor Q(x))$ represents all positive integers are either odd or even, which is true.

(b) $\exists x P(x) \land \exists x Q(x)$ indicates that there is at least an odd number and there is at least an even number, which is true. However, $\exists x(P(x) \land Q(x))$ indicates that there exists an integer that is both odd and even, which is false.

In a compound logic statement consisting of both propositional and predicate logic statements, the **precedence rules** suggest that the universal and existential quantifiers have higher precedence than all logical operators from propositional logic. For instance, the quantified statement $\exists x P(x) \land Q(x)$ means the conjunction of $\exists x P(x)$ and $Q(x)$, that is, we have $\exists x P(x) \land Q(x) \equiv (\exists x P(x)) \land Q(x)$, and not $\exists x(P(x) \land Q(x))$.

Mathematical writing may contain many examples of implicitly quantified statements. For instance, an algebraic identity is an example of implicit universal quantification, say $\forall x((x + 1)^2 = x^2 + 2x + 1)$, and an algebraic equation with at least one solution is an example of implicit existential quantification, say $\exists x((x + 1)^2 = x^2 - 2x + 1)$.

Example 2.7

Provide examples for the following cases.
(a) Implicit universal quantification.
(b) Implicit existential quantification.

Solution

(a) An example of implicit universal quantification is as follows:
  If we have $x > 4$, we then have $x^3 > 64$, that is, $\forall x (x > 4 \rightarrow x^3 > 64)$.
(b) An example of implicit existential quantification is as follows:
  The integer 8 can be the sum of two integers, that is, $\exists x \exists y (x \text{ and } y \text{ integers} \rightarrow x + y = 8)$.

2.3 Negations of Quantified Statements

Consider the statement "Every poor person deserves to live a better life." This statement is a universal quantification, namely $\forall x P(x)$, where $P(x)$ represents "x deserves to live a better life" and the domain consists of all people in the world living in poverty. The negation of this statement is "It is not the case that every poor person deserves to live a better life," or equivalently, "There is at least one poor person who does not deserve to live a better life." This negation is simply the existential quantification of the propositional function $\overline{P(x)}$, that is, $\exists x \overline{P(x)}$. Therefore the negation of a universal statement ("All are") is logically equivalent to an existential statement ("Some are not" or "There is at least one that is not").

Consider the statement "Some people are caring." This statement is an existential quantification, namely, $\exists x P(x)$, where $P(x)$ represents "x is caring," and the domain consists of all people in the world. The negation of this statement is "It is not the case that some people are caring," or equivalently, "No person is caring." This negation is simply the universal quantification of the propositional function $\overline{P(x)}$, that is, $\forall x \overline{P(x)}$. Therefore the negation of an existential statement ("Some are") is logically equivalent to a universal statement ("None is" or "All are not").

The rules for negations of quantified statements, which are called **De Morgan's laws for quantifiers**, are as follows:

$$\begin{cases} \overline{\forall x P(x)} \equiv \exists x \overline{P(x)} \\[2mm] \overline{\exists x P(x)} \equiv \forall x \overline{P(x)} \end{cases}$$

TABLE 2.1 Quantifiers and De Morgan's laws for quantifiers.

Statement	When true?	When false?
$\forall x P(x)$	$P(x)$ is true for every x.	There is an x for which $P(x)$ is false.
$\exists x P(x)$	There is an x for which $P(x)$ is true.	$P(x)$ is false for every x.
$\overline{\forall x P(x)} \equiv \exists x \overline{P(x)}$	There is an x for which $P(x)$ is false.	$P(x)$ is true for every x.
$\overline{\exists x P(x)} \equiv \forall x \overline{P(x)}$	$P(x)$ is false for every x.	There is an x for which $P(x)$ is true.

The meanings of the universal and existential qualifiers as well as those of their negations are summarized in Table 2.1. Note that the quantifier must be changed when negating a quantified proposition in English. For instance, the negation of the quantified statement that "Some people are racist" is not the statement that "Some people are not racist," but it is that "No person is racist."

Example 2.8
Determine the negations of the following statements:
(a) All prime numbers are odd.
(b) There is an honest politician.
(c) Rich people do not have empathy.
(d) Some people do not live to be 100 years old.

Solution
(a) Let $P(x)$ denote "x is an odd number," and the domain consists of all prime numbers. Then, the statement "All prime numbers are odd" is represented by $\forall x P(x)$, and its negation is $\exists x \overline{P(x)}$. This negation can be expressed as "There exists at least one prime number that is not odd."
(b) Let $P(x)$ denote "x is an honest politician," and the domain consists of all politicians. Then, the statement "There is an honest politician" is represented by $\exists x P(x)$, and its negation is $\forall x \overline{P(x)}$. This negation can be expressed as "All politicians are dishonest."
(c) Let $P(x)$ denote "x has empathy," and the domain consists of all rich people. Then, the statement "Rich people do not have empathy" is represented by $\forall x \overline{P(x)}$, and its negation is $\exists x P(x)$. This negation can be expressed as "There exists at least one rich person who has empathy."
(d) Let $P(x)$ denote "x lives to be 100 years old," and the domain consists of all people. Then, the statement "Some people do not live to be 100 years old" is represented by $\exists x \overline{P(x)}$, and its negation is $\forall x P(x)$. This negation can be expressed as "All people live to be 100 years old."

Negations of universal conditional statements are of great importance in mathematics. Noting that in propositional logic, the negation of an implication is logically equivalent to an "and" statement, namely, $\overline{(P(x) \to Q(x))} \equiv P(x) \wedge \overline{Q(x)}$, the negation of a universal conditional statement is thus as follows:

$$\overline{\forall x(P(x) \to Q(x))} \equiv \exists x\left(P(x) \wedge \overline{Q(x)}\right)$$

Example 2.9

Determine the negations of the following statements.
(a) Every person who is a vegetarian is healthy.
(b) Some people weigh more than 100 kg and are not healthy.

Solution

(a) Let $P(x)$ denote "x is a vegetarian," and $Q(x)$ denote "x is healthy," where the domain consists of all people. The statement "Every person who is a vegetarian is healthy" is represented by $\forall x(P(x) \to Q(x))$, and its negation is thus $\exists x\left(P(x) \wedge \overline{Q(x)}\right)$, that is, "There are some people who are vegetarian and not healthy."

(b) Let $P(x)$ denote "x weighs more than 100 kg," and $Q(x)$ denote "x is healthy," where the domain consists of all people. The statement "Some people weigh more than 100 kg and are not healthy" is represented by $\exists x\left(P(x) \wedge \overline{Q(x)}\right)$, and its negation is thus $\forall x(P(x) \to Q(x))$, that is, "Every person who weighs more than 100 kg is healthy."

2.4 Nested Quantifiers

Nested quantifiers are defined where one quantifier is within the scope of another. Quantifications of more than one variable can be viewed as nested loops. A propositional function of n variables has no truth value. However, if it is preceded by a quantifier for each variable, then it denotes a predicate logic statement and has a truth value.

Example 2.10

Determine the truth value of each of the following statements if the domain of each variable consists of all real numbers.
(a) $\forall x_1 \exists x_2((x_1 + x_2 = 2) \wedge (2x_1 - x_2 = 1))$.
(b) $\forall x_1 \forall x_2 \exists x_3(x_3 = 0.5(x_1 + x_2))$.

Solution

(a) The statement says that for every x_1, there exists at least one x_2 that this system of two linear equations is satisfied. It is false. The system has a unique solution, so not any x_1 can then satisfy the simultaneous equations. For example, with $x_1 = 0$, there can be no x_2 satisfying both equations.

(b) The statement says for every x_1 and every x_2, there exists at least one x_3 that the equation is satisfied. It is true, as for every x_1 and every x_2, there exists an $x_3 = 0.5(x_1 + x_2)$.

Table 2.2 presents some logical equivalences in predicate logic, which can prove to be helpful in addressing problems in logic. The order of quantifiers generally matters, as a different ordering of the quantifiers may yield a different statement. However, when they are all universal quantifiers or they are all existential quantifiers, the order of the variables can be changed without affecting the truth value of the proposition. Table 2.3 provides insights into different possible quantifications involving two variables.

TABLE 2.2 Logical equivalences.

$$\forall x P(x) \wedge \forall x Q(x) \equiv \forall x (P(x) \wedge Q(x))$$

$$\forall x P(x) \vee \forall x Q(x) \equiv \forall x \forall y (P(x) \vee Q(y))$$

$$\exists x P(x) \wedge \exists x Q(x) \equiv \exists x \exists y (P(x) \wedge Q(y))$$

$$\exists x P(x) \vee \exists x Q(x) \equiv \exists x (P(x) \vee Q(x))$$

$$\forall x P(x) \wedge \exists x Q(x) \equiv \forall x \exists y (P(x) \wedge Q(y))$$

$$\forall x P(x) \vee \exists x Q(x) \equiv \forall x \exists y (P(x) \vee Q(y))$$

TABLE 2.3 Quantifications of two variables.

Statement	When true?	When false?
$\forall x \forall y P(x, y)$ $\forall y \forall x P(x, y)$	$P(x, y)$ is true for every pair x, y.	There is a pair x, y for which $P(x, y)$ is false.
$\forall x \exists y P(x, y)$	For every x there is a y for which $P(x, y)$ is true.	There is an x such that $P(x, y)$ is false for every y.
$\forall y \exists x P(x, y)$	For every y there is an x for which $P(x, y)$ is true.	There is a y such that $P(x, y)$ is false for every x.
$\exists x \forall y P(x, y)$	There is an x for which $P(x, y)$ is true for every y.	For every x there is a y for which $P(x, y)$ is false.
$\exists y \forall x P(x, y)$	There is a y for which $P(x, y)$ is true for every x.	For every y there is an x for which $P(x, y)$ is false.
$\exists x \exists y P(x, y)$ $\exists y \exists x P(x, y)$	There is a pair x, y for which $P(x, y)$ is true.	$P(x, y)$ is false for every pair x, y.

Example 2.11

Let $P(x, y)$ be the statement $y < x^4$. Determine the truth values of the following quantifications, where the domain for each of the two variables consists of all real numbers $(-\infty, \infty)$. Comment on the results:

(a) $\forall x \forall y P(x, y)$.
(b) $\forall y \forall x P(x, y)$.
(c) $\forall x \exists y P(x, y)$.
(d) $\forall y \exists x P(x, y)$.
(e) $\exists x \forall y P(x, y)$.
(f) $\exists y \forall x P(x, y)$.
(g) $\exists x \exists y P(x, y)$.
(h) $\exists y \exists x P(x, y)$.

Solution

Note that the inequality $y < x^4$ indicates the region of interest in the $x - y$ plane that lies under the graph $y = x^4$.

(a) This quantification denotes the proposition "For all real numbers x and all real numbers y, $y < x^4$." It is false, as no point above the graph satisfies the inequality.

(b) This quantification denotes the proposition "For all real numbers y and all real numbers x, $y < x^4$." It is false, as no point above the graph satisfies the inequality.

(c) This quantification denotes the proposition "For every real number x, there is a real number y, $y < x^4$." It is true, as every vertical line intersecting the region of interest satisfies the inequality.

(d) This quantification denotes the proposition "For every real number y, there is a real number x, $y < x^4$." It is true, as every horizontal line intersecting the region of interest satisfies the inequality.

(e) This quantification denotes the proposition "There is a real number x for all real numbers y, $y < x^4$." It is false, as there is no vertical line wholly within the region of interest satisfying the inequality.

(f) This quantification denotes the proposition "There is a real number y for all real numbers x, $y < x^4$." It is true, as there is a horizontal line wholly within the region of interest satisfying the inequality.

(g) This quantification denotes the proposition "There is a real number x, there is a real number y, $y < x^4$." It is true, as every point in the region of interest satisfies the inequality.

(h) This quantification denotes the proposition "There is a real number y, there is a real number x, $y < x^4$." It is true, as every point in the region of interest satisfies the inequality.

As expected, parts (a) and (b) are the same proposition, simply because they both employ universal quantification, and parts (g) and (h) are the same proposition, as they both employ existential quantification. When the two quantifications are different (i.e., one is universal and the other is existential), the truth values may remain the same, such as parts (c) and (f), or the truth values may differ, such as parts (d) and (e).

Mathematical statements and English sentences can be translated into logical expressions and vice versa.

Example 2.12
(a) Translate the statement "The sum of the squares of two negative numbers is positive, where the domain is all real numbers" into a logical expression.
(b) Translate the statement "Some student has solved at least one exercise in every topic covered in this course" into a logical expression.
(c) Translate the statement "$\forall x(x > 0) \rightarrow \exists y(y = \ln x)$" into English.

Solution
(a) Let x and y represent two negative numbers. The logical expression is thus
$$\forall x \forall y \big(((x < 0) \wedge (y < 0)) \rightarrow (x^2 + y^2) > 0\big).$$
(b) Let $P(x, y)$ mean that student x has solved exercise y, and $Q(y, z)$ mean that exercise y is in topic z in this course. The logical expression is thus
$$\exists x \forall z \exists y (P(x, y) \wedge Q(y, z)).$$
(c) For every positive number x, there exists a real number y such that $y = \ln x$.

Quantifications of more than two variables are also common, and the truth values of such statements can be determined by examining the type of each quantifier and the order of quantifiers.

Example 2.13
Let $P(x, y, z)$ be the statement $x^2 + y^2 = z^2$. Determine the truth values of the following statements, where the domain of each variable consists of all positive numbers.
(a) $\exists x \exists y \exists z P(x, y, z)$.
(b) $\forall x \forall y \forall z P(x, y, z)$.
(c) $\forall x \exists y \exists z P(x, y, z)$.

Solution
(a) The statement means that there exist a positive number x, a positive number y, and a positive number z, for which $x^2 + y^2 = z^2$. The statement is true. For instance, for $x = 3$, $y = 4$, and $z = 5$, the equality holds.

(b) The statement means that for all positive numbers x, for all positive numbers y, and for all positive numbers z, we have $x^2 + y^2 = z^2$. The statement is false. For instance, for $x = 5$, $y = 6$, and $z = 7$, the equality does not hold. In fact, for any positive number x and any positive number y, the value of z cannot be any positive number, as it must have a particular value that is the square root of the sum of the square of x and the square of y.

(c) The statement means that for all positive numbers x, there exist a positive number y and a positive number z for which $x^2 + y^2 = z^2$. The statement is true. For instance, for any positive number x, say $x = 7$, we can choose a positive number y, say $y = \sqrt{15}$, and a positive number z, say $z = 8$, so the equality holds. In fact, for any positive number x, we get to choose a positive number y and a positive number z in such a way that when we subtract the square of y from the square of z, the difference is the square of x.

Quantified statements with more than one variable may be negated by successively applying De Morgan's laws for quantifiers from left to right.

Example 2.14

Consider the statement $\forall x \exists y \exists z (xy > z)$, where the domains of x, y, and z are all real numbers.

(a) Translate the statement into English, and determine its truth value.

(b) Express the negation of the statement, and determine its truth value.

Solution

(a) The English translation of the statement is as follows: "For every x, there exist a y and a z, such that $xy > z$." It is thus true.

(b) As the statement $\forall x \exists y \exists z (xy > z)$ involves nested quantifiers, it can be negated by sequentially applying the rules for negating statements with a single quantifier. The resulting quantification denotes the following proposition:

$$\overline{\forall x \exists y \exists z (xy > z)} \equiv \exists x \overline{\exists y \exists z (xy > z)} \equiv \exists x \forall y \overline{\exists z (xy > z)} \equiv$$
$$\exists x \forall y \forall z \overline{(xy > z)} \equiv \exists x \forall y \forall z (xy \leq z).$$

The negation of the statement is as follows: "There is an x, for all real numbers y and z, such that $xy \leq z$." It is thus false.

Exercises

(2.1)

Determine the truth value of each of the following statements if the domain of each variable consists of all real numbers.

(a) $\exists x (x^4 < -1)$.

(b) $\forall x (x^2 \neq 4x)$.

(2.2)

Determine the truth value of each of the following statements if the domain of each variable consists of positive real numbers.

(a) $\forall x \exists y (xy = 1)$.

(b) $\exists x \forall y (x \leq y^2)$.

(2.3)

Determine the truth values of each of the following statements if the domain of each variable consists of all real numbers.

(a) $\forall x \exists y (x^2 = y)$.

(b) $\forall x \exists y (x = y^2)$.

(c) $\exists x \forall y (xy = 0)$.

(d) $\exists x \forall y (y \neq 0 \rightarrow xy = 1)$.

(e) $\exists z \exists x \exists y (2x + 4y = 7z)$.

(2.4)

Translate the following English statements into logical expressions, where the domains of x and y each consists of all real numbers.

(a) For every x, there exists a y such that $x + y = 100$.

(b) There exists an x such that $x + y = y$ for every y.

(c) For every x and y, $x + y = y + x$.

(d) There exist x and y such that $x + y = 100$.

(2.5)

Negate each of the following statements.

(a) $\exists y \exists x \forall z P(x, y, z)$.

(b) Some people are 90 years old or older.

(c) $\forall y ((\forall x \forall z T(x, y, z)) \wedge (\exists x \exists z U(x, y, z)))$.

(2.6)

Let $P(x, y)$ be the statement "x loves y," where the domains of x and y each consist of all people in the world. Use quantifiers to express each of the following statements.

(a) Everybody loves somebody.

(b) There is somebody whom everybody loves.

(c) Nobody loves everybody.

(d) There is somebody whom no one loves.

(e) Everyone loves himself or herself.

(f) There is someone who loves no one besides himself or herself.

(2.7)

Translate each of the following nested quantifications into an English statement, where the domain of each variable consists of all real numbers.

(a) $\exists x \forall y (xy = y)$.

(b) $\forall x \forall y (((x < 0) \wedge (y < 0)) \rightarrow (xy > 0))$.

(c) $\exists x \exists y ((x^2 > y) \wedge (x < y))$.

(d) $\forall x \forall y \exists z (x + y = z)$.

(2.8)

Use quantifiers to express each of the following statements.

(a) There is a person who has eaten a meal in every restaurant in town.

(b) There does not exist a person who has eaten a meal in every restaurant in town.

(2.9)

Express each of the following statements in predicate logic.

(a) Every positive integer is the sum of the squares of four integers.

(b) A negative real number does not have a square root that is a real number.

(2.10)

Express the meaning of each of the following statements quantifying a predicate with two variables where the domains are real numbers, and determine whether each is true or false.

(a) $\exists x \exists y (x + y = 0)$.

(b) $\forall x \exists y (x + y = 0)$.

(c) $\exists x \forall y (x + y = 0)$.

(d) $\forall x \forall y (x + y = 0)$.

CHAPTER 3

Rules of Inference

Contents

Tautology provides rules of logic that are used in proofs. If the tautology includes an implication, it is often useful to convert it into a statement called a rule of inference. Each step of an extended argument involves drawing intermediate conclusions. When each step of the argument is a valid intermediate conclusion, the argument is then valid. As the rules of inference are the essential building blocks in yielding valid arguments, we will briefly discuss in this chapter the rules of inference in both propositional and predicate logic and also introduce a set of invalid arguments, known as fallacies.

3.1 Valid Arguments

A proof is often called an argument. In everyday life, the word "argument" usually carries a clear connotation of disagreement or controversy. No such negative connotation should be associated with a mathematical argument.

In the context of propositional logic, an ***argument*** is a sequence of propositional statements. All propositions in an argument, except for the final one, are called ***hypotheses***, ***premises***, ***antecedents***, or ***assumptions***, and the final proposition that follows from the hypotheses is called the ***conclusion*** or ***consequence***. A ***valid argument*** is a sequence of propositions where the truth of all the premises implies the truth of the conclusion.

Example 3.1

Consider the two hypotheses that today there are more residents in a small town than the number of days every resident of the town has ever lived, and no one was born today. Show that we can conclude at least two residents of the town have the same age, that is, at least two of them were born on the same day.

Solution

Suppose there are n residents in that small town, where n is obviously a positive integer. The argument contains two hypotheses that we assume they are both true. The first hypothesis suggests that the number of days every resident has ever lived is less than n, and the second hypothesis suggests that the number of days every resident has ever lived is at least 1. If no two residents are of the same age, there must be n positive integers less than n, which is impossible. Therefore at least two residents have been born on the same day. The argument is thus valid because the truth of all the premises implies the truth of the conclusion.

An argument is valid because of its form, not because of its content. An **argument form** is a sequence of compound propositions involving propositional variables. In a **valid argument form**, no matter which particular propositions are substituted for the propositional variables in its premises, the conclusion is true if all the premises are true. To say that an argument is valid means that its form is valid. Therefore an argument is valid if the conjunction of all hypotheses logically implies the conclusion, that is, such an implication is a tautology. Otherwise, the argument is invalid or a fallacy, that is, there is an error in reasoning or, equivalently, a flaw in the argument. One effective way to test an argument form for its validity is to take the following steps:

1. Identify the premises and conclusion of the argument form and construct a truth table showing their truth values, noting that a row of truth table in which all the premises are true is called a **critical row**.
2. Check critical rows. If the conclusion in every critical row is true, the argument form is then valid, otherwise, it is not a valid argument form.

Example 3.2

Determine the validity of this argument form: If $s = q \vee r$ and $t = p \wedge r$, then $r \rightarrow (s \oplus t)$.

Solution

As Table 3.1 reflects, there are critical rows, namely rows 1 and 3, where each row has a false conclusion $(r \rightarrow (s \oplus t))$, but its premises $(s = q \vee r$ and $t = p \wedge r)$ are true. Hence this form of argument is invalid.

TABLE 3.1 Truth table for Example 3.2.

p	q	r	$s = q \vee r$ (premise)	$t = p \wedge r$ (premise)	$s \oplus t$	$r \rightarrow (s \oplus t)$ (conclusion)
T	T	T	T	T	F	F
T	T	F	T	F	T	T
T	F	T	T	T	F	F
T	F	F	F	F	F	T
F	T	T	T	F	T	T
F	T	F	T	F	T	T
F	F	T	T	F	T	T
F	F	F	F	F	F	T

Noting in a valid argument, the truth of all of its premises implies the truth of the conclusion, an argument is called *sound* if and only if it is both valid and all of its premises are true, and as a consequence, its conclusion is true as well. Note that in a sound argument as well as in a valid argument, the conclusion is true. However, in a sound argument, all of its premises are true, whereas in a valid argument, all of its premises are assumed to be true.

Example 3.3
(a) Provide an example of an argument that is valid and sound.
(b) Provide an example of an argument that is valid but not sound.

Solution
(a) All humans need water to survive. My teacher is human. Therefore my teacher needs water to survive. Because of the logical necessity of the conclusion, this argument is valid. The argument is valid and its premises are true; the argument is thus sound.
(b) All birds can fly. Ostriches are birds. Therefore ostriches can fly. This argument is valid because, assuming the premises are true, the conclusion must be true. However, the first premise is false. Not all birds can fly. Hence the argument is valid but not sound.

A *rule of inference* is a valid argument form that can be used in the demonstration that arguments are valid. Rules of inference are the basic tools for establishing the truth of statements.

3.2 Rules of Inference for Propositional Logic

Table 3.2 presents the important rules of inference in propositional logic. They are all tautologies, where the hypotheses are written in a column, followed by a horizontal bar, followed by a line that begins with the therefore symbol "$\therefore$" and ends with the conclusion. Using truth tables, all these rules of inference (tautologies) can be proven. Rules of inference extensively used for propositional logic are as follows:

- *Modus ponens (law of attachment)*: If a conditional statement and its hypothesis are both true, then its conclusion must be true.
- *Modus tollens (law of contrapositive)*: If a conditional statement is true, but its conclusion is false, then its premise is false.
- *Hypothetical syllogism (rule of transitivity)*: If one statement implies a second statement and the second statement implies the third statement, then the first statement implies the third statement.
- *Disjunctive syllogism (rule of elimination)*: When there are two possibilities, and one can be ruled out, the other must be the case.
- *Generalization (rule of addition)*: If a statement is true, then the disjunction of this statement and another statement, regardless of its truth value, is also true.

TABLE 3.2 Rules of inference for propositional statements.

Rule of inference	Tautology	Name
p $p \to q$ $\therefore q$	$(p \wedge (p \to q)) \to q$	Modus ponens (law of attachment)
$p \to q$ $\overline{q}$ $\therefore \overline{p}$	$((p \to q) \wedge \overline{q}) \to \overline{p}$	Modus tollens (law of contrapositive)
$p \to q$ $q \to r$ $\therefore p \to r$	$((p \to q) \wedge (q \to r)) \to (p \to r)$	Hypothetical syllogism (rule of transitivity)
$p \vee q$ $\overline{p}$ $\therefore q$	$((p \vee q) \wedge \overline{p}) \to q$	Disjunctive syllogism (rule of elimination)
p $\therefore p \vee q$	$p \to (p \vee q)$	Generalization (rule of addition)
$p \wedge q$ $\therefore p$	$(p \wedge q) \to p$	Simplification rule
p q $\therefore p \wedge q$	$(p \wedge q) \to (p \wedge q)$	Conjunction rule
$p \vee q$ $\overline{p} \vee r$ $\therefore q \vee r$	$((p \vee q) \wedge (\overline{p} \vee r)) \to (q \vee r)$	Resolution rule

- **Simplification rule**: If the conjunction of two statements is true, then both statements are true.
- **Conjunction rule**: If two statements are true, then their conjunction is also true.
- **Resolution rule**: If two clauses containing complementary propositional variables are true, then a new clause containing all noncomplementary propositional variables is true, where a clause is a disjunction of variables.

Example 3.4
Provide examples illustrating the rules of inference presented in Table 3.2.

Solution
- Modus ponens: Suppose that the conditional statement "If the weather is nice tonight, then I will go for a walk" and its hypothesis "The weather is nice tonight" are both true. Then it follows that the conclusion of the conditional statement "I will go for a walk" is also true.

- Modus tollens: Suppose that the conditional statement "If the weather is nice tonight, then I will go for a walk" is true, but its conclusion, "I will go for a walk," is false. Then it follows that the hypothesis "The weather is nice tonight" is false too.
- Hypothetical syllogism: As the statements "If a number is divisible by 6, then it is divisible by 3" and "If a number is divisible by 3, then the sum of its digits is divisible by 3" are both true, the statement "If a number is divisible by 6, then the sum of its digits is divisible by 3" is also true.
- Disjunctive syllogism: The proposition "Positive integers are either even or odd" is true. Therefore the proposition "If a positive integer is not even, then it must be odd" is true.
- Generalization: Suppose the proposition "It is sunny today" is true. Therefore the proposition "It is sunny today or it is cold today" is true.
- Simplification rule: Suppose the proposition "It is sunny and cold today" is true. Therefore the propositions "It is sunny today" and "It is cold today" are true.
- Conjunction rule: If the propositions "All students in a math course have passed the course" and "All students in a physics course have passed the course" are both true, then the proposition "Any student who took both courses has passed them both" is true.
- Resolution rule: If the propositions "Cyrus is happy or Neda is sad" and "Cyrus is not happy or Bita is happy" are both true, then the proposition "Neda is sad or Bita is happy" is true.

Example 3.5

Consider the following premises:
A. "I am not sad tonight and today is more fun than yesterday."
B. "I will go out only if I am sad."
C. "If I do not go out, then I will watch a basketball game on TV."
D. "If I watch a basketball game on TV, then I sleep late."

Using rules of inference show that these premises lead to the conclusion "I will sleep late."

Solution

We first define these propositions: p as "I am sad tonight," q as "Today is more fun than yesterday," r as "I will go out," s as "I will watch a basketball game on TV," and t as "I will sleep late." We can thus construct the following valid argument:

Step	Reason
(i) $\bar{p} \wedge q$	Premise A
(ii) $\bar{p}$	Simplification using (i)
(iii) $r \rightarrow p$	Premise B

Step	Reason
(iv) $\bar{r}$	Modus tollens using (ii) and (iii)
(v) $\bar{r} \rightarrow s$	Premise C
(vi) s	Modus ponens using (iv) and (v)
(vii) $s \rightarrow t$	Premise D
(viii) t	Modus ponens using (vi) and (vii)

3.3 Rules of Inference for Predicate Logic

Rules of inference for quantified statements, as summarized in Table 3.3, are as follows:

- **Universal instantiation** is the rule of inference that allows us to conclude that $P(a)$ is true, where a is a particular member of the domain, given the hypothesis $\forall x P(x)$. For instance, birds produce offspring by laying eggs, and falcons are birds; we can thus conclude falcons lay eggs.
- **Universal generalization** is the rule of inference that allows us to conclude that $\forall x P(x)$ is true, given the premise that $P(a)$ is true for all elements a in the domain. Note that the element a must be an arbitrary, and not a specific, element of the domain. For instance, every arbitrary university student has a high school diploma; we can therefore conclude that all university students have a high school diploma.
- **Existential instantiation** allows us to conclude that there is a (nonarbitrary) element a in the domain for which $P(a)$ is true, given the premise $\exists x P(x)$. For instance, there is someone who got an A+ in the course, let's call her a and say that a got an A+.
- **Existential generalization** allows us to conclude that $\exists x P(x)$ is true when a particular a with $P(a)$ true is known. For instance, Cyrus got an A+ in the course, therefore someone got an A+ in the course.

TABLE 3.3 Rules of inference for quantified statements.

Name	Rule of inference
Universal instantiation	$\dfrac{\forall x P(x)}{\therefore P(a)}$
Universal generalization	$\dfrac{P(a) \text{ for an arbitrary element } a}{\therefore \forall x P(x)}$
Existential instantiation	$\dfrac{\exists x P(x)}{\therefore P(a) \text{ for some element } a}$
Existential generalization	$\dfrac{P(a) \text{ for some element } a}{\therefore \exists x P(x)}$

Example 3.6

Consider the following premises:

A. "A student writing an exam cheated."

B. "Everyone writing the exam was penalized."

Using rules of inference show that the conclusion "Someone who was penalized had cheated" is implied.

Solution

Let $U(x)$, $V(x)$, and $W(x)$ be propositions "x wrote the exam," "x cheated on the exam," and "x was penalized", respectively. Using rules of inference for both propositional and qualified statements, we can thus construct the following valid argument:

Step	Reason
(i) $\exists x(U(x) \wedge V(x))$	Hypothesis A
(ii) $U(a) \wedge V(a)$	Existential instantiation from (i)
(iii) $U(a)$	Simplification from (ii)
(iv) $\forall x(U(x) \rightarrow W(x))$	Hypothesis B
(v) $U(a) \rightarrow W(a)$	Universal instantiation from (iv)
(vi) $W(a)$	Modus ponens from (iii) and (v)
(vii) $V(a)$	Simplification from (ii)
(viii) $V(a) \wedge W(a)$	Conjunction from (vi) and (vii)
(xi) $\exists x(V(x) \wedge W(x))$	Existential generalization from (viii)

Example 3.7

Using rules of inference, prove $(\forall x(P(x) \rightarrow Q(x)) \wedge \forall xP(x)) \rightarrow \forall xQ(x)$.

Solution

Using rules of inference for both propositional and qualified statements, we can thus construct the following valid argument:

Step	Reason
(i) $\forall x(P(x) \rightarrow Q(x))$	Premise
(ii) $P(x) \rightarrow Q(x)$	Universal instantiation on (i)
(iii) $\forall xP(x)$	Premise
(iv) $P(x)$	Universal instantiation on (iii)
(v) $Q(x)$	Modus ponens on (ii) and (iv)
(vi) $\forall xQ(x)$	Universal generalization on (v)

TABLE 3.4 Rules of inference for propositions and quantified statements.

Name	Rule of inference
Universal modus ponens	$\forall x(P(x) \rightarrow Q(x))$ $\underline{P(a), \text{ where } a \text{ is a particular element}}$ $\therefore Q(a)$
Universal modus tollens	$\underline{}$ $\forall x(P(x) \rightarrow Q(x))$ $\underline{\overline{Q(a)}, \text{ where } a \text{ is a particular element}}$ $\therefore \overline{P(a)}$

The rules of inference for propositional logic and predicate logic can be combined. Two such well-known rules, as summarized in Table 3.4, are universal modus ponens and universal modus tollens:

- **Universal modus ponens** states that if x makes $P(x)$ true, then x makes $Q(x)$ true, and also a makes $P(a)$ true, we can therefore conclude that a makes $Q(a)$ true.
- **Universal modus tollens** states that if x makes $P(x)$ true, then x makes $Q(x)$ true, and also a does not make $Q(a)$ true, we can therefore conclude that a does not make $P(a)$ true.

Example 3.8

Write each of the following arguments in the form of a logical expression using quantifiers and state if each is valid.

(a) All humans want to be healthy, and Bita is a human. We can thus conclude that Bita wants to be healthy.

(b) All humans want to be healthy, and Bita does not want to be healthy. We can thus conclude that Bita is not human.

Solution

(a) Let $P(x)$ and $Q(x)$ be the propositions "x is human" and "x wants to be healthy," respectively, and let a represent Bita. Then the argument $(\forall x(P(x) \rightarrow Q(x))) \wedge P(a) \rightarrow Q(a)$ is in the form of universal modus ponens, and it is therefore valid.

(b) Let $P(x)$ and $Q(x)$ be the propositions "x is human" and "x wants to be healthy," respectively, and let a represent Bita. Then the argument $(\forall x(P(x) \rightarrow Q(x))) \wedge \overline{Q(a)} \rightarrow \overline{P(a)}$ is in the form of universal modus tollens, and it is therefore valid.

3.4 Fallacies

Fallacies arise in invalid arguments where they resemble rules of inference, but they are based on contingencies rather than tautologies. It is important to note that in logic the words "true" and "valid" have totally different meanings. A valid argument may have a false conclusion and an invalid argument may have a true conclusion. In fact, many people often mistake the concept of validity for the concept of truth and vice versa. If they find an argument valid, they accept the conclusion as true, and if they find an argument invalid, they take the conclusion as false. This approach in logic is not correct.

Flawed, yet common, argument forms are known as *fallacies*. Fallacies are statements that might sound reasonable, seemingly plausible, widely agreed, or superficially true, but they are actually defective and deceptive. Fallacies are often psychologically persuasive but logically flawed. Known fallacies are quite many, but a few common fallacies that we come across in life nowadays are briefly highlighted in Table 3.5. Fallacies can be generally divided into two broad categories:

- *Fallacies with irrelevant premises*, such as rejecting a claim by criticizing the person who makes it rather than the claim itself: What she says is totally wrong because she does not have much money.
- *Fallacies with unacceptable premises*, such as incorrectly asserting that only two alternatives exist: Either there should be a reduction in government services or there should be a cut in the social assistance to those in need.

In valid arguments, premises must be both relevant and acceptable, and in a fallacious argument, at least one of these two requirements is not met. Oftentimes a person makes a fallacy either intentionally, usually by one in a position of power to manipulate and persuade by deception, or unintentionally, due to carelessness and ignorance; in either case, it is invalid and appears to be better than it really is. In the age of social media, one must be very careful not to easily fall into accepting fallacies that have been deceptively crafted and masterfully delivered to attract like-minded people.

TABLE 3.5 Common fallacies.

Circular Reasoning: The fallacy of making assertions sufficiently different to obscure the fact that the same proposition occurs as both a premise and a conclusion.
Example: A says "God exists." B says "How do you know that God exists?" A says "The holy book says so." B says "Why should I believe the holy book?" A says "Because it is the word of God."
Hasty Generalization: The fallacy of jumping to conclusions, making assumptions, or reaching results about a group without adequate evidence, such as atypical or just too-small sample size. It is a mistaken use of inductive reasoning.
Examples: (i) Stereotypes about people are a common example of the hasty generalization. Last month, two immigrants who had committed a crime were arrested; I believe all immigrants are criminals. (ii) My neighbor, who is on welfare, is watching TV all the time and doing nothing; I believe social assistance to people should be cut off; we do not want lazy citizens.
Genetic Fallacy: The fallacy of assuming a claim is true or false solely due to its origin or judging something is good or bad on the basis of where it comes from or from whom it comes.

Continued

TABLE 3.5 Common fallacies.—cont'd

Example: Someone appeals to prejudices surrounding someone's background: He is not a good citizen of this country because his parents were not born here.

Composition: The fallacy of assuming that what is true of a part must be true of the whole, that is, the characteristics of the parts are somehow transferable to the whole itself.

Example: If every single tool in a toolbox is lightweight, the toolbox itself must then be light.

Division: The fallacy of assuming that what is true of the whole must be true of individual parts, that is, the properties of the whole must be the same as characteristics of individuals.

Example: He is a government employee, and the government is corrupt; therefore he is a corrupt person.

Equivocation: The fallacy of using a word deliberately in different senses in an argument or altering its definition halfway through a discussion.

Example: Plato says the end of a thing is its perfection; some say that death is the end of life; hence death is the perfection of life. Here the word "end" means goal in Plato's usage, but it means the final event in the second usage.

Appeal to Popularity: The fallacy of asserting a claim must be true simply because a lot of people believe it, that is, using the popularity of a premise or proposition as evidence for its truthfulness. This fallacy, also known as the bandwagon fallacy, is difficult to spot as common sense suggests that if something is popular, it must be good/true/valid/right, but this is not so.

Examples: (i) God exists because most people believe in God. (ii) Capital punishment is the right sentence for a convicted murderer because most people in this country believe it is.

Appeal to Tradition: The fallacy of stating a claim must be true simply because it is part of a tradition or that a premise must be true because people have always believed it or the premise has always worked in the past and will thus work in the future.

Example: A marriage by common law is unacceptable simply because it does not follow the long tradition of a civil or religious ceremony.

Appeal to the Person: The fallacy of rejecting or accepting a claim by attacking or praising the character of the person who makes an argument rather than discussing the substance of the argument itself.

Example: Persuasion comes from irrational psychological transference rather than from an appeal to evidence concerning the issue at hand. She does not deserve healthcare because she is an addict.

Appeal to Hypocrisy: The fallacy of arguing a claim must be true or false just because the claimant is hypocritical by not following it.

Example: Advice given by an obese, inactive father to his children that diet and exercise are important in life is dismissed.

Appeal to Heaven: An extremely dangerous fallacy of asserting that God supports or approves one's own standpoint or actions so it is right, and no further justification is required, and no serious challenge is possible.

Examples: (i) God gave us this land; it is thus ours. (ii) God ordered me to kill my son, and I was just following his orders.

Appeal to Authority: A fallacy in which support for a standpoint is provided by a well-known individual whose authority is in a field unrelated to the argument. This fallacy attempts to capitalize upon feelings of respect or familiarity with the individual. In an appeal to biased authority, the authority is one who truly is knowledgeable on the topic but unfortunately one who may have professional or personal motivations that render that judgment.

TABLE 3.5 Common fallacies.—cont'd

Examples: (i) My dentist, whom I respect a lot, says he will vote for the most conservative candidate in the upcoming election, so I will too. (ii) To determine whether the military budget is big enough, the views of some army generals and the CEOs of military equipment manufacturers were solicited.

Faulty Analogy: The fallacy of reasoning that because two things are similar in some respects, they must be similar in some further respect, that is, relying only on comparisons to prove a point rather than arguing deductively and inductively.

Examples: (i) No one objects to a physician looking up a difficult case in medical books; why, then, shouldn't medical students taking a difficult examination be permitted to use their textbooks? (ii) If knives can kill, and there is a knife in every house, why shouldn't we have guns in every house?

Appeal to Ignorance: The fallacy of arguing a lack of evidence or an absence of knowledge as proof, that is, a claim has to be true (or false) because it has not been proven to be false (or true).

Examples: (i) God exists because it has not been proven that God does not exist. (ii) Scientists cannot positively prove their theory that humans evolved from other creatures because we were not there to see it; therefore it proves the 6-day creation account is literally true as written.

Appeal to Emotion: The fallacy of allowing premises to be based on emotions rather than relevant reasons or manipulating an emotional response in place of a valid argument.

Example: A good citizen must fight in a war for his country and does not question if the war is just, and waging it serves the best interest of the majority of his fellow countrymen.

Red Herring: The fallacy of raising an irrelevant or invalid point deliberately during an argument with the sole purpose of distraction, changing the subject, or diverting the real question at issue.

Example: The president should not be held accountable for cheating on his income tax returns; after all, he was democratically elected.

Straw Man: The fallacy of distorting, weakening, or oversimplifying someone's position so it can be more easily refuted rather than honestly engaging in the real nuances of the debate. It can also refer to attacking one of the opposition's unimportant arguments while ignoring the opposition's best argument.

Examples: (i) One says poor people need the government's financial assistance; the other says you mean people should get a free ride from the tax money of hardworking, honest citizens? (ii) One says vegetarians say animals have feelings too; the other says have you seen a cow ever laugh?

Begging the Question: The fallacy of assuming what is to be proven without having derived it from the premises.

Example: Building a highway in the north of the country that hardly anyone uses is a waste of money; I am therefore against building this highway. It is true that spending money on a useless highway is something that no one wants, but nobody proved this highway was useless.

False Dichotomy: The fallacy of stating that there are binary alternatives when there are more than two possible outcomes.

Examples: (i) Either we go to war with them or our way of life will collapse. (ii) If you do not believe in my religion, you will then go to hell.

Stacking the Deck: A fallacy in which examples that disprove the point are ignored, and examples that support the case are listed.

Continued

TABLE 3.5 Common fallacies.—cont'd

Example: He is a family man who loves his children, goes to church every Sunday, and has been a valued member of his community for decades. There is no mention that he is a racist.

False Cause and Effect: A fallacy that establishes a cause/effect relationship that does not exist. This occurs when one mistakenly assumes that because the first event preceded the second event, it must mean the first event must have caused the second one; sometimes it does, but sometimes it doesn't.

Example: She was scratched by a cat last week, and 2 days later, she came down with a fever. The cat's scratch thus caused the fever.

Othering: A badly corrupted, discriminatory argument where facts, experiences, or objections are arbitrarily disregarded, ignored, or put down without serious consideration because those involved "are not like us" or "don't think like us."

Example: It's OK for those people overseas to earn a buck an hour by our corporations. If it happened here, it is nothing but brutal exploitation and daylight robbery, but over there, the economy is different and they're not like us.

Tiny Percentage Fallacy: An amount or action that is significant in and of itself somehow becomes insignificant simply because it's a tiny percentage of something much larger.

Example: The killings of tens of African Americans by the police every year is a tiny percentage of thousands of African Americans who are arrested by the police every year.

The Big Lie Technique: The contemporary fallacy of repeating a lie, slogan, talking point, nonsense statement, or deceptive half-truth over and over in different forms, particularly in the media, until it becomes part of daily discourse, and people accept it without further proof or evidence.

Examples: (i) The nonexistent "weapons of mass destruction" in Iraq paved the way for the invasion of Iraq in 2003. (ii) The US president-elect in 2016 stated that "millions" of ineligible votes were cast in that year's American presidential election.

Gaslighting: A recently prominent fallacy denying or invalidating a person's own knowledge and experiences by deliberately twisting or distorting known facts, memories, events, and evidence to disorient a vulnerable opponent and to make someone doubt their sanity.

Example: Who are you going to believe? Me or your own eyes? You're crazy! You seriously need to see a shrink.

Example 3.9

Identify the type of each of the following fallacies:

(a) My three best friends failed an introductory calculus course; therefore I conclude that most students who take this course will fail it.

(b) Useless projects like this downtown multimillion-dollar project are a waste of taxpayers' money.

(c) This house is quite large. Therefore all its rooms are large.

(d) Religions have been around for thousands of years. They are thus good for people.

(e) A good citizen of this country will carry out this responsibility.

(f) God exists, and you cannot prove it otherwise.

(g) We must never ban capital punishment in this country; otherwise, committing murder and mass shooting will be as common as jaywalking.

(h) Everyone agrees that an embryo is a human, and all humans have a right to life. Therefore an embryo has a right to life.

(i) Each note of the song sounds great. Therefore the whole song sounds great.

Solution

(a) It is jumping to a conclusion, as from some exceptional cases, one cannot generalize a rule that fits those alone.

(b) It is begging the question, as why this downtown multimillion-dollar project is a useless project.

(c) It is the fallacy of division; what is true of the whole is not true of the parts.

(d) It is appeal to tradition, as how long something has been around brings no validity to it.

(e) It is appeal to emotion, as no evidence is supplied to support if the argument is true; only assertions about people who agree or disagree with the argument are made.

(f) It is appeal to ignorance. It involves the notion of burden of proof, which rests on the side that makes a positive claim.

(g) It is slippery slope, as there is no good reason to believe the assertions.

(h) It is equivocation. Human was first used in the sense of something having human characteristics and was then used in the sense of a person with moral rights.

(i) It is composition, as what is true of the parts is not necessarily true of the whole.

The argument that when an implication and its conclusion are both true, then its hypothesis is true; that is, if $p \rightarrow q$ and q are both true, then p is true is an incorrect reasoning called the **fallacy of affirming the conclusion** or the **fallacy of affirming the consequent**. The proposition $((p \rightarrow q) \wedge q) \rightarrow p$ is not a tautology because it is false when p is false and q is true. This fallacy is also known as **converse error**.

The argument that when an implication and the negation of its hypothesis are both true, then the negation of its conclusion is true, that is, if $p \rightarrow q$ and $\bar{p}$ are both true, then $\bar{q}$ is true is an incorrect reasoning called the **fallacy of denying the hypothesis** or the **fallacy of denying the antecedent**. The proposition $((p \rightarrow q) \wedge \bar{p}) \rightarrow \bar{q}$ is not a tautology because it is false when p is false and q is true. This fallacy is also known as **inverse error**.

Example 3.10

Comment on the following arguments:

(a) The premises "If you do not work hard, then you fail in life" and "You failed in life" are both true. Therefore the statement "you did not work hard" is true.

(b) The premises "If you do not work hard, then you fail in life" and "You worked hard" are both true. Therefore the statement "You did not fail" is true.

Solution

(a) Let p be the proposition "You do not work hard," and q be the proposition "You fail in life." Then the argument is in the form of "if $p \rightarrow q$ and q, then p." This is an example of incorrect argument using the fallacy of affirming the conclusion. It is possible for you to fail in life even if you work hard.

(b) Let p be the proposition "You do not work hard," and q be the proposition "You fail in life." Then the argument is in the form of "if $p \rightarrow q$ and $\bar{p}$, then $\bar{q}$." This is an example of incorrect argument using the fallacy of denying the hypothesis. It is possible for you to work hard, and you still fail.

An argument also in the context of quantified logic can exhibit the converse error or the inverse error. A converse error in the quantified form states that if x makes $P(x)$ true, then x makes $Q(x)$ true and also a makes $Q(a)$ true; we can therefore conclude that a makes $P(a)$ true, which is an invalid conclusion. An inverse error in the quantified form states that if x makes $P(x)$ true, then x makes $Q(x)$ true and also a does not make $P(a)$ true; we can therefore conclude that a does not make $Q(a)$ true, which is an invalid conclusion.

Example 3.11

Write each of the following arguments using quantifiers, variables, and predicate symbols, and state if each is valid:

(a) All humans are mortal, and Rumi is mortal. We can therefore conclude that Rumi is human.

(b) All humans are mortal, and Rumi is not human. We can therefore conclude that Rumi is not mortal.

Solution

(a) Let $P(x)$ and $Q(x)$ be the propositions "x is human" and "x is mortal," respectively, and let a represent Rumi. Then the argument $((\forall x(P(x) \rightarrow Q(x))) \wedge Q(a)) \rightarrow P(a)$ is in the form of converse error for quantified form, and it is therefore invalid. A counterexample is if Rumi is a bird, then it implies that Rumi is mortal but not human.

(b) Let $P(x)$ and $Q(x)$ be the propositions "x is human" and "x is mortal," respectively, and let a represent Rumi. Then the argument $((\forall x(P(x) \rightarrow Q(x))) \wedge \overline{P(a)}) \rightarrow \overline{Q(a)}$ is in the form of inverse error for quantified form, and it is therefore invalid. A counterexample is if Rumi is a bird, then it implies that Rumi is not human but mortal.

In informal language, simple conditional statements are often used to mean biconditional statements. This is the main reason why many people make converse and inverse errors. In fact, if the premise was a biconditional rather than a conditional, the resulting argument would be valid. However, the closer the premise comes to being a biconditional, the more likely the conclusion is to be true. A variation of the converse error, known as abduction, is a powerful reasoning tool, provided it is used prudently. **Abduction** is a reasoning in which the major premise is certain, but the minor premise and therefore the conclusion is only probable; it thus involves forming a conclusion from the information that is known. It goes like this: If x makes $P(x)$ true, then x makes $Q(x)$ true and also $Q(a)$ is true for a particular a. Then check out $P(a)$, as it just might be true.

Abduction is a form of logical reasoning that starts with a set of observations and then seeks to find the simplest and most likely explanation for the observations. This process yields a plausible conclusion but does not positively verify it. Abductive conclusions are thus qualified as having a degree of uncertainty or doubt, which is expressed in terms such as "best available" or "most likely." Abduction is like troubleshooting, which is employed when the symptoms are needed to be identified to the best of knowledge and ability. For instance, it is widely used by doctors to make medical diagnoses, by computer scientists to conduct research in artificial intelligence, and by mechanics to repair cars.

Example 3.12

(a) Provide examples highlighting when people tend to conflate biconditionals and conditionals and the impacts of the resulting arguments.

(b) Provide a real-life example of how abduction can be used.

Solution

(a) (i) All criminals go to the Infamous bar located in the dangerous neighborhood of the city, and Cyrus goes to the Infamous bar. Therefore Cyrus is a criminal. The argument is invalid, as it results from making the converse error, which is due to the fact that the premise is a conditional. (ii) Hardly anyone but criminals go to the Infamous bar in the dangerous neighborhood of the city, and Cyrus also goes to the Infamous bar. Therefore it is likely (though not certain) that Cyrus is a criminal. (iii) Only criminals go to the Infamous bar located in the dangerous neighborhood of the city, and Cyrus goes to the Infamous bar. Therefore Cyrus is certainly a criminal, which is due to the fact that the premise is a biconditional.

(b) A doctor knows that "If a patient has pneumonia, then the patient has a fever and chills, coughs deeply, and feels very tired." The doctor also knows that the patient she is now seeing in her office has the very same symptoms. The doctor thus concludes that a diagnosis of pneumonia is a possibility, and even quite likely, but not a certainty. The doctor, however, will gain further support for the diagnosis through laboratory testing designed to detect pneumonia. As the set of symptoms and lab tests come to being a necessary and sufficient condition for pneumonia, the more certain the doctor can be of her diagnosis.

Exercises

(3.1)

Prove the following rules of inference:

(a) Modus ponens: $(p \wedge (p \rightarrow q)) \rightarrow q$.

(b) Modus tollens: $(\overline{q} \wedge (p \rightarrow q)) \rightarrow \overline{p}$.

(3.2)

For each of the following sets of premises, what relevant conclusion can be drawn?

(a) "If I drink, then I do not sleep well." "I do not sleep well if I have back pain." "I slept well."

(b) "I am either eating and sleeping or working out." "I am not eating and sleeping." "If I am working out, I feel good about myself."

(3.3)

(a) Give an example of a valid argument with false premises and a false conclusion.

(b) Give an example of an invalid argument with true premises and a true conclusion.

(3.4)

Determine what is wrong with each of the following arguments:

(a) Let $P(x)$ be "x is empathetic." Given the premise $\exists x P(x)$, we conclude that $P(\text{Neda})$. Therefore Neda is empathetic.

(b) Let $Q(x, y)$ be "x is more caring than y." Given the premise $\exists w Q(w, \text{Mina})$, it follows that $Q(\text{Mina}, \text{Mina})$. Then by existential generalization, it follows that $\exists x Q(x, x)$, so that someone is more caring than himself.

(3.5)

(a) Prove that if $\forall x(P(x) \rightarrow Q(x))$ and $\forall x(Q(x) \rightarrow R(x))$ are true, then $\forall x(P(x) \rightarrow R(x))$ is true, where the domains of all qualifiers are the same. This valid argument is known as universal transitivity.

(b) Consider these two hypotheses: "No polynomial functions have asymptotes." "This function has an asymptote." Show that "This function is not a polynomial function."

(3.6)

Comment on the following fallacies:

(a) This country needs tougher immigration policies. I have a neighbor who says we should let in many more immigrants. But I must say he is quite lazy, loud, and eats and drinks a lot.

(b) If supporters of the government's gun registry get their way, all recreational and hunting guns will have to be registered, and before you know it, it will be illegal to own a gun for target practice. Eventually, the government will want to know if you own weapons, whether it is a pocket knife or a baseball bat.

(c) My doctor asserts that there is the prevalence of police brutality against visible minorities, including indigenous and black people, while most police brutality goes unreported. But my doctor's husband and her brother are both police officers. I do not think she truly believes there exists a widespread use of excessive and/or unnecessary force by the police.

(3.7)

Consider these two hypotheses: "If an integer is even, then its square is even." "a is a particular integer that is even." Show that "a^2 is even."

(3.8)

Show that the premises $\forall x(\overline{P(x)} \rightarrow Q(x))$ and $\overline{Q(a)}$ for a particular element a in the domain imply $P(a)$.

(3.9)

Consider these two hypotheses: "If Bita comes late, then Bita sits in the back row." "Bita sits in the back row." Can we therefore conclude that "Bita came late"?

(3.10)

Consider these two hypotheses: "If inflation is going up, then employment will go down." "Inflation is not going up." Can we therefore conclude "Employment will not go down"?

CHAPTER 4

Proof Methods

Contents

Proofs are the heart of mathematics, and no mathematical results are accepted as correct unless they are proven using logical reasoning. A proof starts with something that is known. To construct a proof for a theorem, the hypotheses of the theorem, relevant definitions of terms, pertinent identities and axioms, and previously proven theorems along with rules of inference may be employed in intermediate steps leading to the final step of the proof. Each step of a mathematical proof needs to be correct, that is, it must follow logically from the steps preceding it, using relevant assumptions if need be. The statements that need to be proven in mathematics are very diverse and complex, a host of different proof methods are thus needed. The brief focus of this chapter is on some proof methods.

4.1 Terminology

Before embarking on the introduction of some of the methods of proof, it is important to introduce some terminology related to proofs:

Definition: A statement expressing the essential nature of a concept and a set of associated properties that describe the concept.

Axiom: A self-evident true statement, that is, a statement that is accepted on its intrinsic merit without proof. It may also be known as postulate.

Theorem: A mathematical statement that can be shown (proven) to be true.

Corollary: A proposition that can be proven as an immediate consequence of some other theorems.

Lemma: A less important theorem that can help prove a more important theorem.

Discrete Mathematics
ISBN 978-0-12-820656-0, https://doi.org/10.1016/B978-0-12-820656-0.00004-6
55

Conjecture: A statement that is being proposed to be a true statement but is not proven yet.

Example 4.1

Provide specific examples to highlight some terminology related to proofs.

Solution

Definition: A circle is a closed plane curve every point of which is equidistant from a fixed point within the curve.

Axiom: In Euclidean geometry, within a two-dimensional plane, for every given straight line and a point that is not on the line, there exists exactly one straight line passing through the point that is parallel to the line.

Theorem: If two sides of a triangle are equal, then the angles opposite them are equal.

Corollary: If three sides of a triangle are equal, then all three angles of the triangle are equal.

Lemma: If we subtract 1 from a positive integer, then the result is either a positive integer or 0.

Conjecture: If a transformation sends an even integer x to $\frac{x}{2}$ and an odd integer x to $3x + 1$, then for all positive integers x, the repeated application of the transformation will eventually reach integer 1.

A *proof* is a sequence of logically valid statements to demonstrate the validity of some precise statement, simply put, a proof is a derivation of new valid statements from old ones. A *mathematical proof* is an inferential argument for a mathematical statement showing that the stated assumptions methodically and logically lead to guarantee the conclusion. It is imperative to note that every statement that is not an axiom or definition needs to be proven.

Note that in the context of proofs, the phrase "without loss of generality" means that the case being made would not change in any way the validity of the proof, and no additional argument is required to prove other special cases. There are various types of proofs, each of which is appropriate in certain circumstances.

4.2 Proofs of Equivalence

Sometimes a theorem states that a group of $n \geq 2$ propositions p_1, p_2, ..., p_n are equivalent $(p_1 \leftrightarrow p_2 \leftrightarrow ... \leftrightarrow p_n)$, that is, they have the same truth values. To show a *proof of equivalence*, we need to show that the n conditional statements $p_1 \rightarrow p_2$, $p_2 \rightarrow p_3$, ..., $p_n \rightarrow p_1$ are all true, that is, we have the following:

$$(p_1 \leftrightarrow p_2 \leftrightarrow ... \leftrightarrow p_n) \leftrightarrow ((p_1 \rightarrow p_2) \wedge (p_2 \rightarrow p_3) \wedge ... \wedge (p_n \rightarrow p_1)).$$

In other words, instead of biconditional statements (if and only if statements), implications (if, then statements) are employed.

Example 4.2

Prove that for every positive integer n, n is even if and only if $n - 1$ is odd.

Solution

There are two statements p_1: n is even, and p_2: $n - 1$ is odd. To show they are equivalent, that is, $p_1 \leftrightarrow p_2$, we need to prove that $p_1 \rightarrow p_2$ and $p_2 \rightarrow p_1$. We first prove that if n is even, then $n - 1$ is odd. If n is even (i.e., n is an integer that is a multiple of 2), then $n = 2k$ for some positive integer k, and thus $n - 1 = 2k - 1$, which is odd as it is 1 less than the even number $2k$. We then prove that if $n - 1$ is odd, then n is even. If $n - 1$ is odd, then $n - 1 = 2k + 1$ for some positive integer k, and thus $n = 2k + 2 = 2(k + 1)$, which is even as it is a multiple of 2.

4.3 Proof by Counterexample

A counterexample is a form of proof. To prove that a statement of the form $\forall x P(x)$ is false, we need to find an element x such that $P(x)$ is false. In other words, to disprove a statement, we need to find an example in the domain of discourse for which the hypothesis is true and the conclusion is false, such an example is called a *counterexample*. For instance, for the statement that all prime numbers are odd, 2 is a counterexample as it is even. It is imperative to note that a theorem cannot be proven by considering examples unless every possible case in the domain, with no exception, is included.

Example 4.3

(a) Disprove that for every positive integer n, $2^n + n$ is prime.
(b) Disprove that for all real numbers a and b, if $a^2 = b^2$, then $a = b$.

Solution

(a) To disprove it, we need a counterexample. For instance, when $n = 4$, we have $2^4 + 4 = 20$, which is not prime.
(b) Note that if the absolute values of a positive real number and a negative real number are equal, then the statement can be disproved. A counterexample is $a = 5$ and $b = -5$, we thus have $a \neq b$, yet we have $a^2 = b^2 = 25$.

4.4 Vacuous Proofs and Trivial Proofs

The vacuous and trivial proofs are based on the truth table for implication (conditional statement), and they are quite simple. The implication $p \rightarrow q$ can be proven to be true using the truth values of p or q. Vacuous and trivial proofs are often employed to prove special cases of theorems.

If the hypothesis p can be shown to be false, the implication $p \rightarrow q$ is true by default; such a proof is called a *vacuous proof*. Note that in a vacuous proof, the conclusion q is not used.

Example 4.4
Prove the proposition "If $1 = 0$, then $5 = 2$" is true.

Solution
The proposition is vacuously true, as the hypothesis "$1 = 0$" is false. Note that in conclusion, "$5 = 2$" was not used.

If the conclusion q can be shown to be true, the implication $p \rightarrow q$ is true by default, such a proof is called a *trivial proof*. Note that in a trivial proof, the premise p is not used.

Example 4.5
Prove the proposition "If x is a real number with $x^4 + 1 = 0$, then $2 > 1$" is true.

Solution
The proposition is trivially true, as the conclusion "$2 > 1$" is true. Note that the premise "x is a real number with $x^4 + 1 = 0$" was not used.

4.5 Direct Proofs

Direct proof, which is based on using definitions, axioms, theorems, logical equivalences, and the rules of inference, is the most common proof strategy. A *direct proof* of an implication is constructed with the assumption that the premise is true, and a series of intermediate implications eventually leads to the fact that the conclusion of the implication must also be true. In other words, we show that the combination that the premise is true and the conclusion is false never occurs. It is important to note that not all direct proofs are straightforward, as some may require insights.

Example 4.6
Prove the following statements are true.

(a) If n is an odd integer, then n^2 is odd.

(b) If a and b are positive integers, then $a^2 - ab + b^2 > 0$.

(c) The sum of two rational numbers is rational.

Solution

(a) Assuming n is odd, that is, $n = 2k + 1$ for some positive integer k, we have $n^2 = (2k + 1)^2 = 4k^2 + 4k + 1$. Noting that $m = 4k^2 + 4k = 4(k^2 + k)$ is an even integer as it is a multiple of 4, we can thus conclude that $n^2 = m + 1$ is an odd integer.

(b) Assuming $a > 0$ and $b > 0$, we thus have $ab > 0$. Note that $a^2 - ab + b^2 = a^2 - 2ab + b^2 + ab = (a - b)^2 + ab > 0$, as $(a - b)^2$ and ab are both positive.

(c) Suppose s and t are both rational numbers, that is, $s = \frac{a}{b}$ and $t = \frac{c}{d}$ for some integers a, b, c, and d with $b \neq 0$ and $d \neq 0$. We therefore have $s + t = \frac{a}{b} + \frac{c}{d} = \frac{ad + bc}{bd} = \frac{p}{q}$, where p and q are integers because products and sums of integers are integers, and also $q \neq 0$. Hence $s + t$ is rational.

4.6 Proofs by Contraposition and Proofs by Contradiction

When we cannot easily employ a direct proof, we make use of an indirect proof. Indirect proofs do not start with the premises and end with the conclusion. There are two general types of indirect proofs, namely, proofs by contraposition and proofs by contradiction.

A proof by contraposition is based on the law of contrapositive, that is, the conditional statement $p \to q$ is equivalent to its contrapositive $\bar{q} \to \bar{p}$. In other words, in a **proof by contraposition** of $p \to q$, we take $\bar{q}$ as a premise, and we show that $\bar{p}$ must follow.

Example 4.7
Prove the following statements using a proof by contraposition.

(a) If a real number is irrational, then its square root is irrational.

(b) If $r = mn$, where m and n are positive integers, then $m \leq \sqrt{r}$ or $n \leq \sqrt{r}$.

Solution

(a) By letting x be an arbitrary real number, we need to prove that if x is irrational, then $\sqrt{x}$ is irrational. Using a proof by contraposition, we want to prove that if $\sqrt{x}$ is not irrational, then x is not irrational, or equivalently if $\sqrt{x}$ is rational, then x is rational. If $\sqrt{x}$ is rational, then $\sqrt{x} = \frac{m}{n}$ for some integers m and $n \neq 0$. As a result, we have $x = \frac{m^2}{n^2}$, which is the quotient of integers. Hence x is rational. We just showed the negation of the hypothesis of the original conditional statement is true.

(b) Using a proof by contraposition, we want to prove that if $m \leq \sqrt{r}$ or $n \leq \sqrt{r}$ is false, then $r = mn$ is false, or equivalently if both $m > \sqrt{r}$ and $n > \sqrt{r}$ are true, then $r = mn$ is false. However, if $m > \sqrt{r}$ and $n > \sqrt{r}$, then $mn > r$. This shows that $mn \neq r$, which contradicts the premise $mn = r$. We just showed the negation of the hypothesis of the original conditional statement is true.

When a conditional statement is true but the conclusion is false, then the hypothesis is false. In other words, if the implication $\bar{p} \rightarrow q$ is true, but q is false, then $\bar{p}$ is false or, equivalently, p is true. Note that a contradiction is a proposition of the form $r \wedge \bar{r}$, where r may be any proposition, thus it is always false regardless of the truth value of r. Therefore if we show $\bar{p} \rightarrow (r \wedge \bar{r})$ is true, then p is true. The method of **proof by contradiction**, which is an indirect proof, states that if the supposition that statement p is false leads logically to a contradiction, then p is true.

Example 4.8
Prove the following statements using a proof by contradiction.
(a) The earth cannot be flat.
(b) The sum of any rational number and any irrational number is irrational.
(c) $\sqrt{2}$ is irrational.
(d) There is no greatest integer.

Solution
(a) If the earth is flat, then people fall off the edge, which is absurd.
(b) Suppose there is a rational number $t = \frac{a}{b}$ and irrational number u such that $t + u = \frac{c}{d}$ is rational for some integers a, b, c, and d with $b \neq 0$ and $d \neq 0$. We can thus have $u = \frac{bc - ad}{bd}$. Because a, b, c, and d are all integers, $bc - ad$ and $bd \neq 0$ are also integers. Thus by definition, u is rational, which contradicts the supposition that u is irrational.
(c) If $\sqrt{2}$ is not irrational, then it is rational and can be thus written as the fraction of two integers m and n, where we assume $\frac{m}{n}$ is in lowest terms, so m and n are not both even (i.e., we assume $\frac{m}{n}$ is reduced and thus cannot be simplified). Therefore we can reach the following step-by-step conclusions:

$$\sqrt{2} = \frac{m}{n} \rightarrow 2 = \frac{m^2}{n^2} \rightarrow m^2 = 2n^2 \rightarrow m^2 \text{ is even} \rightarrow m \text{ is even} \rightarrow$$

$$m = 2k \rightarrow m^2 = 4k^2 = 2n^2 \rightarrow n^2 = 2k^2 \rightarrow n^2 \text{ is even} \rightarrow n \text{ is even}.$$

We showed that if $\sqrt{2}$ is not irrational, then both m and n are even, which contradicts our assumption that m and n are not both even. Therefore $\sqrt{2}$ is irrational.

(d) Suppose there is a greatest integer N, then $N > n$ for every integer n. Let $M = N + 1$. Therefore M is an integer as it is the sum of integers 1 and N. Thus M is an integer greater than N. Therefore N is not the greatest integer, which is a contradiction to the premise.

4.7 Proof by Cases and Proofs by Exhaustion

Sometimes we need to partition the proof into several disjoint parts whose union is the complete theorem and then prove each part individually. Suppose we must prove $p \rightarrow q$ and that p is equivalent to $p_1 \vee p_2 \vee \ldots \vee p_n$ (where $p_1, p_2, \ldots, p_n$ are the cases). To prove a conditional statement of the form $(p_1 \vee p_2 \vee \ldots \vee p_n) \rightarrow q$, we prove $(p_1 \rightarrow q) \wedge (p_2 \rightarrow q) \wedge \ldots \wedge (p_n \rightarrow q)$, as the two statements are equivalent. Such a proof is called a **proof by cases**, as we have

$$(p \rightarrow q) \leftrightarrow ((p_1 \vee p_2 \vee \ldots \vee p_n) \rightarrow q) \leftrightarrow ((p_1 \rightarrow q) \wedge (p_2 \rightarrow q) \wedge \ldots \wedge (p_n \rightarrow q)).$$

Example 4.9

Assuming k is a positive integer, show that $m = k^3 - k$ is an even integer.

Solution

Using a proof by cases, we consider two mutually exclusive cases for k, that is, k is even, and k is odd, as every positive integer falls into one of these two mutually exclusive cases. Assuming k is even, then for some integer n, we have $k = 2n \rightarrow m = k^3 - k = k(k+1)(k-1) = 2n(2n+1)(2n-1)$, that is, m is even, as it is a multiple of 2. Assuming k is odd, then for some integer n, we have $k = 2n + 1 \rightarrow m = k^3 - k = (k-1)k(k+1) = 2n(2n+1)(2n+2)$, that is, m is even, as it is a multiple of 2.

A proof by cases must check all possible cases that arise in a theorem. However, when each case involves checking an example, such a proof is called a **proof by exhaustion** or an **exhaustive proof**. Note that when the number of cases is infinitely many or just very large, then neither proof by cases nor proof by exhaustion is possible or even feasible.

Example 4.10

Assuming m and n are positive integers, show that $4m^2 + 9n^2 = 36$ has no solutions.

Solution

As m and n are positive integers, $4m^2$ and $9n^2$ are in turn both positive. We can thus conclude that $4m^2 < 36 \rightarrow 2m < 6 \rightarrow m < 3$ and $9n^2 < 36 \rightarrow 3n < 6 \rightarrow n < 2$. Using a proof by exhaustion, this leaves the cases that $m = 1$ or $m = 2$ as well as $n = 1$. Neither when we have $m = 1$ and $n = 1$ nor when we have $m = 2$ and $n = 1$ can $4m^2 + 9n^2$ be equal to 36. It is therefore impossible for $4m^2 + 9n^2 = 36$ to hold when m and n are positive integers.

4.8 Existence Proofs: Constructive Proofs and Nonconstructive Proofs

Some theorems in mathematics are about establishing the existence of a particular object. A proof of a proposition of the form $\exists x P(x)$ is called an **existence proof**. There are two types of existence proofs. If we can find an object a such that $P(a)$ is true, then such an existence proof is called a **constructive existence proof**. If we cannot find an object a such that $P(a)$ is true but rather establish its existence by an indirect proof, usually using a proof by contradiction, then such an existence proof is called a **nonconstructive existence proof**.

Example 4.11
(a) Show that there is a set of three positive integers that the square of one of them is equal to the sum of the squares of the other two integers.
(b) Prove that there exists an even integer that can be written in two ways as a sum of two prime numbers.
(c) Given an integer n, there is an integer m with $m > n$.

Solution
(a) We employ a constructive existence proof. This problem presents in a way a form of Pythagorean theorem. To this effect, there exist the set $\{3, 4, 5\}$ and the set $\{5, 12, 13\}$, which both can satisfy the requirement, as $5^2 = 4^2 + 3^2$ and $13^2 = 5^2 + 12^2$.
(b) Using a constructive existence proof, the integer 24 is an even number and can be written as $24 = 7 + 17$ as well as $24 = 11 + 13$, where 7, 11, 13, and 17 are all prime numbers.
(c) We employ a constructive existence proof. Suppose that n is an integer. Let $m = n + 1$. Then m is an integer and $m > n$. The proof established the existence of the desired integer m by showing that its value can be computed by adding 1 to the value of n.

Example 4.12
(a) Let $\overline{X} = \frac{x_1 + x_2 + \dots + x_n}{n}$ be the average of the n real numbers $x_1, x_2, \dots, x_n$. Show that there exists at least one real number among them, say x_k, where $k \in \{1, 2, \dots, n\}$, which is greater than or equal to $\overline{X}$, that is, $x_k \geq \overline{X}$.
(b) Given a nonnegative integer n, there is always a prime number p that is greater than n.

Solution

(a) We employ a nonconstructive existence proof. Using a proof by contradiction, we assume the negation of the conclusion, that is, we have $\overline{(\exists k(x_k \geq \overline{X}))} \equiv \forall k \overline{(x_k \geq \overline{X})} \equiv \forall k(x_k < \overline{X})$; this in turn means $x_1 < \overline{X}$, $x_2 < \overline{X}, \ldots, x_n < \overline{X}$. This assumption yields first $x_1 + x_2 + \ldots + x_n < n\overline{X}$ and then $\frac{x_1 + x_2 + \ldots + x_n}{n} < \overline{X}$, which contradicts the original hypothesis (the definition of the average of n real numbers).

(b) We employ a nonconstructive existence proof. With n as a nonnegative integer, consider the positive integer $n! + 1$. Then $n! + 1 > 1$ is divisible by some prime number p because every integer greater than 1 is divisible by a prime number. Also, $p > n$ because when $n! + 1$ is divided by any positive integer less than or equal to n, the remainder is 1.

4.9 Proof of a Disjunction

Proof of a disjunction is based on proving $p \rightarrow (q \vee r)$ by proving one of the logical equivalences $(p \wedge \overline{q}) \rightarrow r$ or $(p \wedge \overline{r}) \rightarrow q$ is true.

Example 4.13

Prove that for all integers a and b, if b is prime, then either b is a divisor of a or a is also prime.

Solution

To employ the proof of a disjunction, we define the following propositions:

p: b is prime
q: b is a divisor of a
r: a is prime (i.e., a and b have no common divisor greater than 1)

Noting that there are two ways to prove $p \rightarrow (q \vee r)$, we choose to prove $(p \wedge \overline{q}) \rightarrow r$. Because p is true (i.e., b is prime and its only positive divisors are thus 1 and b), and $\overline{q}$ is true (i.e., b is not a divisor of a, and the only possible positive common divisor of a and b is thus 1), then r is true (i.e., a is prime).

4.10 Uniqueness Proofs

Some theorems state the existence of a unique element with a particular property. To prove such a statement, we need to employ a ***uniqueness proof***, which consists of two distinct parts of existence (i.e., an element with the desired property exists) and uniqueness (i.e., there is no other element with the desired property).

Example 4.14

Show that if m and n are positive integers, then there is a unique integer k such that $m^n - 0.5k = 0$.

Solution

The integer $k = 2m^n$ is a solution of $m^n - 0.5k = 0$. This is the existence part of the proof. If there is an integer such that $m^n - 0.5j = 0$. Then $m^n - 0.5k = m^n - 0.5j$, that is, $k = j$. This means that if $j \neq k$, then $m^n - 0.5j \neq 0$. This establishes the uniqueness part of the proof.

Exercises

(4.1)

(a) Prove "If it is raining, then $1 = 1$" is true.

(b) Prove "If I am dead, then $2 + 2 = 5$" is true.

(4.2)

(a) Consider two different positive real numbers. Prove their arithmetic mean is greater than their geometric mean.

(b) Prove that $x^2 + x^{-2} \geq 2$, if x is a nonzero real number.

(4.3)

(a) Using a proof by contradiction, prove that $\sqrt{3}$ is irrational.

(b) Using a proof by contradiction, prove that $\sqrt{5}$ is irrational.

(4.4)

(a) Using a constructive existence proof, prove that there exists a real number x satisfying $a < x < b$, where a and b are real numbers with $a < b$.

(b) Using a proof by counterexample, disprove that for every prime p, $2^p - 1$ is prime.

(4.5)

(a) Using a proof by contradiction, prove that for all real numbers x and y, if $x + y \geq m$, then either $x \geq \frac{m}{2}$ or $y \geq \frac{m}{2}$, where m is a real number.

(b) Using a proof by contraposition, prove that n is odd if $mn + k$ is odd, k is even, and m is odd.

(4.6)

(a) Using a proof of equivalence, prove that if n is an integer, then n is even if and only if n^2 is even.

(b) Prove that if the product and the sum of two integers are both odd, then both integers are odd.

(4.7)

(a) Suppose that a and b are odd integers with $a \neq b$. Prove there is a unique integer c such that $|a - c| = |b - c|$.

(b) Prove that there are no solutions for $x^3 + y^3 = 64$ with x and y positive integers.

(4.8)

(a) Prove that there exists a pair of consecutive positive integers such that one of these integers is a perfect square and the other is a perfect cube.

(b) Prove that if $n = abc$, where a, b, and c are positive integers, we then have $a < \sqrt[3]{n}$, $b < \sqrt[3]{n}$, or $c < \sqrt[3]{n}$.

(4.9)

(a) Using a proof by exhaustion, show that $(n + 1)^3 \geq 3^n$ if n is a positive integer with $n \leq 4$.

(b) Using a proof by contradiction, show that there are infinitely many primes.

(4.10)

(a) Prove or disprove that if you have an 8-liter jug of water and two empty jugs with capacities of 5 liters and 3 liters, then you can provide 4 liters of water in one of the jars by successively pouring some of all of the water in a jug into another jug.

(b) Prove the pigeonhole principle: If more than n pigeons fly into n pigeon holes, then at least one pigeonhole will contain at least two pigeons.

CHAPTER 5

Sets

Contents

The concept of set is basic to all mathematics and mathematical applications, as almost all mathematical objects can be construed as sets, regardless of any additional properties they may possess. The set is the fundamental discrete structure upon which other discrete structures are built. According to Greg Cantor, the founder of set theory, a set is a *many* that allows itself to be thought of as a *one*. Our focus here is on ***naïve set theory***, which is based on Cantor's intuitive notion of an object and a set as defined informally in natural language, rather than on ***axiomatic set theory***, which is based on the rules of inference provided by formal logic. This chapter briefly highlights fundamental aspects of sets.

5.1 Definitions and Notation

A ***set*** is an unordered collection of distinct objects that are called ***elements*** or ***members*** of the set. It is essential to have a clear and rigorous definition of a set. For instance, "smart children in a town" does not form a set, as the word "smart" does not have a universally agreeable definition, and its membership is debatable, whereas "pregnant women in a town" does form a well-defined set.

It is common to use capital letters, such as A, to denote sets, and lowercase letters, such as x, to refer to set elements. If x is an element of the set A or equivalently x belongs to A, we then use the notation $x \in A$, and if x does not belong to the set A or equivalently x is not an element of A, we then write $x \notin A$. For instance, if A is the set of all capital cities, then Tokyo, denoted by x, is an element of A, that is, $x \in A$, and if B is the set of all European cities, then Tokyo, denoted by x, is not a member of B, because it is a city in Asia, we thus have $x \notin B$.

A set is generally represented by braces (curly brackets), that is, by $\{\}$. One way to specify a set with a finite number of elements is to use the ***set roster method***, by which

Discrete Mathematics
ISBN 978-0-12-820656-0, https://doi.org/10.1016/B978-0-12-820656-0.00005-8

all the elements of the set are listed between curly brackets (i.e., within braces), such as {3, 6, 9}. The order of elements presented in a set is irrelevant, and a set remains the same if its elements are repeated or rearranged. Note that a set of a very large number of elements that follow a recognizable pattern is usually described by listing the first few elements, followed by ellipses "…," which is read as "and so forth," such as {1, 2, 3, 4, …}.

Another way to specify a set is the **set builder notation**, through which some property held only by all members of the set is clearly and completely described, such as $\{x \in N \mid x$ is a multiple of 3, $0 < x < 10\}$, where the vertical line ($\mid$) is read as "such that" and the comma $(,)$ as "and," and N represents the set of all positive integers. Note that the general form $\{x \in S \mid Q(x)\}$, where $Q(x)$ is a predicate indicating the property that the object x of the set S has, is read as "the set of all x in S such that x has the property $Q(x)$."

A set usually presents a group of elements with common properties. However, it is possible for a set to contain any kind of elements whatsoever, and they are not required to be of the same type, such as the set {China, nose, baby, movie, ice cream, π, rainbow, stamp, soccer}.

A **Venn diagram** is a group of simple closed curves arranged in a plane to visually illustrate collections of sets and their logical relationships through geometric intuition so as to help understand set concepts and operations. Fig. 5.1 shows the Venn diagrams for some special sets.

The **universal set**, also known as the **universe of discourse**, denoted by U, is defined to include all elements in a given setting as well as every set under consideration. Thus the universal set varies depending on which objects are of interest. For instance, the universal set may be defined to include all the people living in the world, and the sets under consideration may include people of various nationalities or people with different eye colors. The universal set U is usually represented pictorially as the set of all points within a rectangle, as shown in Fig. 5.1a, whereas the other sets are represented by enclosed areas lying within it, where the interior of each closed curve represents a set.

Two sets A and B are equal if and only if they have exactly the same elements, as shown in Fig. 5.1b. We write $A = B$ if A and B are **equal sets**. For instance, the set $A = \{a, e, i, o, u\}$ and the set $B = \{x \in U \mid x$ is a vowel in the English alphabet, U is the set of all its letters} are equal, that is, $A = B$. If the sets A and B are not equal, then we write $A \neq B$. For instance, the sets $A = \{x \in N \mid x$ is an odd integer} and $B = \{x \in N \mid x$ is a prime number}, where the set N represents all positive integers, are not equal, that is, $A \neq B$.

The set B is a **subset** of the set A, and the set A is a **superset** of the set B, if and only if every member of B is also a member of A. We use the symbol $\subseteq$ to denote subset; $B \subseteq A$ thus implies B is a subset of A, or alternatively B is contained in A or A contains B. In order to show that B is not a subset of A, that is, $B \nsubseteq A$, it is only needed to find one

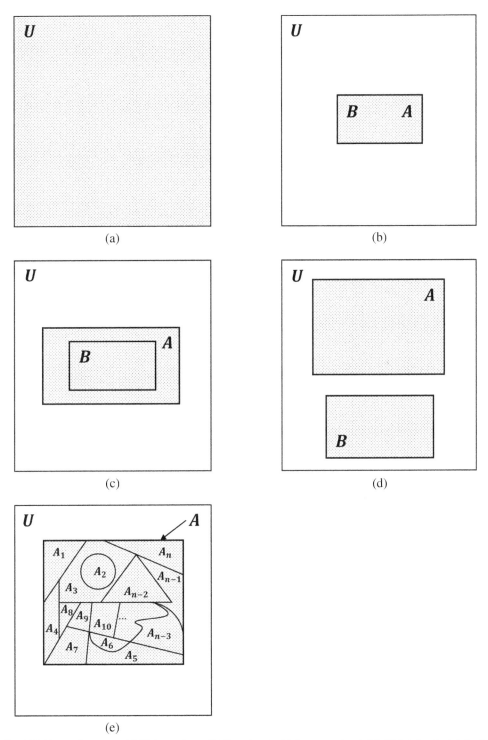

Fig. 5.1 Venn diagrams. (a) Universal set. (b) Equal sets. (c) Proper subset. (d) Disjoint sets. (e) Partitioned set.

element $x \in B$ with $x \notin A$. In other words, find a counterexample x, that is, an x that shows the assertion to be false. For instance, if A is the set of all odd numbers and B is the set of all prime numbers, we have $B \nsubseteq A$, as $2 \in B$, but $2 \notin A$.

Every set is a subset of itself and a subset of the universal set. Note that if $B \subseteq A$ and $A \subseteq C$, then $B \subseteq C$.

The set B is a ***proper subset*** of the set A if every member of B is also a member of A, but there is at least one element of A that is not an element of B. We use the symbol $\subset$ to denote proper subset; $B \subset A$ thus implies B is a proper subset of A, as shown in Fig. 5.1c. Oftentimes the terms "subset" and "proper subset" are interchangeable because it is not important to differentiate them. Fig. 5.2 shows ***special sets of numbers***, where we have $\boldsymbol{P \subset N \subset W \subset Z \subset Q \subset R \subset C}$, and Table 5.1 presents the subsets of real numbers, called ***intervals of real numbers***.

The ***empty set*** or ***null set***, denoted by $\varnothing$, is defined as the set with no elements. For instance, the set of human beings who are 200 years old is the empty set. The empty set is thus a subset of every set. The empty set is unique, that is, there is exactly one empty set. Therefore if A and B are both empty sets, then $A = B$, because they have exactly the same elements, namely, none. To prove a set A is an empty set, we first suppose A has an element x and then derive a contradiction.

A set with one element is called a ***singleton set*** or a ***unit set***. For instance, the set $\{\varnothing\}$ is a singleton set, and its only element is the empty set $\varnothing$. Therefore $\{\varnothing\}$ has one more element than $\varnothing$, so $\{\varnothing\} \neq \varnothing$. Another example of a singleton set is the set of integers that are both prime and even, that is, $\{2\}$.

The sets A and B are known as ***disjoint*** if and only if the sets A and B have no common elements, as shown in Fig. 5.1d. For instance, the set of odd numbers and the set of even numbers are disjoint. In addition, if two sets are disjoint, then neither is a subset of the other unless one is the empty set. The sets A_1, A_2, ..., A_n are ***mutually exclusive***, also known as ***pairwise disjoint***, if and only if no two sets have any element in common. For instance, the sets of birds, cars, books, and trees are all mutually disjoint.

A ***partition*** of a nonempty set A is a finite collection of n nonempty subsets, A_1, A_2, ..., A_n, that are all pairwise disjoint, and every element of the set A belongs to only one of these n mutually exclusive subsets, where $n \geq 2$ is an integer. Note that a set can be arbitrarily partitioned. The nonoverlapping subsets in a partition are called ***cells*** or ***blocks***. Fig. 5.1e shows a partition of the set A. For instance, the set of letters in an English word can be partitioned into two nonempty mutually exclusive subsets of vowels and consonants. Another example is that the population of a country can be partitioned into three nonempty mutually exclusive subsets, children who are younger than 18 years, senior citizens who are at least 65 years old, and those adults who are at least 18 years old but not older than 65 years. Note that the same population can be partitioned into males and females. In principle, the partitioning of a set is done in a way that is most beneficial to help analyze and solve the problem of interest.

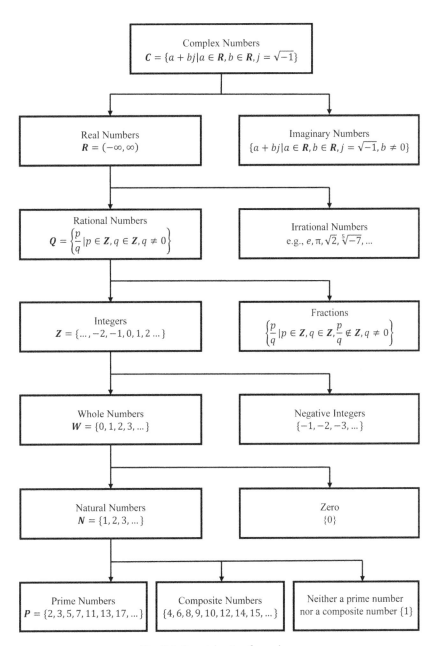

Fig. 5.2 Special sets of numbers.

TABLE 5.1 Intervals of real numbers.

$$[a, b] = \{x \in R \,|\, a \leq x \leq b\}$$
$$[a, b) = \{x \in R \,|\, a \leq x < b\}$$
$$(a, b] = \{x \in R \,|\, a < x \leq b\}$$
$$(a, b) = \{x \in R \,|\, a < x < b\}$$
$$(-\infty, a) = \{x \in R \,|\, -\infty < x < a\}$$
$$(b, \infty) = \{x \in R \,|\, b < x < \infty\}$$
$$(-\infty, a] = \{x \in R \,|\, -\infty < x \leq a\}$$
$$[b, \infty) = \{x \in R \,|\, b \leq x < \infty\}$$
$$(-\infty, \infty) = \{x \in R \,|\, -\infty < x < \infty\}$$

Note: a and b are both real numbers with $a > b$.

Some concepts from both set theory and predicate logic can be tied. A propositional function $P(x)$ defined on a set A has the property that $P(a)$ is true or false for each element a of A. In other words, $P(x)$ becomes a statement with a truth value whenever any element $a \in A$ replaces the variable x. The set A is called the domain of $P(x)$, and the set of all elements of A for which $P(x)$ is true is called the **truth set** of $P(x)$. In general, when A represents a set of numbers, the condition $P(x)$ has the form of an equation or inequality involving the variable x.

Example 5.1
Determine the truth sets of the following predicates, where the domain is the set of all positive integers N:
(a) $T(x)$ is $x^2 = 0$.
(b) $W(x)$ is $x^2 - 3x > 0$.
(c) $V(x)$ is $x^2 - 3x + 2 = 0$.

Solution
(a) The truth set of T, $\{x \in N \,|\, x^2 = 0\}$, is the set of positive integers for which $x^2 = 0$, that is, $x = 0$. As 0 is not a positive integer, the truth set of T is the empty set.
(b) The truth set of W, $\{x \in N \,|\, x^2 - 3x > 0\}$, is the set of positive integers for which $x^2 - 3x > 0$, that is, $x < 0$ or $x > 3$. As the integers less than 0 are not positive integers, the truth set of W is thus the set $\{4, 5, 6, \ldots\}$.
(c) The truth set of V, $\{x \in N \,|\, x^2 - 3x + 2 = 0\}$, is the set of positive integers for which $x^2 - 3x + 2 = 0$, that is, $x = 1$ or $x = 2$. As both 1 and 2 are positive integers, the truth set of V is the set $\{1, 2\}$.

5.2 Set Operations

As propositions can be combined to construct new propositions in various ways, sets can be combined to build a new set, which then has a certain property. There is a close relationship between logic operations and set operations. Fig. 5.3 shows the Venn diagrams for some special set operations.

The **union** of two sets A and B, denoted by $A \cup B$, is the set of all elements that are in A or in B or in both, as shown in Fig. 5.3a, that is, we have

$$A \cup B \triangleq \{x \in U \mid x \in A \text{ or } x \in B\}.$$

Here, "or" within the curly brackets is used in the sense of "and" as well as "or", thus it implies at least in one of the two sets. The **intersection** of two sets A and B, denoted by $A \cap B$, is the set of all elements that exist in both A and B, as shown in Fig. 5.3b, that is, we have

$$A \cap B \triangleq \{x \in U \mid x \in A \text{ and } x \in B\}.$$

The intersection of two disjoint sets A and B is thus the empty set, that is, $A \cap B = \emptyset$. The **difference** of sets A and B (or the **relative complement** of B with respect to A), denoted by $A - B$ or $A \backslash B$, is the set of elements in A that are not in B, as shown in Fig. 5.3c, that is, we have

$$A - B \triangleq \{x \in U \mid x \in A \text{ and } x \notin B\}.$$

Note that the set $A - B$, read as "A minus B," is different from the set $B - A$. The **absolute complement** or, simply, the **complement** of a set A, with respect to the universal set U, denoted by A^c or $\overline{A}$, is the set of all elements that are not in A, as shown in Fig. 5.3d, that is, we have

$$A^c = \overline{A} \triangleq \{x \in U \mid x \notin A\}.$$

Note that the complement of the universal set is the empty set and vice versa, the union of a set and its complement is the universal set, that is, $A \cup A^c = U$, and the intersection of a set and its complement is the empty set, that is, $A \cap A^c = \emptyset$. The **symmetric difference** of sets A and B, denoted by $A \oplus B$ or $A \Delta B$, consists of those elements that belong to A or B but not to both, as shown in Fig. 5.3e, that is, we have

$$A \oplus B = A \Delta B \triangleq \{x \in U \mid (x \in A, x \notin B) \text{ or } (x \notin A, x \in B)\}.$$

We thus have

$$A \oplus B = (A \cup B) - (A \cap B).$$

The **precedence rules**, which can reduce the number of parentheses required, must be performed in the following order: (i) operations from left to right, (ii) operations between

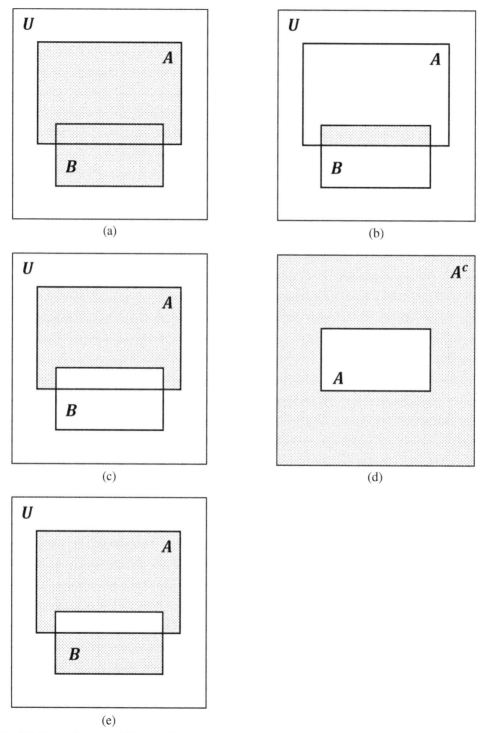

Fig. 5.3 Venn diagrams. (a) Union of two sets. (b) Intersection of two sets. (c) Difference between two sets. (d) Complement of a set. (e) Symmetric difference.

parentheses first with the innermost of nested parentheses, (iii) complementation, (iv) intersections, and (v) unions. Note that the set difference or symmetric difference must always use parentheses.

Example 5.2

Suppose the universal set U represents all possible outcomes when a typical six-sided cube-shaped die is rolled, that is, $U = \{1, 2, 3, 4, 5, 6\}$. We also define the set A representing the odd outcomes and the set B representing the outcomes that are prime. Determine the following sets: $A \cup B$, $A \cap B$, $A - B$, A^c, and $A \oplus B$.

Solution

We first identify the elements of the sets A and B from the universal set U and then perform the required set operations:

$$U = \{1,2,3,4,5,6\} \rightarrow \begin{cases} A = \{1,3,5\} \\ B = \{2,3,5\} \end{cases} \rightarrow \begin{cases} A \cup B = \{1,2,3,5\} \\ A \cap B = \{3,5\} \\ A - B = \{1\} \\ A^c = \{2,4,6\} \\ A \oplus B = \{1,2\}. \end{cases}$$

The union and intersection operations can be repeated for an arbitrary number of sets. Thus the union of n sets is the set of all elements that are in at least one of the n sets, and the intersection of n sets is the set of all elements that are shared by all n sets, where $n \geq 2$ is an integer. Note that the intersection of any n sets, B_1, B_2, ..., B_n, is a subset of each of the n sets, and in turn, each of the n sets is a subset of the union of the n sets, that is, we have:

$$\left(\bigcap_{i=1}^{n} B_i \right) \subseteq B_i \subseteq \left(\bigcup_{i=1}^{n} B_i \right) \quad i = 2, 3, ..., n.$$

Example 5.3

Suppose the universal set U represents all positive integers less than or equal to 20, that is, we have $U = \{1, 2, ..., 20\}$. We also define the set A representing the positive integers that are divisible by 3, the set B representing the positive integers that are divisible by 4, and the set C representing the positive integers that are divisible by 6. Determine the following sets: $A \cup B \cup C$, $A \cap B \cap C$, $(A \cup B) \cap C$, and $A \cup (B \cap C)$. Comment on the results.

Solution

We first identify the elements of the sets A, B, and C and then perform the required set operations:

$$U = \{1, 2, ..., 20\} \rightarrow \begin{cases} A = \{3, 6, 9, 12, 15, 18\} \\ B = \{4, 8, 12, 16, 20\} \\ C = \{6, 12, 18\} \end{cases}$$

$$\rightarrow \begin{cases} A \cup B \cup C = \{3, 4, 6, 8, 9, 12, 15, 16, 18, 20\} \\ A \cap B \cap C = \{12\} \\ (A \cup B) \cap C = \{6, 12, 18\} \\ A \cup (B \cap C) = \{3, 6, 9, 12, 15, 18\}. \end{cases}$$

When the set operations are all unions or all intersections, their order does not matter. However, when they are a mix of unions and intersections, the order of unions and intersections does matter. Hence we have $(A \cup B) \cap C \neq A \cup (B \cap C)$.

Suppose there are n distinct sets, X_1, X_2, ..., X_n. A **fundamental product** of these n sets is a set defined as $Y_1 \cap Y_2 \cap ... \cap Y_n$, where Y_i is either the set X_i or its complement, that is, X_i^c, for $i = 1, 2, ..., n$, where $n \geq 1$ is an integer. There are therefore 2^n such fundamental products, as there are two choices for each Y_i. There is a geometric description for each fundamental product. It can be shown that all 2^n fundamental products are disjoint and their union is the universal set U, that is, the universal set U is partitioned by the 2^n sets representing the fundamental products of the n sets.

Example 5.4

Consider three sets A, B, and C. Identify all fundamental products and show them in a Venn diagram.

Solution

There are three sets, the number of fundamental products is then $8 \ (= 2^3)$. The fundamental products are as follows:

$$\begin{aligned} S_1 &= A \cap B \cap C \\ S_2 &= A \cap B \cap C^c \\ S_3 &= A \cap B^c \cap C \\ S_4 &= A^c \cap B \cap C \\ S_5 &= A \cap B^c \cap C^c \\ S_6 &= A^c \cap B \cap C^c \\ S_7 &= A^c \cap B^c \cap C \\ S_8 &= A^c \cap B^c \cap C^c. \end{aligned}$$

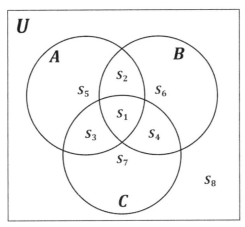

Fig. 5.4 Venn diagram: Fundamental products for Example 5.4.

These eight fundamental products correspond to eight disjoint sets S_1, S_2, ..., S_8 through which the universal set U is partitioned, as shown in Fig. 5.4.

5.3 Set Identities and Methods of Proof

A *set identity* is an equality between two set expressions that is true for all elements of the sets involved in the identity. In a set identity, some basic set operations are combined to form another set. Table 5.2 presents some important set identities. These identities in set theory are similar to the logical equivalences in logic.

It is a fact of set algebra, called the *principle of duality*, that the dual of an identity is also an identity, where the dual of an identity can be obtained by replacing each occurrence of $\cup$, $\cap$, U, and $\varnothing$ in the identity by $\cap$, $\cup$, $\varnothing$, and U, respectively. Many of the identities in Table 5.2 arranged in pairs reflect the principle of duality.

De Morgan's laws are prominent set identities that provide a pair of transformation rules. The laws can be expressed as the complement of the union of two sets is the same as the intersection of their complements, and the complement of the intersection of two sets is the same as the union of their complements. De Morgan's laws are used when the complements of sets are easier to define than the sets themselves.

To prove set identities, membership tables can be used. A table that displays the membership of elements in sets is called a *membership table*, also known as a *truth table*. The columns of a membership table must represent the original basic sets and the two sets on both sides of the set identity, where 1 is used to indicate an element that is in the set, and 0 is used to indicate an element that is not in the set. Note that there is a great similarity between membership tables in set theory and truth tables in propositional logic.

TABLE 5.2 Set identities.

Identity	Name
$A \cup B = B \cup A$ $A \cap B = B \cap A$	Commutative laws
$(A \cup B) \cup C = A \cup (B \cup C) = A \cup B \cup C$ $(A \cap B) \cap C = A \cap (B \cap C) = A \cap B \cap C$	Associative laws
$A \cup (B \cap C) = (A \cup B) \cap (A \cup C)$ $A \cap (B \cup C) = (A \cap B) \cup (A \cap C)$	Distributive laws
$A \cup \varnothing = A$ $A \cap U = A$	Identity laws
$A \cup U = U$ $A \cap \varnothing = \varnothing$	Domination laws
$A \cup A = A$ $A \cap A = A$	Idempotent laws
$(A^c)^c = A$	Complementation law
$A \cup A^c = U$ $A \cap A^c = \varnothing$	Complement laws
$A - B = A \cap B^c$	Relative complement law
$A \cup (A \cap B) = A$ $A \cap (A \cup B) = A$	Absorption laws
$(A \cup B)^c = A^c \cap B^c$ $(A \cap B)^c = A^c \cup B^c$	De Morgan's laws
$A \subseteq B$ iff $A \cup B = B$ $A \subseteq B$ iff $A \cap B = A$	Consistency laws
$|A \cup B| = |A| + |B| - |A \cap B|$	Inclusion-exclusion principle

Table 5.3 presents a membership table for the union, intersection, difference, and symmetric difference of two sets, as well as the complements of the two sets.

Insights regarding set identities can be obtained from Venn diagrams, but Venn diagrams cannot be used when proving theorems unless special attention is paid to make sure that the diagrams are sufficiently general to encompass all possible cases, and that is a difficult task. As the role of Venn diagrams is not to provide formal proofs, we need formal methods of proving set identities. Here are three distinct methods to prove a set identity:
1. Show each side of the identity is a subset of the other side. This method of proof is known as the **element argument** or **containment proof**. In other words, to prove

TABLE 5.3 Membership tables for basic set operations.

A	B	A∩B	A∪B	A^c	B^c	A − B	B − A	A⊕B
1	1	1	1	0	0	0	0	0
1	0	0	1	0	1	1	0	1
0	1	0	1	1	0	0	1	1
0	0	0	0	1	1	0	0	0

$M = K$, we need to prove $M \subseteq K$ and $K \subseteq M$. This powerful method brings insight into the proof, but in some cases, this proof method may prove to be rather complex.

2. Transform one side into the other side step by step by employing the other known set identities. This method of proof is known as the **algebraic proof**. This is usually the shortest method, provided that there are relevant set identities that can be applied to simplify the set expressions.

3. Build a membership table step by step for each side of the set identity, and show the columns corresponding to the both sides of the identity are identical. This method, known as **proof by membership table**, does not provide any insight into the proof. However, it is a straightforward method if the number of the original sets in the identity is just a few, otherwise, a computer should be used to build the membership table of interest.

Example 5.5

Prove that $(A-C) \cap (B-C) = (A \cap B) \cap C^c$, using all three methods of proof.

Solution

The first method is to show that each side of the identity is a subset of the other side. The first step is to show $(A-C) \cap (B-C) \subseteq (A \cap B) \cap C^c$. When $x \in (A-C) \cap (B-C)$, then by definition of intersection, $x \in (A-C)$ and $x \in (B-C)$. When $x \in (A-C)$, then by definition of difference, $x \in A$ and $x \notin C$, and when $x \in (B-C)$, then by definition of difference, $x \in B$ and $x \notin C$. When $x \in A$ and $x \in B$ and $x \notin C$, then by definition of complement, $x \in A$ and $x \in B$ and $x \in C^c$. Hence $x \in (A \cap B) \cap C^c$ by definition of intersection. The second step is to show $(A \cap B) \cap C^c \subseteq (A-C) \cap (B-C)$. When $x \in (A \cap B) \cap C^c$, then $x \in (A \cap B)$ and $x \in C^c$ by definition of intersection. When $x \in (A \cap B)$, then $x \in A$ and $x \in B$ by definition of intersection, and when $x \in C^c$, then $x \notin C$ by definition of complement. When $x \in A$ and $x \in B$ and $x \notin C$, then $x \in (A-C)$ and $x \in (B-C)$ by definition of difference. Hence $(A-C) \cap (B-C)$ by definition of intersection.

The second method is to apply the relevant set identities step by step to make one side equal to the other side. We thus have

$$
\begin{aligned}
(A-C) \cap (B-C) &= (A \cap C^c) \cap (B \cap C^c) \quad \text{by the difference equivalence} \\
&= (A \cap B) \cap (C^c \cap C^c) \quad \text{by the associative law} \\
&= (A \cap B) \cap C^c \quad \text{by the idempotent law.}
\end{aligned}
$$

The third method is to build a membership table. There are three basic sets A, B, and C that have been combined to create the set identity of interest. The presence of an element to any one of these three sets is denoted by 1 and its absence by 0. We thus have a membership table consisting of 8 $(= 2^3)$ rows, as shown in Table 5.4. After forming the three columns associated with the three sets A, B, and

TABLE 5.4 Membership table for Example 5.5.

A	B	C	A − C	B − C	(A −C)∩(B −C)	A∩B	C^c	(A∩B)∩C^c
1	1	1	0	0	0	1	0	0
1	1	0	1	1	1	1	1	1
1	0	1	0	0	0	0	0	0
1	0	0	1	0	0	0	1	0
0	1	1	0	0	0	0	0	0
0	1	0	0	1	0	0	1	0
0	0	1	0	0	0	0	0	0
0	0	0	0	0	0	0	1	0

C, which are the building sets of the identity, we first form the columns corresponding to the sets $(A-C)$ and $(B-C)$ to construct the column associated with the set $(A -C)\cap(B -C)$, which is the left-hand side of the identity. We then form the columns corresponding to the sets $(A -B)$ and C^c to construct the column associated with the set $(A\cap B)\cap C^c$, which is the right-hand side of the identity. We notice that because the columns for the sets $(A-C)\cap(B-C)$ and $(A\cap B)\cap C^c$ are the same, the identity is valid.

Example 5.6

Let A, B, and C be sets, where the sets A^c, B^c, and C^c are complements of A, B, and C, respectively. Using a membership table, show the following identities:
(a) $(A\cup B^c\cup C^c)^c \equiv A^c\cap B\cap C.$
(b) $(A^c\cup B\cup C)^c \equiv A\cap B^c\cap C^c.$
(c) $(A\cup B\cup C^c)^c \equiv A^c\cap B^c\cap C.$

Solution
Table 5.5 shows the membership tables for all three cases.
(a) Columns 7 and 8 are identical.
(b) Columns 9 and 10 are identical.
(c) Columns 11 and 12 are identical.

5.4 Cardinality of Sets

The number of distinct elements in a set A is called the **cardinality** of A, written as $|A|$. The cardinality of a set (i.e., the size of a set) may be finite or infinite. For instance, we have $|\varnothing| = 0$ because the empty set has no elements. A set with a finite number of elements is defined as a **finite set**, and it is thus countable. The exact number of elements in a finite set can be known, such as the set of cards in a deck of playing cards, or unknown,

TABLE 5.5 Membership table for Example 5.6.

A	A^c	B	B^c	C	C^c	$(A\cup B^c\cup C^c)^c$	$A^c\cap B\cap C$	$(A^c\cup B\cup C)^c$	$A\cap B^c\cap C^c$	$(A\cup B\cup C)^c$	$A^c\cap B^c\cap C$
1	0	1	0	1	0	0	0	0	0	0	0
1	0	1	0	0	1	0	0	0	0	0	0
1	0	0	1	1	0	0	0	0	0	0	0
1	0	0	1	0	1	0	0	1	1	0	0
0	1	1	0	1	0	1	1	0	0	0	0
0	1	1	0	0	1	0	0	0	0	0	0
0	1	0	1	1	0	0	0	0	0	0	1
0	1	0	1	0	1	0	0	0	0	1	0

such as the set of fish in the world. A set that is not finite is **infinite**, an infinite set is either countable or uncountable. In a **countably infinite set**, it is possible to list the elements of the set in a sequence indexed by positive integers, such as the set of all prime numbers. On the other hand, in an **uncountably infinite set**, it is not possible to list the elements of the set in a sequence indexed by positive integers, such as the set of all real numbers between 0 and 1.

A set can have other sets as members. The set of all subsets of a set A, which also includes the empty set $\varnothing$ and the set A itself, is called the **power set** of A and is denoted by $P(A)$. If A is a finite set, then we have

$$|P(A)| = 2^{|A|}$$

which in turn implies $|A| < |P(A)|$. Given two sets of A and B, the **Cartesian product** of A and B, denoted by $A \times B$ and read as "A cross B," is the set of all **ordered pairs** (a, b), where $a \in A$ and $b \in B$. The number of ordered pairs in the Cartesian product of A and B is equal to the product of the number of elements in the set A and the number of elements in the set B, that is, $|A \times B| = |A| \, |B|$. The Cartesian product of more than two sets can also be defined. The Cartesian product of n sets A_1, A_2, ..., A_n is the set of all **ordered n-tuples**, and symbolically is shown as follows:

$$A_1 \times A_2 \times ... \times A_n = \{(a_1, a_2, ..., a_n) \mid a_1 \in A_1, \, a_2 \in A_2, \, ..., \, a_n \in A_n\}.$$

The notation for an ordered n-tuple is a generalization of the notation for an ordered pair, and it takes both order and multiplicity into account.

A subset R of the Cartesian product $A \times B$ is called a **relation** from the set A to the set B. The elements of R are ordered pairs, where the first element belongs to A and the second to B. In general, we have $A \times B \neq B \times A$, unless $A = \varnothing$, $B = \varnothing$, or $A = B$.

Example 5.7

(a) Suppose $X = \{3, 6, 9\}$; determine the power set of set X.

(b) Suppose $A = \{1, 2, 3\}$, $B = \{a, b\}$, and $C = \{1, \#\}$; determine the Cartesian products $A \times B \times C$ and $A \times C \times B$. Comments on the results.

Solution

(a) Because we have $|X| = 3$, we have $|P(X)| = 2^3 = 8$. $P(X)$ is thus a set with the following eight subsets:

$$P(X) = \{\varnothing, \{3\}, \{6\}, \{9\}, \{3,6\}, \{3,9\}, \{6,9\}, \{3,6,9\}\}.$$

(b) We have

$$A \times B \times C = \{(1, a, 1), (1, a, \#), (1, b, 1), (1, b, \#),$$
$$(2, a, 1), (2, a, \#), (2, b, 1), (2, b, \#),$$
$$(3, a, 1), (3, a, \#), (3, b, 1), (3, b, \#)\}.$$

and

$$A \times C \times B = \{(1,1,a),\ (1,1,b),\ (1,\#,a),\ (1,\#,b),$$
$$(2,1,a),\ (2,1,b),\ (2,\#,a),\ (2,\#,b),$$
$$(3,1,a),(3,1,b),(3,\#,a),(3,\#,b)\}.$$

Each of these two Cartesian products is a set consisting of 12 ($= 3 \times 2 \times 2$) ordered triples. However, the resulting two sets are different, that is, $A \times B \times C \neq A \times C \times B$, because both order and multiplicity matter.

The cardinality of the union of two finite sets A and B can be found using the **principle of inclusion–exclusion**, that is, we have

$$|A \cup B| = |A| + |B| - |A \cap B|$$

Note that $|A| + |B|$ counts each element that is in set A but not in set B once and in set B but not in set A once, and each element that is in both sets A and B exactly twice. The number of elements that are in both A and B, that is, $|A \cap B|$, is then subtracted from $|A| + |B|$ so as to count the elements in the intersection only once. If the sets A and B are disjoint, then we have $|A \cup B| \triangleq |A| + |B|$. The principle of inclusion–exclusion can be extended to n sets $A_1, A_2, \ldots, A_n$, thus we can have

$$|A_1 \cup A_2 \cup \ldots \cup A_n| = \sum_{i=1}^{n} |A_i| - \sum_{i=1}^{n-1} \sum_{j=i+1}^{n} |A_i \cap A_j| + \sum_{i=1}^{n-2} \sum_{j=i+1}^{n-1} \sum_{k=j+1}^{n} |A_i \cap A_j \cap A_k|$$
$$- \ldots + (-1)^{n+1} |A_1 \cap A_2 \cap \ldots \cap A_n|$$

In general, for n sets, where n is a positive integer, the principle of inclusion–exclusion has a maximum of $2^n - 1$ terms. However, some of these terms may be zero because it is possible that some of the n sets are mutually exclusive.

Example 5.8
Give formulas for the number of elements in the union of three sets as well as in the union of four sets.

Solution
The formula for three sets contains 7 ($= 2^3 - 1$) different terms, and that for four sets contains 15 ($= 2^4 - 1$) different terms:

$$|A_1 \cup A_2 \cup A_3| = |A_1| + |A_2| + |A_3| - |A_1 \cap A_2| - |A_1 \cap A_3| - |A_2 \cap A_3|$$
$$+ |A_1 \cap A_2 \cap A_3|$$

and

$$|A_1 \cup A_2 \cup A_3 \cup A_4| = |A_1| + |A_2| + |A_3| + |A_4| - |A_1 \cap A_2| - |A_1 \cap A_3|$$
$$- |A_1 \cap A_4| - |A_2 \cap A_3| - |A_2 \cap A_4| - |A_3 \cap A_4|$$
$$+ |A_1 \cap A_2 \cap A_3| + |A_1 \cap A_2 \cap A_4| + |A_1 \cap A_3 \cap A_4|$$
$$+ |A_2 \cap A_3 \cap A_4| - |A_1 \cap A_2 \cap A_3 \cap A_4|$$

5.5 Computer Representation of Sets

Various set operations can be implemented using a computer. Although the members of a set have inherently no order, an order is imposed to make computing combinations of sets easy. Suppose that the universal set U is finite, whose number of elements n is not larger than the memory size of the computer being used. In computer representation of sets, the elements are represented by the bits 0 and 1, where the universal set U is an array with n bits, that is, a_1, a_2, ..., a_n, each containing a 1, and a subset A of U is an array with n bits, where the ith bit in this string is 1 if a_i belongs to A and is 0 if a_i does not belong to A. The sets are represented by arrays of bits.

Example 5.9

With the universal set U consisting of positive integers less than 9, consider the set A consisting of even integers less than 9 and the set B consisting of integers less than 9 that are divisible by 3. Determine the bit sequences representing the union, intersection, difference, and symmetric difference of the sets A and B.

Solution

We first determine the sets U, A, and B, and then obtain their computer representations as follows:

$$U = \{1, 2, 3, 4, 5, 6, 7, 8\} \rightarrow U = \{1, 1, 1, 1, 1, 1, 1, 1\}$$
$$A = \{2, 4, 6, 8\} \qquad\quad \rightarrow A = \{0, 1, 0, 1, 0, 1, 0, 1\}$$
$$B = \{3, 6\} \qquad\qquad\quad \rightarrow B = \{0, 0, 1, 0, 0, 1, 0, 0\}$$

The bit sequences representing the union, intersection, difference, and symmetric difference of sets A and B are thus as follows:

$$A \cup B = \{2, 3, 4, 6, 8\} \rightarrow A \cup B = \{0, 1, 1, 1, 0, 1, 0, 1\}$$
$$A \cap B = \{6\} \qquad\quad \rightarrow A \cap B = \{0, 0, 0, 0, 0, 1, 0, 0\}$$
$$A - B = \{2, 4, 8\} \quad\; \rightarrow A - B = \{0, 1, 0, 1, 0, 0, 0, 1\}$$
$$A \oplus B = \{2, 3, 4, 8\} \; \rightarrow A \oplus B = \{0, 1, 1, 1, 0, 0, 0, 1\}$$

TABLE 5.6 Representation of subsets by bit strings.

Subset	Bit string
$\varnothing$	0000
$\{a\}$	1000
$\{b\}$	0100
$\{c\}$	0010
$\{d\}$	0001
$\{a, b\}$	1100
$\{a, c\}$	1010
$\{a, d\}$	1001
$\{b, c\}$	0110
$\{b, d\}$	0101
$\{c, d\}$	0011
$\{a, b, c\}$	1110
$\{a, b, d\}$	1101
$\{a, c, d\}$	1011
$\{b, c, d\}$	0111
$\{a, b, c, d\}$	1111

It is important to note that there is a close relationship between sets and bit strings. For instance, Table 5.6 lists the bit sequences representing all subsets of the set $\{a, b, c, d\}$.

5.6 Multisets

As defined earlier, a set is an unordered collection of objects, where the multiplicity of objects is ignored, and the membership of an object has a binary status, that is, either an element belongs to the set or it does not. We now deviate from this general definition of a set to briefly introduce multisets, where the multiplicity of an object is explicitly significant, and later present fuzzy sets, where membership of an object is not binary but a continuum of values.

A ***multiset*** (short form for multiple-membership set), also known as a ***bag***, is an unordered collection of objects where an object can occur as a member of a set more than once, that is, repeated occurrences of objects are allowed. For instance, multisets $\{7, 8, 9\}$ and $\{9, 8, 7\}$ are the same, but multisets $\{7, 8, 9\}$ and $\{7, 8, 7, 9\}$ are different. The number of occurrences, given for each element, is called the ***multiplicity*** of the element in the multiset. A multiset corresponds to an ordinary set if the multiplicity of every element is one.

Example of multisets may include the multiset of prime factors of an integer, such as the integer 360 that has the prime factorization $360 = 2^3 \times 3^2 \times 5^1$, which gives the multiset $\{2, 2, 2, 3, 3, 5\}$. The sets of distinct letters forming the words "are," "era," "ear," and "rear" are the same, which is $\{r, a, e\}$; however, their multisets of letters forming these words are different, as the multiset of the words "are," "era," and "ear" is $\{r,\ a,\ e\}$, whereas that for the word "rear" is $\{r, r,\ a,\ e\}$.

The multiset A is a **subbag** of the multiset B, that is, $A \subseteq B$, if the number of occurrences of each element x in A is less than or equal to the number of occurrences of x in B. For instance, if $A = \{a,\ b,\ c, b\}$ and $B = \{a,\ b,\ c, a, b\}$, then A is a subbag of B, but B is not a subbag of A. Two bags A and B are equal if and only if A is a subbag of B and B is a subbag of A.

The notation $\{m_1 \cdot a_1,\ m_2 \cdot a_2,\ \dots,\ m_n \cdot a_n\}$ denotes the multiset with the element a_1 occurring m_1 times, the element a_2 occurring m_2 times, and so on. The numbers m_i, $i = 1,\ \dots,\ n$ are called the multiplicities of the elements a_i, $i = 1,\ \dots,\ n$, where elements not in a multiset are assigned 0 as their multiplicity. The **cardinality of a multiset** is determined by summing up the multiplicities of all its elements, that is, $m_1 + m_2 + \dots + m_n$. For example, in the multiset $\{c,\ a,\ n,\ a,\ d,\ i,\ a,\ n\}$, the multiplicities of the distinct members $c,\ a,\ n,\ d,$ and i are respectively 1, 3, 2, 1, and 1, and therefore the cardinality of this multiset is 8 $(= 1 + 3 + 2 + 1 + 1)$.

The **union** or **intersection** of two multisets is the multiset in which the multiplicity of an element is the maximum or the minimum of its multiplicities in those two multisets, respectively. The **difference of two multisets** is the multiset in which the multiplicity of an element is the difference between the multiplicities of the element in these two multisets, unless the difference is negative, in which case the multiplicity is 0. The **sum of two multisets** is the multiset in which the multiplicity of an element is the sum of multiplicities in those two multisets.

Example 5.10

Suppose that H and K are the multisets $\{4 \cdot a,\ 3 \cdot b,\ 2 \cdot c,\ 1 \cdot d\}$ and $\{2 \cdot a,\ 3 \cdot b,\ 4 \cdot c,\ 1 \cdot e\}$, respectively. Determine their union $(H \cup K)$, intersection $(H \cap K)$, difference $(H - K)$, and sum $(H + K)$.

Solution

We thus have the following multisets sets:

$$\begin{aligned}
H \cup K &= \{\max(4, 2) \cdot a, \max(3, 3) \cdot b, \max(2, 4) \cdot c, \max(1, 0) \cdot d, \max(0, 1) \cdot e\} \\
&= \{4 \cdot a, 3 \cdot b, 4 \cdot c, 1 \cdot d, 1 \cdot e\}. \\
H \cap K &= \{\min(4, 2) \cdot a, \min(3, 3) \cdot b, \min(2, 4) \cdot c, \min(1, 0) \cdot d, \min(0, 1) \cdot e\} \\
&= \{2 \cdot a, 3 \cdot b, 2 \cdot c\}. \\
H - K &= \{\max(4 - 2, 0) \cdot a, \max(3 - 3, 0) \cdot b, \max(2 - 4, 0) \cdot c, \max(1 - 0, 0) \cdot d, \\
&\qquad \max(0 - 1, 0) \cdot e\} \\
&= \{2 \cdot a, 1 \cdot d\}. \\
H + K &= \{(4 + 2) \cdot a, (3 + 3) \cdot b, (2 + 4) \cdot c, (1 + 0) \cdot d, (0 + 1) \cdot e\} \\
&= \{6 \cdot a, 6 \cdot b, 6 \cdot c, 1 \cdot d, 1 \cdot e\}.
\end{aligned}$$

5.7 Fuzzy Sets

In a world of many shades of gray, a black-white dichotomy is an unnecessary artificial imposition. The concept of fuzzy sets is an important and practical generalization of the notion of classical sets. For instance, if the universe of discourse consists of knowledgeable people, then in fuzzy set theory, members of a set can have varying degrees of knowledge. **Fuzzy sets**, introduced by Lotfi Zadeh, where each member of the set is defined by the degree of fuzziness, have an array of applications in modeling, control systems, linguistics, information retrieval, decision-making, and of course artificial intelligence, where information is incomplete or imprecise.

In **classical set theory,** a set A is defined in terms of its **characteristic function** $\mu_A(x)$, a mapping from the universal set U to the binary set $\{0, 1\}$, where x belongs to A if and only if $\mu_A(x) = 1$ and x does not belong to A if and only if $\mu_A(x) = 0$. In **fuzzy set theory**, a set A is defined in terms of its **membership function** $\mu_A(x)$, a mapping from the universal set U to the unit interval $[0, 1]$, where x in the fuzzy set A has a certain **degree of membership**. Therefore the fuzzy set A is denoted by listing the elements with their degrees of membership.

Classical sets are special cases of fuzzy sets, in which the membership functions of fuzzy sets only take values 0 or 1. In the context of fuzzy sets, classical sets are usually called **crisp sets**. For instance, the membership functions for fuzzy and crisp sets of tall people reflecting their degrees of tallness are shown in Fig. 5.5. The crisp set assigns a number from the binary set $\{0, 1\}$ to indicate whether a person is considered tall or not (e.g., whether the person's height is greater than or less than 180 cm), whereas the fuzzy set assigns a real number in the interval $[0, 1]$ to indicate the extent to which a person is a member of the set of tall people (e.g., the person's height ranges between 170 cm and 190 cm).

The degree of fuzziness for each member of the fuzzy set needs to be always specifically stated, noting that elements with 0 degree of membership are not listed. As an example, the fuzzy set A of healthy people consists of a, b, c, d, and e, whose degrees

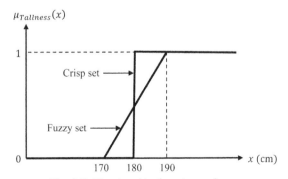

Fig. 5.5 Membership functions of sets.

of membership (i.e., degrees of healthiness) are as follows: $\mu_A(a) = 0.99$, $\mu_A(b) = 0.9$, $\mu_A(c) = 0.5$, $\mu_A(d) = 0.05$, and $\mu_A(e) = 0.001$. In turn, this points to a being the healthiest and e having the poorest health in the fuzzy set A. As another example, the fuzzy set B of wealthy people consists of a, b, c, and d, whose degrees of membership (i.e., degrees of wealthiness) are as follows: $\mu_B(a) = 0.999$, $\mu_B(b) = 0.95$, $\mu_B(c) = 0.2$, and $\mu_B(d) = 0.001$. This in turn indicates that a is the wealthiest and d is the poorest in the fuzzy set B.

The concepts of set inclusion and equality can also be extended to fuzzy sets. Assuming A and B are fuzzy sets, we have $A \subset B$, that is, A is a proper subset of B, if and only if for every element x, we have $\mu_A(x) < \mu_B(x)$, and we have $A = B$ if and only if for every element x, we have $\mu_A(x) = \mu_B(x)$.

Set operations in classical sets can be extended to fuzzy sets in terms of membership function, namely, we have

- The complement of fuzzy set A is A^c, where $\mu_{A^c}(x) = 1 - \mu_A(x)$.
- The union of fuzzy sets A and B is $A \cup B$, where $\mu_{A \cup B}(x) = \max\{\mu_A(x), \mu_B(x)\}$.
- The intersection of fuzzy sets A and B is $A \cap B$, where $\mu_{A \cap B}(x) = \min\{\mu_A(x), \mu_B(x)\}$.

Example 5.11

Suppose the fuzzy set I of three intelligent students a, b, and c has the degrees of membership $\mu_I(a) = 0.9$, $\mu_I(b) = 0.7$, and $\mu_I(c) = 0.3$, respectively, and the fuzzy set D of three diligent students a, b, and c has the degrees of membership $\mu_D(a) = 0.5$, $\mu_D(b) = 0.6$, and $\mu_D(c) = 0.8$, respectively. Determine the fuzzy sets $I \cup D$ and $I \cap D$.

Solution

We thus have the following fuzzy sets:

$$I \cup D = \{a | \mu_{I \cup D}(a) = 0.9, b | \mu_{I \cup D}(b) = 0.7, c | \mu_{I \cup D}(c) = 0.8\}$$

and

$$I \cap D = \{a | \mu_{I \cap D}(a) = 0.5, b | \mu_{I \cap D}(b) = 0.6, c | \mu_{I \cap D}(c) = 0.3\}.$$

5.8 Paradoxes in Set Theory

A *paradox* is defined as a self-contradictory statement that at first seems true. There are paradoxes in naïve set theory, where any property whatever (i.e., without restrictions) as the defining property of a set can lead to paradoxes (logical inconsistencies). We now introduce just a few well-known paradoxes in set theory.

Suppose S is the set of all sets, then every subset of S is also a member of S. The power set $P(S)$ is therefore a subset of S, that is, we have $P(S) \subseteq S$, which in turn means $|P(S)| \leq |S|$. However, we know that we always have $|S| < |P(S)|$. Thus the concept of the set of all sets leads to a contradiction, known as **Cantor's paradox**. Most sets are not members of themselves, yet a few are. For instance, the set of all countries is not a country, therefore such a set does not belong to the set of countries. On the other hand, the set of all sets each having at least one member is a set with at least one member, therefore such a set does belong to the set of all sets each having at least one member (i.e., such a set belongs to itself).

Suppose S is the set of all sets that are not members of themselves, that is, we have $\{S \mid S \text{ is a set, } S \notin S\}$. Is S a member of itself, that is, do we have $S \in S$? On the one hand, if $S \in S$, then by definition, $S \notin S$ but on the other hand, if $S \notin S$, then again by definition, $S \in S$. Therefore in either case, there is a contradiction. This paradox, known as **Russel's paradox**, shows that not every predicate defines a set, that is there is no set consisting of all sets that do not contain themselves. Russel devised a puzzle, known as the **barber puzzle**, to help explain his paradox. In a certain town, there is a male barber who shaves all those men, and only those men, who do not shave themselves. The question is, who shaves the barber? The answer is neither yes nor no.

The **liar's paradox**, also known as **Epimenides' paradox**, reveals a problem with self-reference. A person says, "I am lying." If the person is lying, then the sentence "I am lying" is false. Hence the person is telling the truth. If the person is telling the truth, then the sentence "I am lying" is true. Hence the person is lying.

It is interesting to note that there may be cases in which countably infinite sets, vis-à-vis finite sets, can address certain issues. This is best illustrated by **Hilbert's paradox**. Hilbert imagined a grand hotel that has a countably infinite number of rooms, each occupied by a guest. In a hotel with a finite number of rooms where all rooms are occupied, a new guest cannot be accommodated without evicting a current guest. However, in the grand hotel, a new guest can always be accommodated, even when all rooms are occupied. Moving the guest in room 1 to room 2, the guest in room 2 to room 3, and so forth, frees up room 1, which we assign to the new guest, and all the current guests have rooms.

Exercises

(5.1)

Using the truth table, prove the following set identities:

(a) $A \oplus B = (A \cup B) - (A \cap B)$.

(b) $A \oplus B = (A - B) \cup (B - A)$.

(c) $A - B = A \cap B^c$.

(5.2)

(a) Prove the set identity $(A \cup B) - (C - A) = A \cup (B - C)$.

(b) Write the dual of the set identity $(U \cap A) \cup (B \cap A) = A$.

(5.3)

(a) Using set identities, verify that $(X - Y) - Z = X - (Y \cup Z)$.

(b) Simplify the set expression $(A \cap B^c) \cup (A^c \cap B) \cup (A^c \cap B^c)$.

(5.4)

(a) Find the Cartesian product of sets $A = \{x\}$, $B = \{y, z\}$, and $C = \{1, 2, 3\}$

(b) Partition the set of nonnegative integers into four blocks of integers.

(5.5)

Assuming $A_i = \{.., -4, -3, -2, -1, 0, 1, 2, 3, 4, ..., i\}$, determine the following set expressions:

(a) $\bigcup_{i=1}^{n} A_i$.

(b) $\bigcap_{i=1}^{n} A_i$.

(5.6)

In a survey of 1200 people, it was found that 650 have shares of AA stock, 450 have shares of GG stock, and 420 have shares of ZZ stock. It was also found that 200 have shares in both AA and GG stocks, 250 have shares in both AA and ZZ stocks, 150 have shares in both GG and ZZ stocks, and 80 have shares in all three AA, GG, and ZZ stocks.

(a) Determine the number of people who have shares at least in one of the three stocks.

(b) Determine the number of people who have shares exactly in one stock.

(5.7)

Consider the set $A = \{a, b, c, d\}$.

(a) Determine the power set of A.

(b) Find all partitions of A.

(5.8)

Let A, B, and C be sets. Prove $(A \cup (B \cap C))^c = (C^c \cup B^c) \cap A^c$ using the following methods:

(a) Membership tables.

(b) Set identities.

(5.9)

Let $U = \{a, b, c, d, e, f, g, h\}$, $A = \{b, e, g, h\}$, and $B = \{a, b, c, e, g\}$. Using bit representations, find the following sets as eight-bit words:

(a) $A \cap B$.

(b) $A \cup B$.

(c) $A \oplus B$.

(d) A^c.

(e) $B - A$.

(5.10)

Five people, a, b, c, d, and e, are rated based on how rich they are, yielding the following fuzzy set:

$$R = \{a|\mu_R(a) = 0.9, b|\mu_R(b) = 0.7, c|\mu_R(c) = 0.5, d|\mu_R(d) = 0.3, e|\mu_R(e) = 0.05\}$$

and based on their conservatism in their political beliefs, yielding the following fuzzy set:

$$C = \{a|\mu_C(a) = 0.99, b|\mu_C(b) = 0.88, c|\mu_C(c) = 0.7, d|\mu_C(d) = 0.1, e|\mu_C(e) = 0.01\}.$$

Determine the fuzzy intersection reflecting their wealth and conservatism.

CHAPTER 6

Matrices

Contents

Matrices arise in representing discrete structures. In almost all fields, from science and engineering to social and medical sciences, it is necessary to express and use data in rectangular arrays. However, when an array is very large, which is almost always the case, analytical methods fail, and it is required to obtain numerical solutions. With the advent of computers and rapid advances in computer processing power and speed in recent decades, matrix formulation of problems as well as efficient methods of matrix manipulation have been incorporated into all widely used software packages. In this chapter, the basic aspects of matrices are introduced, and some applications of matrices will be briefly discussed.

6.1 Definitions and Special Matrices

A *matrix* is a rectangular array of numbers, and the numbers in the array are called the *entries* or *elements*. A matrix is denoted by a boldface capital letter, such as A, and its entries by a regular lowercase letter with two subscripts, such as a_{ij}. In the context of matrix algebra, a quantity described by a real number and represented by a regular lowercase letter, such as k, is referred to as a *scalar*.

Matrices vary in size, and the *size* of a matrix is specified by its number of *rows* (horizontal sets of numbers) and its number of *columns* (vertical sets of numbers). Noting that m and n are positive integers, a matrix A with m rows and n columns is called an $m \times n$ *matrix*, and thus it has a total of $m \times n$ entries. Every entry is denoted by a variable with two subscripts reflecting its row number and column number. For instance, the entry a_{ij} of the matrix A appears in the ith row and the jth column, where $i \in \{1, 2, \ldots, m\}$ and

Discrete Mathematics
ISBN 978-0-12-820656-0, https://doi.org/10.1016/B978-0-12-820656-0.00006-X

$j \in \{1, 2, \ldots, n\}$. Generally, the entries of a matrix are real numbers. A common short-hand notation to express the matrix A is to write (a_{ij}); we thus have the following:

$$A = \begin{pmatrix} a_{11} & \cdots & a_{1n} \\ \vdots & \ddots & \vdots \\ a_{m1} & \cdots & a_{mn} \end{pmatrix} \leftrightarrow A = (a_{ij}), \; i = 1, 2, \ldots, m \; \& \; j = 1, 2, \ldots, n.$$

Two matrices are said to be **equal** if they have the same size (i.e., the same number of rows and the same number of columns) and the corresponding entries in every position in the two matrices are equal. Therefore the equality of two $m \times n$ matrices is equivalent to a system of $m \times n$ equalities, one for each corresponding pair of entries.

If a matrix has only one row (i.e., $m = 1$), then it is called a **row vector**, and if a matrix has only one column (i.e., $n = 1$), then it is called a **column vector**. Vectors are generally denoted by boldface lowercase letters, such as u and v. Moreover, if $m = n = 1$, then the matrix is referred to as a scalar.

A matrix A with the same number of rows as columns (i.e., when $m = n$) is a **square matrix**. The entries $a_{11}, a_{22}, \ldots, a_{nn}$ are said to be on the **main diagonal** of the square matrix A. The sum of entries on the main diagonal is called the **trace**. For instance, the trace of the square matrix A is as follows:

$$\text{trace } A = a_{11} + a_{22} + \ldots + a_{nn}.$$

If all entries off the main diagonal of a square matrix are zero, then it is called a **diagonal matrix**. If all entries of the square matrix A below the main diagonal are zero, then A is called an **upper triangular matrix**. If all entries of the square matrix A above the main diagonal are zero, then A is called a **lower triangular matrix**. Triangular matrices arise in solving systems of linear equations.

A square matrix with 1s on the main diagonal and 0s off the main diagonal is called an **identity matrix**. An identity matrix of size n is denoted by I_n, where $n \geq 2$ is an integer. Some examples of identity matrices are as follows:

$$I_2 = \begin{pmatrix} 1 & 0 \\ 0 & 1 \end{pmatrix}, \quad I_3 = \begin{pmatrix} 1 & 0 & 0 \\ 0 & 1 & 0 \\ 0 & 0 & 1 \end{pmatrix}, \quad \ldots \quad I_n = \begin{pmatrix} 1 & 0 & \cdots & 0 & 0 \\ 0 & 1 & 0 & \vdots & 0 \\ \vdots & 0 & 1 & 0 & \vdots \\ 0 & \vdots & 0 & 1 & 0 \\ 0 & 0 & \cdots & 0 & 1 \end{pmatrix}.$$

Note that an identity matrix is a diagonal matrix. A matrix whose entries are all 0s is called a **null matrix** or a **zero matrix**. A null matrix of size $m \times n$ is denoted by $\mathbf{0}_{m \times n}$, where $m \geq 2$ and $n \geq 2$ are both integers. Some examples of null matrices are as follows:

$$\mathbf{0}_{2 \times 3} = \begin{pmatrix} 0 & 0 & 0 \\ 0 & 0 & 0 \end{pmatrix}, \ \mathbf{0}_{3 \times 3} = \begin{pmatrix} 1 & 0 & 0 \\ 0 & 1 & 0 \\ 0 & 0 & 1 \end{pmatrix}, \ \mathbf{0}_{m \times n} = \begin{pmatrix} 0 & 0 & \cdots & 0 & 0 \\ 0 & 0 & 0 & \vdots & 0 \\ \vdots & 0 & 0 & 0 & \vdots \\ 0 & \vdots & 0 & 0 & 0 \\ 0 & 0 & \cdots & 0 & 0 \end{pmatrix}.$$

If a matrix has relatively few nonzero entries, it is then called a **sparse matrix**. Sparse matrices frequently arise in solving large systems of linear equations because in many physical models, a given variable typically interacts with relatively few others. Linear systems derived from sparse matrices require less storage space and can be solved more efficiently than those derived from a **dense matrix**, a matrix where most of its entries are nonzero.

If the m rows and n columns of the matrix $\boldsymbol{A}$ are interchanged, the resulting matrix is then called the **transpose** of the matrix $\boldsymbol{A}$, denoted by $\boldsymbol{A}^{\mathrm{T}}$, which has n rows and m columns. In other words, we have

$$\boldsymbol{A} = \left(a_{ij} \right) \leftrightarrow \boldsymbol{A}^{\mathrm{T}} = \left(a_{ji} \right), i = 1, 2, \ldots, m \ \& \ j = 1, 2, \ldots, n.$$

For instance, the following two matrices $\boldsymbol{A}$ and $\boldsymbol{B}$ are the transpose of one another:

$$\boldsymbol{A} = \boldsymbol{B}^{\mathrm{T}} = \begin{pmatrix} -1 & 2 & 4 \\ -7 & 0 & 5 \end{pmatrix} \ \ \& \ \ \boldsymbol{B} = \boldsymbol{A}^{\mathrm{T}} = \begin{pmatrix} -1 & -7 \\ 2 & 0 \\ 4 & 5 \end{pmatrix}.$$

Note that the transpose of a row vector is a column vector, and vice versa. If for the square matrix $\boldsymbol{A}$, we have $\boldsymbol{A} = \boldsymbol{A}^{\mathrm{T}}$, then we have $a_{ij} = a_{ji}$. Such a matrix is called a **symmetric matrix**. For instance, the identity matrix is a symmetric matrix. Symmetric matrices are found in many applications, such as control theory, statistical analyses, and optimization, and they play important roles in many computations.

Example 6.1

Consider the matrix $\boldsymbol{A} = \begin{pmatrix} x & w \\ z & y \end{pmatrix}$. Specify the relationship among x, y, z, and w, and determine their values for each of the following cases.

(a) A is an upper triangular matrix.
(b) A is a lower triangular matrix.
(c) A is an identity matrix.
(d) A is a null matrix.
(e) A is a symmetric matrix.

Solution

(a) $z = 0$, where x, y, and w can have any values.
(b) $w = 0$, where x, y, and z can have any values.
(c) $x = y = 1$ and $z = w = 0$.
(d) $x = y = z = w = 0$.
(e) $z = w$, where x and y can have any values.

6.2 Matrix Addition and Scalar Multiplication

It is of great importance to highlight the fact that only matrices of the same size can be added. The sum of any two matrices of the same size is obtained by adding entries in the corresponding positions. **Matrix addition** is thus defined as follows:

$$A = \left(a_{ij}\right) \; \& \; B = \left(b_{ij}\right) \; \rightarrow \; A + B = \left(a_{ij} + b_{ij}\right), i = 1, 2, \dots, m$$
$$\& \, j = 1, 2, \dots, n.$$

Note that the zero matrix in the context of matrix addition plays much the same role as the number 0 plays in the arithmetic addition of real numbers. In other words, if the matrix A and the zero matrix $\mathbf{0}$ are of the same size, then their sum is A.

Scalar multiplication refers to the product of a matrix A and a scalar k, where every entry of A is multiplied by the constant k. A matrix of any size can be multiplied by a scalar. Thus we have

$$A = \left(a_{ij}\right) \; \rightarrow \; kA = \left(ka_{ij}\right), \; i = 1, 2, \dots, m \; \& \; j = 1, 2, \dots, n.$$

Note that if $k = -1$, then the matrix $-A$ is called the negative of A. If for the square matrix A, we have $A = -A^{\mathsf{T}}$, then we have $a_{ij} = -a_{ji}$. Such a matrix is called a **skew-symmetric matrix** or **antisymmetric matrix**. Note that the diagonal entries of a skew-symmetric matrix are all zero.

The difference between any two matrices of the same size is obtained by subtracting entries in the corresponding positions. **Matrix subtraction** is thus defined as follows:

$$A = \left(a_{ij}\right) \; \& \; B = \left(b_{ij}\right) \; \rightarrow \; A - B = A + (-1)B = \left(a_{ij} + (-1)b_{ij}\right), i = 1, \dots, m$$
$$\& \, j = 1, \dots, n.$$

Note that matrices of different sizes cannot be added or subtracted. Thus we can conclude that a linear combination of any number of matrices can be determined as long as all matrices are of the same size.

Example 6.2

Suppose the matrices A, B, and C are as follows:

$$A = \begin{pmatrix} x & 1 \\ 1 & y \end{pmatrix}, B = \begin{pmatrix} 1 & z \\ 4 & 2 \end{pmatrix} \text{ \& } C = 2A - 3B^T = \begin{pmatrix} 3 & t \\ w & 2 \end{pmatrix}.$$

Noting C is a symmetric matrix, determine the values of the unknown variables x, y, z, w, and t, as well as the matrices A, B, and C.

Solution

Because C is a symmetric matrix, we have $t = w$. Having B, its transpose is then

$B^T = \begin{pmatrix} 1 & 4 \\ z & 2 \end{pmatrix}$. Thus we can obtain the following:

$$C = 2A - 3B^T = 2\begin{pmatrix} x & 1 \\ 1 & y \end{pmatrix} + (-3)\begin{pmatrix} 1 & 4 \\ z & 2 \end{pmatrix}$$

$$= \begin{pmatrix} 2x & 2 \\ 2 & 2y \end{pmatrix} + \begin{pmatrix} -3 & -12 \\ -3z & -6 \end{pmatrix} = \begin{pmatrix} 2x-3 & 2-12 \\ 2-3z & 2y-6 \end{pmatrix}$$

$$= \begin{pmatrix} 3 & w \\ w & 2 \end{pmatrix}.$$

The equality of the two 2×2 matrices warrants a system of $4\ (= 2\times2)$ equations, one for each corresponding pair of entries. Thus we have

$$\begin{cases} 2x - 3 = 3 \\ 2 - 12 = w \\ 2 - 3z = w \\ 2y - 6 = 2 \end{cases} \rightarrow \begin{cases} x = 3 \\ w = -10 \\ z = 4 \\ y = 4 \end{cases} \rightarrow$$

$$A = \begin{pmatrix} 3 & 1 \\ 1 & 4 \end{pmatrix}, B = \begin{pmatrix} 1 & 4 \\ 4 & 2 \end{pmatrix} \text{ \& } C = \begin{pmatrix} 3 & -10 \\ -10 & 2 \end{pmatrix}.$$

6.3 Matrix Multiplication

It is important to note that the product of two matrices is defined only when the number of columns in the first matrix and the number of rows in the second matrix are the same. If the matrix A is an $m \times n$ matrix and the matrix B is an $n \times r$ matrix, then the product of A and B, i.e., AB, is the $m \times r$ matrix whose entry in the ith row and the kth column is the sum of the product of the corresponding entries from the ith row of A and the kth column of B. **Matrix multiplication** is defined as follows:

$$A = (a_{ij}) \ \& \ B = (b_{jk}) \ \to \ C = AB = (c_{ik}) = \left(\sum_{j=1}^{n} a_{ij} b_{jk} \right), i = 1, \ldots, m$$

$\& \ k = 1, \ldots, r.$

A convenient way to determine whether a product of two matrices is defined is to write down the size of the first matrix, i.e., $m \times n$, and to the right of it, write down the size of the second matrix, i.e., $n \times r$. If the inside integers are the same, then the product is defined. The outside integers, i.e., $m \times r$, then give the size of the product. In other words, we have

$$A_{m \times n} \times B_{n \times r} = C_{m \times r}.$$

Note that the identity matrix in the context of matrix multiplication plays much the same role as the number 1 plays in the arithmetic multiplication of real numbers. In other words, if the matrix A and the identity matrix I can be multiplied, then their product is A.

In general, the commutative law for multiplication does not hold in matrix algebra. In other words, AB and BA need not be equal. One reason can be that AB is defined, but BA is undefined, say A is a 3×3 matrix and B is a 3×2 matrix. Another reason can be that AB and BA are both defined but have different sizes, say A is a 2×3 matrix and B is a 3×2 matrix. Finally, AB and BA are both defined and have the same size (A and B are both square matrices), but the end results are different, that is, $AB \neq BA$.

Example 6.3

Consider the row vector $A = (1 \quad 3 \quad 2)$ and the row vector $B^{T} = (3 \quad 2 \quad 4)$. Determine AB and BA.

Solution

We have the following matrix products:

$$AB = (1 \quad 3 \quad 2) \begin{pmatrix} 3 \\ 2 \\ 4 \end{pmatrix} = 17.$$

and

$$BA = \begin{pmatrix} 3 \\ 2 \\ 4 \end{pmatrix} (1 \quad 3 \quad 2) = \begin{pmatrix} 3 & 9 & 6 \\ 2 & 6 & 4 \\ 4 & 12 & 8 \end{pmatrix}.$$

Both AB and BA are defined, yet they have different sizes.

Example 6.4

Consider the following four matrices:

$$A = \begin{pmatrix} 0 & 2 \\ 0 & 4 \end{pmatrix}, B = \begin{pmatrix} 2 & 2 \\ 6 & 8 \end{pmatrix}, C = \begin{pmatrix} 4 & 10 \\ 6 & 8 \end{pmatrix}, \& D = \begin{pmatrix} 9 & 21 \\ 0 & 0 \end{pmatrix}.$$

Determine AB, BA, AC, and AD. Comment on the results.

Solution

We obtain the following matrix products:

$$AB = \begin{pmatrix} 12 & 16 \\ 24 & 32 \end{pmatrix}, BA = \begin{pmatrix} 0 & 12 \\ 0 & 44 \end{pmatrix}, AC = \begin{pmatrix} 12 & 16 \\ 24 & 32 \end{pmatrix} \& AD = \begin{pmatrix} 0 & 0 \\ 0 & 0 \end{pmatrix}.$$

Both AB and BA are defined, but they are not equal. We have $AB = AC$ and $A \neq 0$; nevertheless, we cannot employ the cancellation law that exists in the arithmetic of real numbers. More specifically, it is incorrect to cancel A from both sides of $AB = AC$ and write $B = C$ because we know B and C are not equal. Also, we have $AD = 0$, but neither A nor D is a zero matrix; this is another rule of the arithmetic of real numbers that does not hold for matrix algebra.

Multiplication by diagonal matrices has the effect of scaling the rows or columns of a matrix. More specifically, premultiplication by a diagonal matrix scales the rows, and postmultiplication by a diagonal matrix scales the columns.

Example 6.5

Consider the following two matrices:

$$A = \begin{pmatrix} 4 & 0 \\ 0 & 7 \end{pmatrix} \& B = \begin{pmatrix} 1 & 2 \\ 3 & 4 \end{pmatrix}.$$

Determine AB and BA. Comment on the results.

Solution

We have the following matrix products:

$$AB = \begin{pmatrix} 4 & 8 \\ 21 & 28 \end{pmatrix} \quad \& \quad BA = \begin{pmatrix} 4 & 14 \\ 12 & 28 \end{pmatrix}.$$

Note that the nonzero entries of the diagonal matrix A are 4 and 7. By comparing AB to B, we realize that the first row of B has been multiplied by 4 ($= a_{11}$) and its second row by 7 ($= a_{22}$) to get AB. By comparing BA to B, we realize that the first column of B has been multiplied by 4 ($= a_{11}$) and its second column by 7 ($= a_{22}$) to get BA.

Powers of square matrices can be defined, as matrix multiplication is associative. Assuming A is an $n \times n$ matrix, we can then define A^r, meaning the matrix A is multiplied by itself r times, where r is a positive integer. For instance, for $r = 2$, we have $A^2 = AA$, and for $r = 1$, we have $A^1 = A$. Note that $A^0 \triangleq I_n$.

6.4 Matrix Inversion

A square matrix A is said to be ***invertible*** if there exists a square matrix B such that

$$AB = BA = I \rightarrow B = A^{-1} \quad \& \quad A = B^{-1},$$

where the matrix B is called the ***inverse*** of the matrix A and denoted by A^{-1}. If B is the inverse of A, then A is the inverse of B. A square matrix that is not invertible is called ***singular***. Note that if a matrix is not square, then it has no inverse. In addition, a product of invertible matrices is always invertible, and the inverse of the product is the product of the inverses in the reverse order, that is, we have $(AB)^{-1} = B^{-1}A^{-1}$. The inverse of the matrix A plays much the same role in matrix algebra that the reciprocal of a number plays in the arithmetic of real numbers.

An effective method to find the inverse of a matrix is to employ the elementary row operations. The ***elementary row operations*** consist of the following suboperations:

- Interchange two rows.
- Multiply all entries in a row by a nonzero number.
- Add a multiple of a row to another row.

To find the inverse of the matrix A of size n using the elementary row operations, we take the following steps:

1. Form the $n \times 2n$ matrix (A, I), that is, the matrix A is in the left half of it and the identity matrix I is in its right half.

2. Play around with the rows by row switching, row multiplication, and row addition, in no particular order, to change (A, I) step by step to form (I, B), where the identity matrix I has then replaced A in the left half. It is of great importance that there is no unique set of steps to make (A, I) into (I, B), as the process highly depends on the matrix A. It is imperative to note that if the process generates a zero row in the left half, then A has no inverse.

3. Set $A^{-1} = B$, where B is in the right half of the resulting matrix.

Example 6.6

Consider the matrix $A = \begin{pmatrix} 1 & 1 & 2 \\ 2 & 1 & 0 \\ 1 & 2 & 2 \end{pmatrix}$. Determine A^{-1}.

Solution

We first form the matrix (A, I) and then use the elementary row operations to create the matrix (I, B) starting with the leftmost column of (A, I), noting that $R1$, $R2$, and $R3$ refer to row 1, row 2, and row 3, respectively. Thus we have

$$(A, I) = \begin{pmatrix} 1 & 1 & 2 & 1 & 0 & 0 \\ 2 & 1 & 0 & 0 & 1 & 0 \\ 1 & 2 & 2 & 0 & 0 & 1 \end{pmatrix}.$$

$$\begin{array}{l} R1 = R1 \\ \text{Use } R2 = R2 - 2 \times R1 \text{ to obtain} \\ R3 = R3 - R1 \end{array} \begin{pmatrix} 1 & 1 & 2 & 1 & 0 & 0 \\ 0 & -1 & -4 & -2 & 1 & 0 \\ 0 & 1 & 0 & -1 & 0 & 1 \end{pmatrix}.$$

$$\begin{array}{l} R1 = R1 + R2 \\ \text{Use } R2 = -R2 \quad \text{to obtain} \\ R3 = R3 + R2 \end{array} \begin{pmatrix} 1 & 0 & -2 & -1 & 1 & 0 \\ 0 & 1 & 4 & 2 & -1 & 0 \\ 0 & 0 & -4 & -3 & 1 & 1 \end{pmatrix}.$$

$$\begin{array}{l} R1 = R1 \\ \text{Use } R2 = R2 \quad \text{to obtain} \\ R3 = -0.25 \times R3 \end{array} \begin{pmatrix} 1 & 0 & -2 & -1 & 1 & 0 \\ 0 & 1 & 4 & 2 & -1 & 0 \\ 0 & 0 & 1 & 0.75 & -0.25 & -0.25 \end{pmatrix}.$$

$$R1 = R1 + 2 \times R3$$

Use $R2 = R2 - 4 \times R3$ to obtain

$$R3 = R3$$

$$\begin{pmatrix} 1 & 0 & 0 & 0.5 & 0.5 & -0.5 \\ 0 & 1 & 0 & -1 & 0 & 1 \\ 0 & 0 & 1 & 0.75 & -0.25 & -0.25 \end{pmatrix} = (I, B).$$

Thus we have

$$B = A^{-1} = \begin{pmatrix} 0.5 & 0.5 & -0.5 \\ -1 & 0 & 1 \\ 0.75 & -0.25 & -0.25 \end{pmatrix} = \frac{1}{4} \begin{pmatrix} 2 & 2 & -2 \\ -4 & 0 & 4 \\ 3 & -1 & -1 \end{pmatrix}.$$

Example 6.7

Consider the matrix $C = \begin{pmatrix} 1 & 3 & -4 \\ 1 & 5 & -1 \\ 3 & 13 & -6 \end{pmatrix}$. Determine C^{-1}.

Solution

We first form the matrix (C, I) and then use the elementary row operations to create the matrix (I, D) starting with the leftmost column of (C, I), noting that $R1$, $R2$, and $R3$ refer to row 1, row 2, and row 3, respectively. Thus we have

$$(C, I) = \begin{pmatrix} 1 & 3 & -4 & 1 & 0 & 0 \\ 1 & 5 & -1 & 0 & 1 & 0 \\ 3 & 13 & -6 & 0 & 0 & 1 \end{pmatrix}.$$

$$\begin{matrix} R1 = R1 \\ \text{Use } R2 = R2 - R1 \\ R3 = R3 - 3 \times R1 \end{matrix} \quad \text{to obtain} \quad \begin{pmatrix} 1 & 3 & -4 & 1 & 0 & 0 \\ 0 & 2 & 3 & -1 & 1 & 0 \\ 0 & 4 & 6 & -3 & 0 & 1 \end{pmatrix}.$$

$$\begin{matrix} R1 = R1 \\ \text{Use } R2 = R2 \\ R3 = R3 - 2 \times R2 \end{matrix} \quad \text{to obtain} \quad \begin{pmatrix} 1 & 3 & -4 & 1 & 0 & 0 \\ 0 & 2 & 3 & -1 & 1 & 0 \\ 0 & 0 & 0 & -1 & -2 & 1 \end{pmatrix}.$$

A zero row in its left half of the matrix indicates that C has no inverse.

If the inverse of a real square matrix A is equal to its transpose, that is, $A^{-1} = A^{\mathrm{T}}$, then the matrix is called an **orthogonal matrix**.

Example 6.8

Show the matrix $A = \frac{1}{9} \begin{pmatrix} 4 & 8 & 1 \\ 7 & -4 & 4 \\ -4 & 1 & 8 \end{pmatrix}$ is orthogonal.

Solution

Using the elementary row operations, the inverse of the matrix A is as follows:

$$A^{-1} = \frac{1}{9} \begin{pmatrix} 4 & 7 & -4 \\ 8 & -4 & 1 \\ 1 & 4 & 8 \end{pmatrix}.$$

Noting that $A^{-1} = A^{\mathrm{T}}$, A is an orthogonal matrix.

Table 6.1 presents a list of matrix properties, where a, b, and c are some scalars, n is a positive integer, and A, B, and C as well as I and 0 have sizes that allow the matrix operations to be performed.

TABLE 6.1 Matrix identities.

$$A + B = B + A$$
$$A + (B+C) = (A+B) + C$$
$$A(BC) = (AB)C$$
$$A(B\pm C) = AB \pm AC$$
$$(B\pm C)A = BA \pm CA$$
$$a(B\pm C) = aB \pm aC$$
$$(a\pm b)C = aC \pm bC$$
$$(ab)C = a(bC)$$
$$a(BC) = (aB)C = B(aC)$$
$$A + 0 = 0 + A = A$$
$$A - A = 0$$
$$0 - A = -A$$
$$A0 = 0A = 0$$
$$IA = A$$
$$AI = A$$
$$(A^{\mathrm{T}})^{\mathrm{T}} = A$$
$$(A + B)^{\mathrm{T}} = A^{\mathrm{T}} + B^{\mathrm{T}}$$
$$(aA)^{\mathrm{T}} = aA^{\mathrm{T}}$$
$$(AB)^{\mathrm{T}} = B^{\mathrm{T}}A^{\mathrm{T}}$$
$$AA^{-1} = A^{-1}A = I$$
$$(AB)^{-1} = B^{-1}A^{-1}$$
$$(A^{-1})^{-1} = A$$
$$(A^{n})^{-1} = (A^{-1})^{n}$$

6.5 Zero-One Matrix

Assuming a and b are binary digits, also known as bits (0 or 1), the **Boolean operations** $\vee$ and $\wedge$ are defined as follows:

$$a \vee b = \begin{cases} 0 & \text{if } a = b = 0 \\ 1 & \text{otherwise} \end{cases}$$

and

$$a \wedge b = \begin{cases} 1 & \text{if } a = b = 1 \\ 0 & \text{otherwise} \end{cases}$$

A matrix whose entries are either 0 or 1 and subject to the Boolean operations is called a **zero-one matrix**, **Boolean matrix**, or **logical matrix**. Let A and B be zero-one matrices of the same size. The **join** of A and B, denoted by $A \vee B$, and the **meet** of A and B, denoted by $A \wedge B$, are defined respectively as follows:

$$A \vee B = \left(a_{ij} \vee b_{ij} \right)$$

and

$$A \wedge B = \left(a_{ij} \wedge b_{ij} \right).$$

Let $A = \left(a_{ij} \right)$ be an $m \times n$ zero-one matrix and $B = \left(b_{ij} \right)$ be an $n \times r$ zero-one matrix. The **Boolean product** of A and B, denoted by $A \odot B$, is the $m \times r$ zero-one matrix $C = \left(c_{ij} \right)$, where we have the following:

$$A \odot B = C \rightarrow \left(c_{ik} \right) = \left(\left(a_{i1} \wedge b_{1k} \right) \vee \ldots \vee \left(a_{in} \wedge b_{nk} \right) \right), \quad i = 1, \ldots, m \; \& \; k = 1, \ldots, r.$$

The Boolean product is obtained in the same fashion as the ordinary product of matrices where addition and multiplication are replaced with the operations $\vee$ and $\wedge$ respectively, noting that the Boolean product can be obtained by finding the usual product of matrices and then replacing any nonzero integer by 1.

Example 6.9

Suppose we have the following zero-one matrix:

$$A = \begin{pmatrix} 1 & 1 & 0 \\ 0 & 1 & 1 \\ 1 & 0 & 0 \end{pmatrix}.$$

Determine the Boolean power of A^r, where $r \geq 2$ is an integer.

Solution

We find A^r, $r = 2, 3, \ldots$, until we see a pattern.

$$A^2 = \begin{pmatrix} 1 & 1 & 0 \\ 0 & 1 & 1 \\ 1 & 0 & 0 \end{pmatrix} \begin{pmatrix} 1 & 1 & 0 \\ 0 & 1 & 1 \\ 1 & 0 & 0 \end{pmatrix} = \begin{pmatrix} 1 & 1 & 1 \\ 1 & 1 & 1 \\ 1 & 1 & 0 \end{pmatrix}$$

$$A^3 = AA^2 = \begin{pmatrix} 1 & 1 & 0 \\ 0 & 1 & 1 \\ 1 & 0 & 0 \end{pmatrix} \begin{pmatrix} 1 & 1 & 1 \\ 1 & 1 & 1 \\ 1 & 1 & 0 \end{pmatrix} = \begin{pmatrix} 1 & 1 & 1 \\ 1 & 1 & 1 \\ 1 & 1 & 1 \end{pmatrix}$$

and

$$A^4 = AA^3 = \begin{pmatrix} 1 & 1 & 0 \\ 0 & 1 & 1 \\ 1 & 0 & 0 \end{pmatrix} \begin{pmatrix} 1 & 1 & 1 \\ 1 & 1 & 1 \\ 1 & 1 & 1 \end{pmatrix} = \begin{pmatrix} 1 & 1 & 1 \\ 1 & 1 & 1 \\ 1 & 1 & 1 \end{pmatrix}.$$

Because $A^3 = A^4$, we can conclude that $A^r = A^3$ for every integer $r \geq 3$.

6.6 Applications of Matrices

There are applications of matrices in multitudes of fields, as reflected in Table 6.2. However, the brief focus of this section is only on three applications, namely, solving systems of linear equations, the best fitting of a linear function, and linear transformations.

TABLE 6.2 Applications of matrices.

Systems of linear equations
Mesh and node analysis in circuits
Markov chains in stochastic processes
Multivariate normal distribution
Leontief models in economics
Optimal strategy in game theory
Inheritance of traits in genetics
Simplex method in linear programming
Least squares fitting of a straight line
Cryptography
Geometrical optics
Computer graphics
Equilibrium of rigid bodies
Finite methods
Network modeling
Channel coding in digital systems
Linear transformations

One of the important applications of matrices is to help solve **systems of linear equations** efficiently. Matrices can be used to compactly write and work with n linear equations with n unknowns, as follows:

$$\begin{cases} a_{11}x_1 + a_{12}x_2 + \ldots + a_{1n}x_n = b_1 \\ a_{21}x_1 + a_{22}x_2 + \ldots + a_{2n}x_n = b_2 \\ \quad\quad\quad \vdots \quad = \quad \vdots \\ a_{n1}x_1 + a_{n2}x_2 + \ldots + a_{nn}x_n = b_n \end{cases} \rightarrow$$

$$\begin{pmatrix} a_{11} & a_{12} & \ldots & a_{1n} \\ a_{21} & a_{22} & \ldots & a_{2n} \\ \vdots & \vdots & \vdots & \vdots \\ a_{n1} & a_{n2} & \ldots & a_{nn} \end{pmatrix} \begin{pmatrix} x_1 \\ x_2 \\ \vdots \\ x_n \end{pmatrix} = \begin{pmatrix} b_1 \\ b_2 \\ \vdots \\ b_n \end{pmatrix} \rightarrow Ax = b,$$

where A is the matrix of coefficients, x is the vector of unknowns, and b is the vector of constants. If $b = 0$, the system is called **homogeneous**, and if $b \neq 0$, the system is called **nonhomogeneous**.

When a system of linear equation has at least one solution, it is said to be **consistent**; otherwise, it is called **inconsistent**. Table 6.3 provides all possible cases arising in finding a solution to a system of linear equations.

TABLE 6.3 Possible solutions to a system of linear equations.

	$b = 0$	$b \neq 0$
A^{-1} exists	The trivial solution: $x = 0$	A unique solution: $x = A^{-1}b$, $x \neq 0$
A^{-1} does not exist	Infinitely many solutions	Either no solution or infinitely many solutions

---

Example 6.10

Solve the following systems of linear equations:

(a) $\begin{cases} 2x_1 - x_2 = 0 \\ 3x_1 + x_2 = 5 \end{cases}$

(b) $\begin{cases} 2x_1 - x_2 = 0 \\ 3x_1 + x_2 = 0 \end{cases}$

(c) $\begin{cases} 2x_1 - x_2 = 0 \\ 4x_1 - 2x_2 = 5 \end{cases}$

(d) $\begin{cases} 2x_1 - x_2 = 2.5 \\ 4x_1 - 2x_2 = 5 \end{cases}$

(e) $\begin{cases} 2x_1 - x_2 = 0 \\ 4x_1 - 2x_2 = 0 \end{cases}$

Solution

(a) $\begin{cases} 2x_1 - x_2 = 0 \\ 3x_1 + x_2 = 5 \end{cases} \rightarrow b = \begin{pmatrix} 0 \\ 5 \end{pmatrix} \neq 0 \quad \& \quad A = \begin{pmatrix} 2 & -1 \\ 3 & 1 \end{pmatrix} \rightarrow$

$A^{-1} = \begin{pmatrix} 0.2 & 0.2 \\ -0.6 & 0.4 \end{pmatrix} \rightarrow x = \begin{pmatrix} 0.2 & 0.2 \\ -0.6 & 0.4 \end{pmatrix}\begin{pmatrix} 0 \\ 5 \end{pmatrix} = \begin{pmatrix} 1 \\ 2 \end{pmatrix} \rightarrow$

A unique solution $(x_1 = 1, x_2 = 2)$.

(b) $\begin{cases} 2x_1 - x_2 = 0 \\ 3x_1 + x_2 = 0 \end{cases} \rightarrow b = \begin{pmatrix} 0 \\ 0 \end{pmatrix} = 0 \quad \& \quad A = \begin{pmatrix} 2 & -1 \\ 3 & 1 \end{pmatrix} \rightarrow$

$A^{-1} = \begin{pmatrix} 0.2 & 0.2 \\ -0.6 & 0.4 \end{pmatrix} \rightarrow x = \begin{pmatrix} 0.2 & 0.2 \\ -0.6 & 0.4 \end{pmatrix}\begin{pmatrix} 0 \\ 0 \end{pmatrix} = \begin{pmatrix} 0 \\ 0 \end{pmatrix} \rightarrow$

The trivial solution $(x_1 = x_2 = 0)$.

(c) $\begin{cases} 2x_1 - x_2 = 0 \\ 4x_1 - 2x_2 = 5 \end{cases} \rightarrow b = \begin{pmatrix} 0 \\ 5 \end{pmatrix} \neq 0 \quad \& \quad A = \begin{pmatrix} 2 & -1 \\ 4 & -2 \end{pmatrix} \rightarrow$

A^{-1} does not exist $\rightarrow \begin{cases} 4x_1 - 2x_2 = 0 \\ 4x_1 - 2x_2 = 5 \end{cases} \rightarrow$

Inconsistent equations $\rightarrow$ No solution

(d) $\begin{cases} 2x_1 - x_2 = 2.5 \\ 4x_1 - 2x_2 = 5 \end{cases} \rightarrow b = \begin{pmatrix} 2.5 \\ 5 \end{pmatrix} \neq 0 \ \& \ A = \begin{pmatrix} 2 & -1 \\ 4 & -2 \end{pmatrix} \rightarrow$

A^{-1} does not exist $\rightarrow \begin{cases} 4x_1 - 2x_2 = 5 \\ 4x_1 - 2x_2 = 5 \end{cases} \rightarrow$

Only one equation (# of equations $<$ # of unknowns) $\rightarrow$
Infinitely many solutions $(x_1 = t \ \& \ x_2 = 2t - 2.5$, the trivial solution not included).

(e) $\begin{cases} 2x_1 - x_2 = 0 \\ 4x_1 - 2x_2 = 0 \end{cases} \rightarrow b = \begin{pmatrix} 0 \\ 0 \end{pmatrix} = 0 \ \& \ A = \begin{pmatrix} 2 & -1 \\ 4 & -2 \end{pmatrix} \rightarrow$

A^{-1} does not exist $\rightarrow \begin{cases} 4x_1 - 2x_2 = 0 \\ 4x_1 - 2x_2 = 0 \end{cases} \rightarrow$

Only one equation (# of equations $<$ # of unknowns) $\rightarrow$
Infinitely many solutions $(x_1 = t \ \& \ x_2 = 2t \ \forall \, t$, including the trivial solution).

Another important application of matrices lies in the **best fitting of a polynomial function** to a set of points in the plane. Our brief focus here is the least squares fitting of a straight line to bivariate data, which is known as **linear regression**. It is commonly available in statistical software packages.

Given $n > 1$ points $(x_1, y_1), (x_2, y_2), \ldots, (x_n, y_n)$, the least squares straight line fit is $y = mx + b$, where the unknown constants m and b can be found as follows:

$$X = \begin{pmatrix} 1 & x_1 \\ 1 & x_2 \\ \vdots & \vdots \\ 1 & x_n \end{pmatrix} \quad \& \quad y = \begin{pmatrix} y_1 \\ y_2 \\ \vdots \\ y_n \end{pmatrix} \rightarrow v = \left(X^T X\right)^{-1} X^T y = \begin{pmatrix} b \\ m \end{pmatrix}.$$

Example 6.11
Find the least squares straight line fit to the four points $(0, 1)$, $(1, 3)$, $(2, 4)$, and $(3, 4)$.

Solution
We have

$$X = \begin{pmatrix} 1 & 0 \\ 1 & 1 \\ 1 & 2 \\ 1 & 3 \end{pmatrix} \rightarrow X^T = \begin{pmatrix} 1 & 1 & 1 & 1 \\ 0 & 1 & 2 & 3 \end{pmatrix} \rightarrow$$

$$X^T X = \begin{pmatrix} 1 & 1 & 1 & 1 \\ 0 & 1 & 2 & 3 \end{pmatrix} \begin{pmatrix} 1 & 0 \\ 1 & 1 \\ 1 & 2 \\ 1 & 3 \end{pmatrix} = \begin{pmatrix} 4 & 6 \\ 6 & 14 \end{pmatrix} \rightarrow$$

$$\left(X^T X\right)^{-1} = \begin{pmatrix} 0.7 & -0.3 \\ -0.3 & 0.2 \end{pmatrix} \quad \& \quad y = \begin{pmatrix} 1 \\ 3 \\ 4 \\ 4 \end{pmatrix}.$$

We can now find the unknowns m and b as follows:

$$v = \begin{pmatrix} 0.7 & -0.3 \\ -0.3 & 0.2 \end{pmatrix} \begin{pmatrix} 1 & 1 & 1 & 1 \\ 0 & 1 & 2 & 3 \end{pmatrix} \begin{pmatrix} 1 \\ 3 \\ 4 \\ 4 \end{pmatrix} = \begin{pmatrix} 1.5 \\ 1 \end{pmatrix} = \begin{pmatrix} b \\ m \end{pmatrix} \rightarrow y = x + 1.5.$$

Linear transformations, also known as ***linear maps***, can be performed by using matrices. A real $m \times n$ transformation matrix A gives rise to a linear transformation $R^n \rightarrow R^m$, mapping each column vector x in R^n to the matrix product Ax, which is a column vector in R^m. In a two-dimensional system, linear transformations can be represented using a 2×2 transformation matrix. The most common geometric transformations in R^2 that keep the origin fixed are linear, and they are as follows:

- ***Rotation*** (by an angle about the origin).
- ***Scaling*** (stretching or compression along the x-axis or the y-axis).
- ***Shearing*** (a transformation in which all points along a given line remain fixed, whereas other points are shifted parallel the line by a distance proportional to their perpendicular distance from the line).
- ***Reflection*** (with respect to the x-axis or the y-axis).
- ***Squeezing*** (stretching in one axis and compression in the other axis).

Example 6.12

Suppose we have a unit square whose vertices are $(0, 0)$, $(1, 0)$, $(0, 1)$, and $(1, 1)$. Suppose we use the following matrix A to transform this square to a quadrilateral shape.

$$A = \begin{pmatrix} a & b \\ c & d \end{pmatrix}.$$

Determine the type of the quadrilateral shape. What should be the relationship among a, b, c, and d that the quadrilateral shape becomes a rhombus whose every side is 1?

Solution

The vertices of the quadrilateral shape are as follows:

$$
\left\{
\begin{aligned}
\begin{pmatrix} x_1 \\ y_1 \end{pmatrix} &= \begin{pmatrix} a & b \\ c & d \end{pmatrix} \begin{pmatrix} 0 \\ 0 \end{pmatrix} = \begin{pmatrix} 0 \\ 0 \end{pmatrix} \\
\begin{pmatrix} x_2 \\ y_2 \end{pmatrix} &= \begin{pmatrix} a & b \\ c & d \end{pmatrix} \begin{pmatrix} 1 \\ 0 \end{pmatrix} = \begin{pmatrix} a \\ c \end{pmatrix} \\
\begin{pmatrix} x_3 \\ y_3 \end{pmatrix} &= \begin{pmatrix} a & b \\ c & d \end{pmatrix} \begin{pmatrix} 0 \\ 1 \end{pmatrix} = \begin{pmatrix} b \\ d \end{pmatrix} \\
\begin{pmatrix} x_4 \\ y_4 \end{pmatrix} &= \begin{pmatrix} a & b \\ c & d \end{pmatrix} \begin{pmatrix} 1 \\ 1 \end{pmatrix} = \begin{pmatrix} a+b \\ c+d \end{pmatrix}
\end{aligned}
\right.
$$

The line connecting $(0, 0)$ and (a, c) and the line connecting (b, d) and $(a+b, c+d)$ have the same slope of $\frac{c}{a}$, that is, they are parallel. Also, the line connecting $(0, 0)$ and (b, d) and the line connecting (a, c) and $(a+b, c+d)$ have the same slope of $\frac{d}{b}$, that is, they are parallel. Thus the quadrilateral shape is a parallelogram. To be a rhombus, two adjacent sides of a parallelogram must be equal, so we should have $\sqrt{a^2 + c^2} = \sqrt{b^2 + d^2} = 1$.

Exercises

(6.1)

Let A and B be two matrices given by

$$
A = \begin{bmatrix} x+y & 6 \\ 2x-3 & 2-y \end{bmatrix} \quad \& \quad B = \begin{bmatrix} 5 & 5x+2 \\ y & x-y \end{bmatrix}.
$$

Determine if there are values of x and y so that A and B are equal.

(6.2)

Let A and B be two matrices given by

$$A = \begin{bmatrix} 3 & 1 & 5 \\ -2 & 0 & 6 \end{bmatrix} \quad \& \quad B = \begin{bmatrix} 4 & 1 & 0 \\ 8 & 1 & -3 \end{bmatrix}.$$

Determine $C = 3A - 2B$.

(6.3)

Let A and B be two matrices given by

$$A = \begin{bmatrix} x & y \\ 2 & w \end{bmatrix} \quad \& \quad B = \begin{bmatrix} 1 & 3 \\ 2 & 4 \end{bmatrix}.$$

Determine the values of x, y, z, and w so that we have $AB = BA$.

(6.4)

Find the inverse of the following matrix:

$$A = \begin{bmatrix} 4 & 2 \\ 3 & 1 \end{bmatrix}.$$

(6.5)

Suppose we have

$$A = \begin{bmatrix} 1 & 2 \\ 3 & -4 \end{bmatrix}.$$

Determine $B = A^2 + 3A - 10I$.

(6.6)

Suppose we have

$$A = \begin{bmatrix} 1 & 0 & 2 \\ 2 & -1 & 3 \\ 4 & 1 & 8 \end{bmatrix}.$$

Find the inverse of matrix A.

(6.7)

Suppose we have

$$A = \begin{bmatrix} 1 & 3 \\ 4 & -3 \end{bmatrix}.$$

Determine the column vector $u = \begin{bmatrix} x \\ y \end{bmatrix}$ such that $Au = 3u$.

(6.8)

Find the Boolean product of A and B, where we have

$$A = \begin{bmatrix} 1 & 0 & 0 & 1 \\ 0 & 1 & 0 & 1 \\ 1 & 1 & 1 & 1 \end{bmatrix} \quad \& \quad B = \begin{bmatrix} 1 & 0 \\ 0 & 1 \\ 1 & 1 \\ 1 & 0 \end{bmatrix}.$$

(6.9)

Determine $A \vee B$, $A \wedge B$, and $A \odot B$ if A and B are as follows:

$$A = \begin{bmatrix} 1 & 1 \\ 0 & 1 \end{bmatrix} \quad \& \quad B = \begin{bmatrix} 0 & 1 \\ 1 & 0 \end{bmatrix}.$$

(6.10)

Solve the following system of linear equations using elementary row operations:

$$\begin{cases} 7x_1 - 8x_2 + 5x_3 = 5 \\ -4x_1 + 5x_2 - 3x_3 = -3 \\ x_1 - x_2 + x_3 = 0 \end{cases}$$

CHAPTER 7

Functions

Contents

The word *function* indicates the dependence of one varying quantity on another. The concept of function is one of the very important concepts in mathematics. Functions were first defined by Leonard Euler and later formulated by Peter Dirichlet. In discrete mathematics, functions have various applications of great importance. For instance, functions are used in the definition of discrete structures, such as sequences and strings, or they are employed to represent how long it takes for a computer to solve a problem of a given size. The focus of this chapter is on basic aspects of functions and also some special functions.

7.1 Basic Definitions

A function associates each member in one set, say the set of all people living in the world, with a member in another set, say the set of all positive real numbers representing their heights. Let X and Y be two nonempty sets of real numbers. A ***function*** from X to Y, denoted by $f: X \rightarrow Y$, is a relation from X to Y, a subset of $X \times Y$, that must satisfy both of the following two requirements:

1. Every element in X is related to some element in Y.

2. No element in X is related to more than one element in Y.

A relation from X to Y that contains only one ordered pair (x, y) for every element $x \in X$ defines a function f from X to Y where $y \in Y$, as shown in Fig. 7.1. For instance, assuming X and Y are the set of real numbers, $y = x^2$ is a function, as it meets both requirements. However, $y^2 = x$ is not a function because there is an element $x > 0$ in X that is related to the two elements $-\sqrt{x}$ and $\sqrt{x}$ in Y. Note that $f(x)$ does not mean f times x; it simply means f is a function of x, and $f(x)$, read as "f of x," is just the value of f at x. Although it is incorrect, we may call $f(x)$ the function. $f(x)$ is the short form for a function of x. Note that functions may also be called ***maps***, ***mappings***, or ***transformations***.

Discrete Mathematics
ISBN 978-0-12-820656-0, https://doi.org/10.1016/B978-0-12-820656-0.00007-1

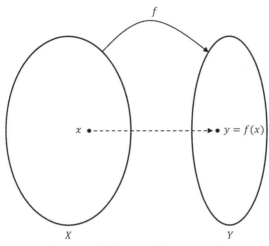

Fig. 7.1 The function f maps X to Y.

With f as a function from X to Y, the set X is the **domain** of the function f, and Y is the **codomain** of f. Moreover, y is the **image** of x, and x is a **preimage** or an **inverse image** of y. The **range** of f is the set of all images of elements of X. Note that the codomain of a function is the set of all possible values (i.e., all elements of Y), and the range is the set of all values of $f(x)$ for $x \in X$; therefore the range is a subset of the codomain.

Example 7.1
Determine whether or not the correspondences in Fig. 7.2 are functions. If an assignment is a function, determine its domain, codomain, and range.

Solution
(a) The correspondence in Fig. 7.2a is not a function because it maps an element in X to two distinct elements in Y.
(b) The correspondence in Fig. 7.2b is a function because each member in X is associated with exactly one member in Y. Its domain is the set $\{a, b, c, d\}$, codomain is the set $\{1, 2, 3, 4\}$, and range is the set $\{1, 3, 4\}$.
(c) The correspondence in Fig. 7.2c is not a function because there is an element of X, namely b that is not mapped to any element of Y.

For a function of $y = f(x)$, the variable x is called the **independent variable**, as it can have any value from its domain, and the variable y is called the **dependent variable** because its value solely depends on the value of x. If f represents a system, then y is the **output** corresponding to the **input** x.

As a function is defined by its domain, codomain, and the mapping of elements of the domain to elements in the codomain, two functions are **equal** when they have the same

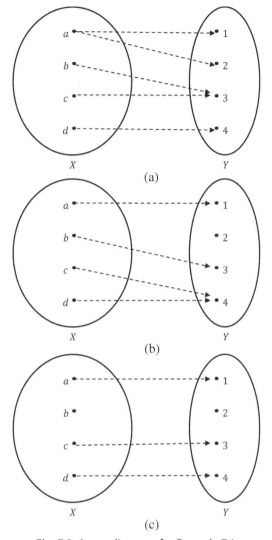

Fig. 7.2 Arrow diagrams for Example 7.1.

domain, the same codomain, and the same mapping of elements in the domain. A function is called ***real valued*** if its codomain is the set of real numbers **R**, and it is called ***integer valued*** if its codomain is the set of integers **Z**.

Example 7.2

Let $f(x) = 2x$, where the domain is the set of real numbers. Determine the ranges for the following domains:

(a) $f(\mathbf{Z})$, where **Z** is the set of integers.

(b) $f(N)$, where N is the set of natural numbers (positive integers).
(c) $f(R)$, where R is the set of real numbers.

Solution
(a) The range consists of the set of even integers.
(b) The range consists of the set of positive even integers.
(c) The range consists of the set of real numbers.

A function is uniquely represented by the set of all pairs $(x, f(x))$. When the domain X and codomain Y of a function are sets of real numbers, each such pair may be considered as the Cartesian coordinates of a point in the plane. The **graph** of the function f is the set of ordered pairs $\{(x,\ y)\,|\,x \in X,\ f(x) = y \in Y\}$.

Let $f: X \to R$ and $g: X \to R$. The sum and product of the functions f and g denoted by $f + g$ and fg, respectively, are also functions from X to R defined for all $x \in X$ by

$$(f + g)(x) = f(x) + g(x)$$

and

$$(fg)(x) = f(x) \times g(x).$$

The functions $f + g$ and fg are defined only wherever both f and g are defined. The domain of the function $f + g$ or the domain of the function fg is the intersection of the domain of f and the domain of g.

Example 7.3
Let $f(x) = \sqrt{4 - x^2}$ and $g(x) = \sqrt{x - 1}$, where $f(x)$ and $g(x)$ are both real-valued functions. Determine the functions $(f + g)(x)$ and $(fg)(x)$ and their domains.

Solution
We first find the domains of $f(x)$ and $g(x)$. The domain of $f(x)$ is $-2 \leq x \leq 2$ and the domain of $g(x)$ is $x \geq 1$. Therefore both $(f + g)(x)$ and $(fg)(x)$ are defined only when $1 \leq x \leq 2$. We can now have the sum and product of the two functions as follows:

$$(f + g)(x) = \sqrt{4 - x^2} + \sqrt{x - 1}$$

and

$$(fg)(x) = \sqrt{4 - x^2} \times \sqrt{x - 1} = \sqrt{(4 - x^2)(x - 1)}.$$

7.2 Special Functions

Let the domain and codomain of the function f be subsets of the set of real numbers. f is called a **nondecreasing function** if $f(x_1) \leq f(x_2)$ and an **increasing function** if $f(x_1) < f(x_2)$, whenever $x_1 < x_2$ and both x_1 and x_2 are in the domain of f. f is called a **nonincreasing function** if $f(x_1) \geq f(x_2)$ and a **decreasing function** if $f(x_1) > f(x_2)$, whenever $x_1 < x_2$ and both x_1 and x_2 are in the domain of f. Using quantifiers and assuming x_1 and x_2 are in the domain of the function, and $f(x_1)$ and $f(x_2)$ are in the codomain of the function, we have the following:

- Nondecreasing function: $\forall x_1 \forall x_2 (x_1 < x_2 \rightarrow f(x_1) \leq f(x_2))$.
- Increasing function: $\forall x_1 \forall x_2 (x_1 < x_2 \rightarrow f(x_1) < f(x_2))$.
- Nonincreasing function: $\forall x_1 \forall x_2 (x_1 < x_2 \rightarrow f(x_1) \geq f(x_2))$.
- Decreasing function: $\forall x_1 \forall x_2 (x_1 < x_2 \rightarrow f(x_1) > f(x_2))$.

Example 7.4

Consider the following four functions of the form $f : \mathbf{R} \rightarrow \mathbf{R}$. Determine the intervals for which they are nondecreasing, decreasing, nonincreasing, or increasing functions.

(a) $f(x) = x^2$.
(b) $g(x) = e^{-x}$.
(c) $h(x) = x^3 - x$.
(d) $k(x) = x^3$.

Solution

The following categorizes each function over some intervals:

(a) It is decreasing over the interval $(-\infty, 0)$ and increasing over the interval $(0, \infty)$.

(b) It is decreasing over the interval $(-\infty, \infty)$.

(c) It is increasing over the intervals $\left(-\infty, -\frac{1}{\sqrt{3}}\right)$ and $\left(\frac{1}{\sqrt{3}}, \infty\right)$ and decreasing over the interval $\left(-\frac{1}{\sqrt{3}}, \frac{1}{\sqrt{3}}\right)$.

(d) It is increasing over the interval $(-\infty, \infty)$.

The definition of a function in mathematics generally consists of just one formula. However, many practical functions reflecting real-world applications may consist of more than one formula, depending on the values of x. Such a function is called a **piecewise defined function**. Functions that are defined piecewise are written as if-then-else statements in most programming languages.

The **unit step function**, used often in modeling physical systems, is piecewise defined as follows:

$$f(x) = 0, \text{ if } x < 0, \text{ and } f(x) = 1, \text{ if } x \geq 0.$$

The **absolute-value function** is piecewise defined as follows:

$$f(x) = |x| \rightarrow f(x) = x, \text{ if } x \geq 0, \text{ and } f(x) = -x, \text{ if } x < 0.$$

A function that assigns each element $x \in X$ to itself is called the **identity function** on X. For an identity function $f: X \rightarrow X$, we have $f(x) = x$ (i.e., it leaves every input unchanged). The graph of the identity function on $\mathbf{R}$ is the straight line $y = x$.

The **floor function**, also known as the **greatest integer function**, assigns to the real number x the largest integer that is less than or equal to x. The floor of x rounds down x and is denoted by $\lfloor x \rfloor$. The **ceiling function**, also known as the **least integer function**, assigns to the real number x the smallest integer that is greater than or equal to x. The ceiling of x rounds up x and is denoted by $\lceil x \rceil$. For instance, if $x = 6.4$, then $\lfloor x \rfloor = 6$ and $\lceil x \rceil = 7$, and if $x = -6.4$, then $\lfloor x \rfloor = -7$ and $\lceil x \rceil = -6$. Table 7.1 presents some of the properties of the floor and ceiling functions. The floor and ceiling functions are often used in the analysis of the number of steps required by algorithms and can thus help provide measures of complexities of algorithms. Note that all programming languages provide the floor and ceiling functions as built-in functions.

TABLE 7.1 Properties of the floor and ceiling functions.

$$\lfloor x \rfloor = n \quad \text{if and only if } n \leq x < n+1$$
$$\lceil x \rceil = n \quad \text{if and only if } n - 1 < x \leq n$$
$$\lfloor x \rfloor = n \quad \text{if and only if } x - 1 < n \leq x$$
$$\lceil x \rceil = n \quad \text{if and only if } x \leq n < x+1$$
$$\lfloor x + n \rfloor = \lfloor x \rfloor + n$$
$$\lceil x + n \rceil = \lceil x \rceil + n$$
$$\left\lfloor \frac{n}{2} \right\rfloor = \frac{(n-1)}{2} \text{ if } n \text{ is odd \& } \left\lfloor \frac{n}{2} \right\rfloor = \frac{n}{2} \text{ if } n \text{ is even}$$
$$\left\lceil \frac{n}{2} \right\rceil = \frac{(n+1)}{2} \text{ if } n \text{ is odd \& } \left\lceil \frac{n}{2} \right\rceil = \frac{n}{2} \text{ if } n \text{ is even}$$
$$\lfloor -x \rfloor = -\lceil x \rceil$$
$$\lceil -x \rceil = -\lfloor x \rfloor$$
$$x - 1 < \lfloor x \rfloor \leq x \leq \lceil x \rceil < x + 1$$

Note: n is an integer and x is a real number.

Example 7.5

Determine the maximum number of identical cube boxes that can be fit inside a storage room, where the size of a box is 0.5 m × 0.5 m × 0.5 m and the size of the room is 5.6 m × 5 m × 3.75 m.

Solution

As the number of boxes that can be put in any direction is an integer, the maximum number of boxes that can be fit inside the storage room is as follows:

$$\left\lfloor \frac{5.6}{0.5} \right\rfloor \times \left\lfloor \frac{5}{0.5} \right\rfloor \times \left\lfloor \frac{3.75}{0.5} \right\rfloor = 11 \times 10 \times 7 = 770.$$

Example 7.6

Suppose a and b are both positive integers. Show that the number of positive integers no greater than a and divisible by b is $\left\lfloor \frac{a}{b} \right\rfloor$.

Solution

Suppose there are k positive integers no greater than a and divisible by b. Therefore the largest multiple of b is $kb \leq a$ (i.e., $k \leq \frac{a}{b}$). In addition, $(k+1)b > a$ (i.e., $k+1 > \frac{a}{b}$ or $\frac{a}{b} - 1 < k$). We thus have

$$\frac{a}{b} - 1 < k \leq \frac{a}{b} \quad \rightarrow \quad k = \left\lfloor \frac{a}{b} \right\rfloor.$$

A function $f: \mathbf{R} \to \mathbf{R}$ defined as $f(x) = a_n x^n + a_{n-1} x^{n-1} + \ldots + a_1 x + a_0$, where the coefficients $a_0, a_1 \ldots, a_{n-1}$ and $a_n \neq 0$ are all real numbers and n is a positive integer, is called a **polynomial function of degree n**. When $n = 1$, $f(x)$ is called a **linear function**, and when $n = 2$, it is called a **quadratic function**. Polynomial functions have numerous features, including being continuous for all values of x, where $x \in (-\infty, \infty)$, and having no asymptotic lines. Moreover, the term $a_n x^n$ in a polynomial function, regardless of the value of a_n, becomes the dominant term as the value of x becomes very large.

Noting $m \in \mathbf{W} = \{0, 1, 2, 3, \ldots\}$ and $f(m) \in \mathbf{N} = \{1, 2, 3, \ldots\}$, the function $f: \mathbf{W} \to \mathbf{N}$, denoted by $f(m) = m!$, where $m! \triangleq m \times (m-1) \times (m-2) \times \ldots \times 2 \times 1$ and $f(0) = 0! \triangleq 1$, is called the **factorial function**. For instance, if $m = 6$, then $f(6) = 6! = 6 \times 5 \times 4 \times 3 \times 2 \times 1 = 720$.

As a very useful tool in the calculation of m factorial when m is very large, the factorial of m can be asymptotically approximated by the **Stirling's formula** $m! \approx \sqrt{2\pi m} \left(\frac{m}{e} \right)^m$, as $m \to \infty$, where $\pi = 3.141592\ldots$ and $e = 2.718281\ldots$.

An important use for the factorial notation is in calculating values of quantities that occur in the study of counting methods. The symbol $\begin{pmatrix} n \\ r \end{pmatrix} = \frac{n!}{r!(n-r)!}$, read as n *choose* r and called a *binomial coefficient*, represents the number of subsets of size r that can be chosen from a set with n elements, where $0 \leq r \leq n$.

The function $f : R \rightarrow R^+$, defined by $f(x) = a^x$, where $a \in R^+$ and $a \neq 1$, is called the *exponential function* to the base a. Note that $x \in R$ (i.e., x belongs to the set of all real numbers), $f(x) \in R^+$ (i.e., $f(x)$ belongs to the set of all positive real numbers), and $a^0 \triangleq 1$. Table 7.2 presents some properties of exponential functions.

The function $f : R^+ \rightarrow R$, defined by $f(x) = \log_a x$, where $a \in R^+$ and $a \neq 1$, is called the *logarithmic function* to the base a. Note that $x \in R^+$ (i.e., x belongs to the set of all positive real numbers) and $f(x) \in R$ (i.e., $f(x)$ belongs to the set of all real numbers). Table 7.3 presents some properties of logarithmic functions.

Exponential and logarithmic functions are thus related as follows:

$$x = \log_a f(x) \leftrightarrow f(x) = a^x.$$

As an example, with $a = 2$, we have $f(8) = 2^8 = 256$ as well as $8 = \log_2 256$. Compared to the linear function $y = x$, for $a > 1$, the logarithmic function $y = \log_a x$ grows very slowly and the exponential function $y = a^x$ grows very quickly. The most frequently used bases for logarithmic functions are as follows:

- If $a = \lim_{n \to \infty} \left(1 + \frac{1}{n}\right)^n \approx 2.718281828459$, which is known as e and referred to as

  Euler's number or *Napier's constant*, the logarithmic function is then called the *natural logarithm* and denoted by $\ln f(x)$, rather than by $\log_e f(x)$.

TABLE 7.2 Properties of exponential functions.

$$a^{x_1} a^{x_2} = a^{x_1 + x_2}$$
$$\left(a^{x_1}\right)^{x_2} = a^{x_1 x_2}$$
$$\frac{a^{x_1}}{a^{x_2}} = a^{x_1 - x_2}$$
$$(ab)^{x_1} = a^{x_1} b^{x_1}$$

Note: $a > 0$, $b > 0$, x_1, and x_2 are real numbers.

TABLE 7.3 Properties of logarithmic functions.

$$\log_a x_1 + \log_a x_2 = \log_a(x_1 x_2)$$
$$\log_a x_1 - \log_a x_2 = \log_a\left(\frac{x_1}{x_2}\right)$$
$$\log_a(x_1)^{x_2} = x_2 \log_a x_1$$
$$\log_a x_1 = \frac{\log_b x_1}{\log_b a}$$

Note: $a \neq 1$, $b \neq 1$, x_1, and x_2 are positive real numbers.

- If $a = 10$, the logarithmic function is then called the ***common logarithm*** and denoted by $\log f(x)$ rather than by $\log_{10} f(x)$.

There are other important functions, such as the mod and div functions, Boolean and hashing functions, and recursively defined functions, which will be discussed in other chapters in broader contexts.

7.3 One-to-One and Onto Functions

A function $f: X \rightarrow Y$ is said to be ***one-to-one*** or ***injection*** if and only if $f(x_1) = f(x_2)$ implies that $x_1 = x_2$ for all elements in X. In other words, if at least two different elements in the domain of a function can be found that have the same element in the codomain, then the function is not one-to-one. Using quantifiers, a function f is one-to-one if $\forall x_1 \forall x_2 (f(x_1) = f(x_2) \rightarrow x_1 = x_2)$ or equivalently, $\forall x_1 \forall x_2 (f(x_1) \neq f(x_2) \rightarrow x_1 \neq x_2)$, where x_1 are x_2 are in the domain of the function and $f(x_1)$ and $f(x_2)$ are in the codomain of the function.

A function $f: X \rightarrow Y$ is said to be ***onto*** or ***surjection*** if and only if, for every element $y \in Y$, there is at least one element $x \in X$ with $f(x) = y$. In other words, if the range and codomain are not the same, then the function is not onto. Using quantifiers, a function f is onto if $\forall y \exists x (f(x) = y)$, where x and y are in the domain and codomain of the function, respectively.

A function $f: X \rightarrow Y$ is said to be ***one-to-one correspondence*** or ***bijection*** if and only if it is both one-to-one and onto. When a function is a one-to-one correspondence, the elements of its domain and codomain match up perfectly.

Example 7.7

Consider the four arrow diagrams in Fig. 7.3, where each represents a function. Identify the functions that are one-to-one and those that are onto.

Solution

Only the arrow diagrams in (a) and (c) are one-to-one functions, because in each of them, different elements of the domain have distinct images (i.e., no two values in the domain are assigned to the same function value). Only the arrow diagrams in (b) and (c) are onto functions, because in each of them, all elements in the codomain are images of elements in the domain. In summary, the function in (a) is one-to-one, but not onto, the function in (b) is onto, but not one-to-one, the function in (c) is both one-to-one and onto, i.e., one-to-one correspondence, and the function in (d) is neither one-to-one nor onto.

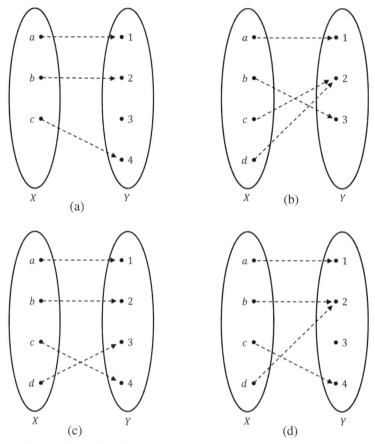

Fig. 7.3 Arrow diagrams for various functions in Example 7.7: (a) One-to-one, but not onto. (b) Onto, but not one-to-one. (c) Both one-to-one and onto. (d) Neither one-to-one nor onto.

Geometrical characterization of one-to-one and onto functions can bring about meaningful insights. Consider functions of the form $f : \mathbf{R} \to \mathbf{R}$. The graphs of such functions can be plotted in the Cartesian plane using the set of ordered pairs $((a, b) | a \in \mathbf{R}$ and $f(a) = b)$, where the graph of a function f is an aid in understanding the behavior of the function. The concepts of being one-to-one, onto, and one-to-one correspondence have some geometrical meaning, which are as follows:

- If the function $f : \mathbf{R} \to \mathbf{R}$ is one-to-one, then each horizontal line intersects the graph of the function f in at most one point (i.e., the number of intersection points ≤ 1).
- If the function $f : \mathbf{R} \to \mathbf{R}$ is onto, then each horizontal line intersects the graph of the function f in at one or more points (i.e., the number of intersection points ≥ 1).
- If the function $f : \mathbf{R} \to \mathbf{R}$ is a one-to-one correspondence, then each horizontal line intersects the graph of the function f in exactly one point (i.e., the number of intersection points $= 1$).

Example 7.8

Consider the following four functions of the form $f: R \rightarrow R$. Using their geometric characterizations, identify the functions that are one-to-one and those that are onto:

(a) $f(x) = x^2$.
(b) $g(x) = e^{-x}$.
(c) $h(x) = x^3 - x$.
(d) $k(x) = x^3$.

Solution

The graphs of these functions are shown in Fig. 7.4. The functions $g(x)$ and $k(x)$ are both one-to-one, because in the graphs of $g(x)$ and $k(x)$, no horizontal line intersects the graph at more than one point. The functions $h(x)$ and $k(x)$ are both onto, because in the graphs of $h(x)$ and $k(x)$, each horizontal line intersects the graph at one or more points. In summary, the function $g(x)$ is one-to-one, but not onto, the function $h(x)$ is onto, but not one-to-one, the function $k(x)$ is both one-to-one and onto (i.e., one-to-one correspondence), and the function $f(x)$ is neither one-to-one nor onto.

7.4 Compositions of Functions

In addition to simple operations on functions, such as addition and multiplication, there is a fundamentally different way, called composition, to combine two functions so as to construct a new function.

Consider the function $f: X \rightarrow Y$ and the function $g: Y \rightarrow Z$. The **composition** of the functions f and g, denoted by $g \circ f$ and read as "g circle f," is a function from X to Z, defined as follows:

$$(g \circ f)(x) = g(f(x)).$$

In order to find $(g \circ f)(x)$, we first apply the function f to x to obtain $f(x)$, and then we apply the function g to $f(x)$ to obtain $(g \circ f)(x) = g(f(x))$. Fig. 7.5 shows the composition of functions.

In general, the domain of the function g need not be the same as the codomain of the function f. The composition of $g \circ f$ cannot be defined unless the range of the function f is a subset of the domain of the function g. For instance, suppose the domain of the function g is the set of positive real numbers if the range of the function f is the set of positive integers, then $g \circ f$ can be defined; however, if the range of the function f is the set of all integers, then $g \circ f$ cannot be defined.

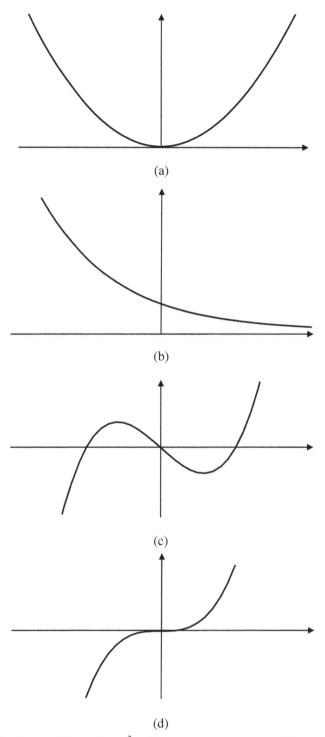

(a)

(b)

(c)

(d)

Fig. 7.4 Graphs for Example 7.8: (a) $f(x) = x^2$: neither one-to-one nor onto; (b) $g(x) = e^{-x}$: one-to-one, but not onto; (c) $h(x) = x^3 - x$: onto, but not one-to-one; (d) $k(x) = x^3$: both one-to-one and onto.

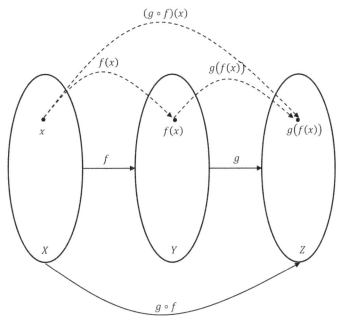

Fig. 7.5 The composition of the functions g and f.

At a system level, the composite function can be viewed as a system with two sub-systems in series, where the output of the first subsystem forms the input of the second subsystem, and the composite function represents the system output $(g \circ f)(x)$ for the system input x.

Note that the order of the functions matters in a composition. Even if both $g \circ f$ and $f \circ g$ are defined, though, in general, we have $g \circ f \neq f \circ g$. In other words, the commutative law does not hold for the composition of functions.

Example 7.9

Consider the functions $f: \mathbf{R} \to \mathbf{R}$ and $g: \mathbf{R} \to \mathbf{R}$, where $\mathbf{R}$ represents the set of real numbers. Assuming $f(x) = x + 1$ and $g(x) = x^3 + x$, determine $(g \circ f)(x)$ and $(f \circ g)(x)$. Comment on the results.

Solution

We form the compositions of functions:

$$(g \circ f)(x) = g(f(x)) = g(x+1) = (x+1)^3 + (x+1) = x^3 + 3x^2 + 4x + 2$$

and

$$(f \circ g)(x) = f(g(x)) = f(x^3 + x) = (x^3 + x) + 1 = x^3 + x + 1.$$

Because we have $x^3 + 3x^2 + 4x + 2 \neq x^3 + x + 1$, we can conclude $g \circ f \neq f \circ g$.

It is worth noting that with the functions $f: X \rightarrow Y$ and $g: Y \rightarrow Z$, we can make the following statements:

- If the functions f and g are one-to-one, then $(g \circ f)(x) = g(f(x))$ is one-to-one.
- If the functions f and g are onto, then $(g \circ f)(x) = g(f(x))$ is onto.
- If the functions f and g are one-to-one correspondence, then $(g \circ f)(x) = g(f(x))$ is a one-to-one correspondence.

Suppose $f: X \rightarrow Y$ is a one-to-one correspondence. The **inverse function** of the function f, denoted by f^{-1}, is the function that assigns to an element y belonging to Y the unique element x in X such that $f(x) = y$. Hence we have

$$f^{-1}(y) = x \leftrightarrow y = f(x).$$

Fig. 7.6 shows the concept of an inverse function. Note that the domain of f becomes the codomain of f^{-1} and the codomain of f becomes the domain of f^{-1}. The independent variable x for f acts as the dependent variable for f^{-1}, and correspondingly the dependent variable y for f becomes the independent variable for f^{-1}. For instance, if we have $f(x) = e^x$, we then have $\ln f(x) = x$. By changing the roles of the independent and dependent variables, the inverse function of $f(x) = e^x$ is then $f^{-1}(x) = \ln x$.

A function is **invertible** if and only if it is a one-to-one correspondence, consequently, a function is not invertible if it is not a one-to-one correspondence because the inverse of such a function does not exist.

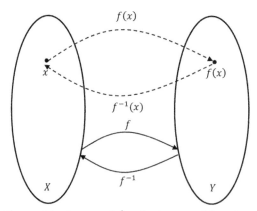

Fig. 7.6 The function f^{-1} is the inverse of function f.

Example 7.10

Determine if the following functions are invertible, where their domains and codomains are the set of real numbers:

(a) $h(x) = (x - 1)^4 + 2$.
(b) $k(x) = (x - 7)^5$.

Solution

(a) The function $h(x)$ is not a one-to-one correspondence because it is not even one-to-one. Therefore it is not invertible. However, the inverse is obtained as follows:

$$h(x) = (x - 1)^4 + 2 \rightarrow x = \sqrt[4]{h(x) - 2} + 1 \rightarrow h^{-1}(x) = \pm\sqrt[4]{x - 2} + 1.$$

Its inverse, $h^{-1}(x) = \pm\sqrt[4]{x - 2} + 1$, is not a function because for a given $x > 2$, we have two distinct values for $h^{-1}(x)$.

(b) The function $k(x)$ is a one-to-one correspondence. Therefore it is invertible. The inverse is obtained as follows:

$$k(x) = (x - 7)^5 \rightarrow x = \sqrt[5]{k(x)} + 7 \rightarrow k^{-1}(x) = \sqrt[5]{x} + 7.$$

As expected, its inverse, $k^{-1}(x) = \sqrt[5]{x} + 7$, is also a function.

It is important to note that if $f: X \rightarrow Y$ and $g: Y \rightarrow X$ are both invertible, the function $g \circ f$ is then invertible, and we have $(g \circ f)^{-1} = f^{-1} \circ g^{-1}$.

Example 7.11

Show $(g \circ f)^{-1} = f^{-1} \circ g^{-1}$, if $f(x) = x^3 + 2$ and $g(x) = x + 4$.

Solution

We have

$$(g \circ f) = g(f(x)) = (x^3 + 2) + 4 = x^3 + 6 \rightarrow (g \circ f)^{-1} = \sqrt[3]{x - 6}$$

and also

$$\begin{cases} f(x) = x^3 + 2 \rightarrow f^{-1} = \sqrt[3]{x - 2} \\ g(x) = x + 4 \rightarrow g^{-1} = x - 4 \end{cases} \rightarrow f^{-1} \circ g^{-1} = \sqrt[3]{(x - 4) - 2} = \sqrt[3]{x - 6}$$

We can thus conclude

$$(g \circ f)^{-1} = f^{-1} \circ g^{-1} = \sqrt[3]{x - 6}$$

Exercises

(7.1)
Let $X = \{a, b, c\}$ and $Y = \{1, 2, 3, 4\}$, and also let the function $f: X \rightarrow Y$ be defined as $\{(a, 2), (b, 4), (c, 2)\}$.
(a) Determine the domain, codomain, and range of the function f.
(b) Determine $f(a)$ and inverse images of 1, 2, 3, 4.

(7.2)
Let $f(x) = 2x + 1$. Determine the range of $f(x)$ if the domain is as follows:
(a) The set of integers.
(b) The set of positive integers.
(c) The set of real numbers.

(7.3)
Prove or give a counterexample for each of the following statements:
(a) An increasing function from R to R is one-to-one.
(b) A nondecreasing function from R to R is one-to-one.
(c) A decreasing function from R to R is one-to-one.
(d) A nonincreasing function from R to R is one-to-one.

(7.4)
(a) Round down $e = 2.718281...$ to two decimal places.
(b) Round up $\pi = 3.141592...$ to three decimal places.

(7.5)
Suppose a and b are both real numbers where $a < b$. Use the floor and/or ceiling functions to express the number of integers n that satisfy the following cases:
(a) $a \le n \le b$.
(b) $a < n < b$.
(c) Assuming $a = 4.56$ and $b = 9$, determine n for the above two parts.

(7.6)
Let $f: Z \rightarrow Z$, noting Z is the set of integers. Assuming $f(n) = 5n - 2$, determine if it is (a) one-to-one, (b) onto, and (c) one-to-one correspondence. Prove or give a counterexample.

(7.7)
Let $f: R^+ \rightarrow R$, where R is the set of real numbers and R^+ is the set of positive real numbers. Assuming $f(x) = \log x$, determine if it is (a) one-to-one, (b) onto, and (c) one-to-one correspondence. Prove or give a counterexample.

(7.8)

Let $A = \{a, b, c\}$, $B = \{x, y, z\}$, and $C = \{r, s, t\}$. Let $f: A \rightarrow B$ and $g: B \rightarrow C$ be defined by the following: $f = \{(a, y), (b, x), (c, y)\}$ and $g = \{(x, s), (y, t), (z, r)\}$.

(a) Determine composition function $g \circ f: A \rightarrow C$.

(b) Determine images of f, g, and $g \circ f$.

(7.9)

Let $f(x) = ax + b$ and $g(x) = cx + d$, where a, b, c, and d are constants. Determine the necessary and sufficient conditions on the constants a, b, c, and d so that $g \circ f = f \circ g$.

(7.10)

Show that the function $f(x) = e^x$ from the set of real numbers to the set of real numbers is not invertible, but if the codomain is restricted to the set of positive real numbers, the function is then invertible.

CHAPTER 8

Boolean Algebra

Contents

Boolean algebra was developed by the mathematician George Boole. His contribution was the development of a theory of logic using symbols instead of words. The laws that are used to define an abstract mathematical structure called Boolean algebra are similar to those for sets and propositional logic. Later, Claude Shannon, known as the father of information theory, used Boolean algebra to analyze digital circuits. His contributions allowed Boolean algebra to become an indispensable tool for the analysis and design of electronic circuits, including all digital devices, systems, and networks. In this chapter, we develop basic properties of Boolean algebra and briefly discuss its applications to digital circuits.

8.1 Basic Definitions

A Boolean algebra is a mathematical system. A ***Boolean algebra*** consists of a nonempty set B together with two binary operations of addition "$+$" and multiplication "$\cdot$", where they map elements of $B \times B$ to elements of B (i.e., if x, $y \in B$, then $x + y$ and $x \cdot y$ are also in B), a unary operation of complementation "$'$" where it maps elements of B to elements of B, two distinct elements "1" and "0", and the axioms for all elements x, y, and z in B, as summarized in Table 8.1.

In symbols, a Boolean algebra is designated by its six parts $\{B, +, \cdot, ', 0, 1\}$. The operators "$+$", "$\cdot$", and "$'$" are called ***sum***, ***product***, and ***complement***, respectively. Note that "$+$" and "$\cdot$" are not the usual arithmetic operators "plus" and "times." The symbols "0" and "1", which are called the ***zero element*** and the ***unit element***, respectively, do not represent numbers on the real number line. However, the names "plus," "times," "complement," "zero," and "one" are commonly used informally when discussing Boolean algebras. For convenience, "$x \cdot y$" is shown as "xy", and it is also common to replace the complement operation "$'$" by a bar "$^-$" (e.g., $x' = \overline{x}$).

Discrete Mathematics
ISBN 978-0-12-820656-0, https://doi.org/10.1016/B978-0-12-820656-0.00008-3

TABLE 8.1 Boolean algebra axioms.

Identity laws:
For every $x \in B$, there exist distinct elements 0 and 1 in B such that
$$x + 0 = x \quad \text{and} \quad x \cdot 1 = x$$

Complement laws:
For every $x \in B$, there exists a unique element $\overline{x} \in B$ such that
$$x + \overline{x} = 1 \quad \text{and} \quad x \cdot \overline{x} = 0$$

Commutative laws:
For every pair of (not necessarily distinct) elements $x, y \in B$,
$$x + y = y + x \quad \text{and} \quad x \cdot y = y \cdot x$$

Distributive laws:
For every three (not necessarily distinct) elements $x, y, z \in B$,
$$x \cdot (y + z) = x \cdot y + x \cdot z \quad \text{and} \quad x + y \cdot z = (x + y) \cdot (x + z)$$

It is important to follow the **precedence rules** governing Boolean algebra. No parentheses appear when there is no possibility of confusion, and parentheses have the highest precedence. The complement operator "¯" has the second-highest precedence, followed by the product operator "$\cdot$," and the sum operator "+" with the lowest precedence. For instance, we have $xy + \overline{x}\,\overline{y} = (xy) + ((\overline{x})(\overline{y}))$.

Example 8.1
Provide examples for sets and propositional logic as Boolean algebras.

Solution
(i) In sets as Boolean algebra, we can let S be any nonempty set and B be the power set of S (i.e., the elements of Boolean algebra will be all the subsets of S), 1 be the set S, and 0 be the null set $\varnothing$, plus be union, times be intersection, and the Boolean complement be set complement. For instance, assuming $C, D \in S$, if the Boolean notation is $(0 + C)(\overline{D} + 1)$, then the corresponding set notation is $(\varnothing \cup C) \cap (\overline{D} \cup S)$.

(ii) In propositional logic as Boolean algebra, we can let $B = \{F, T\}$, plus be disjunction, times be conjunction, the Boolean complement be negation, equality be logical equivalence, and the Boolean variables be propositions. For instance, assuming P and Q are propositions, if the Boolean notation is $(0 + \overline{P}) \cdot (Q + 1)$, then the corresponding propositional notation is $(F \vee \overline{P}) \wedge (Q \vee T)$.

According to the definition of Boolean algebra, a Boolean algebra contains the zero element and the unit element. Consequently, the simplest Boolean algebra contains exactly two elements of 0 and 1. Let us now focus on such a Boolean algebra. Let $B = \{0, 1\}$, the set of **bits** (binary digits). The complement operation is defined as $\overline{0} = 1$ and

$\overline{1} = 0$. The binary operations "$+$" (Boolean sum) and "$\cdot$" (Boolean product) for all cases are defined as follows:

$$
\begin{cases}
1 + 1 = 1 \\
1 + 0 = 1 \\
0 + 1 = 1 \\
0 + 0 = 0
\end{cases}
\quad \& \quad
\begin{cases}
1 \cdot 1 = 1 \\
1 \cdot 0 = 0 \\
0 \cdot 1 = 0 \\
0 \cdot 0 = 0
\end{cases}
$$

Notice the unusual definition for $1 + 1 = 1$, and note that when $x \in B$, we have $x + x = x$ and $x \cdot x = x$.

Example 8.2

Let x, y, and z be variables whose values are in the binary set $B = \{0, 1\}$. Determine the values of the following Boolean expressions, if $x = 0$, $y = 1$, and $z = 1$:

(a) $1 \cdot x + (0 + y) + z$.

(b) $(x \cdot y) + (z \cdot z) \cdot \overline{y}$.

(c) $\overline{((x + y) \cdot z)}$.

Solution

(a) $1 \cdot x + (0 + y) + z = 1 \cdot 0 + (0 + 1) + 1 = 0 + 1 + 1 = 1$.

(b) $(x \cdot y) + (z \cdot z) \cdot \overline{y} = (0 \cdot 1) + (1 \cdot 1) \cdot \overline{1} = 0 + 1 \cdot 0 = 0 + 0 = 0$.

(c) $\overline{((x + y) \cdot z)} = \overline{((0 + 1) \cdot 1)} = \overline{(1 \cdot 1)} = \overline{1} = 0$.

8.2 Boolean Expressions and Boolean Functions

Let $B = \{0, 1\}$. The variable x is called a **Boolean variable** if it assumes values only from B (i.e., if its only possible values are 0 and 1). Then, $B^n = \{x_1, x_2, ..., x_n | x_i \in B$ for $1 \le i \le n\}$ is the set of all possible n-tuples of 0s and 1s and has 2^n elements. A function from B^n to B is called a **Boolean function of degree n**. Boolean functions can be defined by Boolean tables. Because a Boolean function is an assignment of 0 or 1 to each of these 2^n different n-tuples, there are $(2)^{(2^n)}$ different Boolean functions of degree n, labeled $F_1, F_2, ..., F_{2^{2^n}}$. Table 8.2 presents all Boolean functions of degree two (i.e., $n = 2$), labeled $F_1, F_2, ..., F_{16}$.

A Boolean expression consists of Boolean variables and Boolean operators. The **Boolean expression** in the Boolean variables $x_1, x_2, ..., x_n$ are defined recursively through the **basic clause** that states that 0, 1, $x_1, x_2, ..., x_n$ are Boolean expressions, and the **recursive clause** that states that the sum and the product of any two Boolean expressions as well as the complement of any Boolean expression are also Boolean expressions. Each Boolean

TABLE 8.2 All Boolean functions of degree 2.

x	y	F_1	F_2	F_3	F_4	F_5	F_6	F_7	F_8	F_9	F_{10}	F_{11}	F_{12}	F_{13}	F_{14}	F_{15}	F_{16}
1	1	1	1	1	1	1	1	1	1	0	0	0	0	0	0	0	0
1	0	1	1	1	1	0	0	0	0	1	1	1	1	0	0	0	0
0	1	1	1	0	0	1	1	0	0	1	1	0	0	1	1	0	0
0	0	1	0	1	0	1	0	1	0	1	0	1	0	1	0	1	0

expression represents a Boolean function. The values of a Boolean function are obtained by substituting 0 and 1 for the Boolean variables in the Boolean expression.

Example 8.3
Verify that $xy + \bar{x}\bar{y}$ is a Boolean expression, where the variables x and y are Boolean variables.

Solution
Because x and y are Boolean variables, they are Boolean expressions by the basic clause. Consequently, xy and $\bar{x}\bar{y}$ are Boolean expressions by the recursive clause. Therefore by the recursive clause, $xy + \bar{x}\bar{y}$ is a Boolean expression.

Every Boolean function $f: B^n \rightarrow B$ can be represented by a Boolean expression in n Boolean variables. Two Boolean expressions that represent the same function are called **equivalent**. For instance, the Boolean expressions $\overline{xy}$ and $\bar{x} + \bar{y}$ are equivalent. The complement of the Boolean function $F(x_1, x_2, \ldots, x_n)$ is the function $\overline{F(x_1, x_2, \ldots, x_n)}$. The Boolean sum and product of two Boolean functions of degree n are as follows:

$$
\begin{cases}
S(x_1, x_2, \ldots, x_n) = F(x_1, x_2, \ldots, x_n) + G(x_1, x_2, \ldots, x_n) \\
\\
P(x_1, x_2, \ldots, x_n) = F(x_1, x_2, \ldots, x_n) \cdot G(x_1, x_2, \ldots, x_n)
\end{cases}
$$

For notational simplicity, we often write the elements of B^n as an n-bit sequence without commas.

Example 8.4
Assume $x = 11100$ and $y = 01010$ belong to B^n, where $n = 5$. Determine the sum of their complements and the complement of their product.

Solution
The operations in Boolean algebra are done bit by bit in the 5-bit sequences. We thus have $\bar{x} = 00011$ and $\bar{y} = 10101$, and consequently, obtain $\bar{x} + \bar{y} = 00011 + 10101 = 10111$ and $\overline{xy} = \overline{(11100)\,(01010)} = \overline{01000} = 10111$.

8.3 Identities of Boolean Algebra

There is an array of identities in Boolean algebra. The important identities that are widely used in the analysis and design of digital circuits are shown in Table 8.3. Note that any identity can be proven using a Boolean table, as both sides of a Boolean identity are the same in a Boolean table.

Example 8.5

Prove the distributive identity $x + yz = (x + y)(x + z)$ using a Boolean table.

Solution

As there are three Boolean variables, the Boolean table shown in Table 8.4 has $8\ (= 2^3)$ rows. The identity holds because the last two columns representing both sides of the identity are identical.

The Boolean identities, the logical equivalences, and the set identities are all the special cases of the same identities. Each collection of identities can be obtained by making the appropriate changes.

In order to determine the **dual** of a Boolean expression, Boolean sums and Boolean products should be interchanged, and 0s and 1s should also be interchanged. For instance,

TABLE 8.3 Boolean identities.

Identity	Name
$x + x = x$	Idempotent laws
$xx = x$	
$(x + y) + z = x + (y + z)$	Associativity laws
$(xy)z = x(yz)$	
$\overline{\overline{x}} = x$	Involution law
$x + 1 = 1$	Domination laws
$x \cdot 0 = 0$	
$\overline{x + y} = \overline{x}\,\overline{y}$	De Morgan's laws
$\overline{xy} = \overline{x} + \overline{y}$	
$x + xy = x$	Absorption laws
$x \cdot (x + y) = x$	
$x + 0 = x$	Identity laws
$x \cdot 1 = x$	
$x + y = y + x$	Commutative laws
$xy = yx$	
$x + yz = (x + y)(x + z)$	Distributive laws
$x(y + z) = xy + xz$	
$x + \overline{x} = 1$	Unit property
$x\overline{x} = 0$	Zero property

TABLE 8.4 Boolean table for Example 8.5.

x	y	z	yz	$x + y$	$x + z$	$x + yz$	$(x + y)(x + z)$
1	1	1	1	1	1	1	1
1	1	0	0	1	1	1	1
1	0	1	0	1	1	1	1
1	0	0	0	1	1	1	1
0	1	1	1	1	1	1	1
0	1	0	0	1	0	0	0
0	0	1	0	0	1	0	0
0	0	0	0	0	0	0	0

the dual of the Boolean expression $x + 1(y + 0)$ is $x \cdot (0 + (y \cdot 1))$. The **principle of duality** states that when the duals of both sides of an identity are taken, another identity is obtained. For instance, the dual of the identity $\overline{xy} = \overline{x} + \overline{y}$ is another identity $\overline{x + y} = \overline{x}\,\overline{y}$. Due to the principle of duality, all identities in Table 8.3, except the law of the double complement (involution), come in pairs.

8.4 Representing Boolean Functions

We often need to determine a Boolean expression that represents a given Boolean function. Any Boolean function can be represented by a Boolean sum of Boolean products of the Boolean variables and their complements. A **literal** is a Boolean variable or its complement. A **minterm** of the Boolean variables x_1, x_2, ..., x_n is a Boolean product $y_1 y_2 \ldots y_n$, where $y_i = x_i$ or $y_i = \overline{x}_i$, and $i = 1, 2, \ldots, n$. Hence a minterm is a product of n literals, with one literal for each of the n variables. A minterm $y_1 y_2 \ldots y_n$ has the value 1 if and only if each y_i is 1 for $i = 1, 2, \ldots, n$. This occurs when $y_i = x_i$ for $x_i = 1$, and $y_i = \overline{x}_i$ for $x_i = 0$. For instance, with 3 literals x_1, x_2, and x_3, there are 8 $(= 2^3)$ minterms and each has value 1, as reflected in Table 8.5.

TABLE 8.5 Minterms for three literals.

x_1	x_2	x_3	Minterm
1	1	1	$x_1 x_2 x_3$
1	1	0	$x_1 x_2 \overline{x}_3$
1	0	1	$x_1 \overline{x}_2 x_3$
1	0	0	$x_1 \overline{x}_2 \overline{x}_3$
0	1	1	$\overline{x}_1 x_2 x_3$
0	1	0	$\overline{x}_1 x_2 \overline{x}_3$
0	0	1	$\overline{x}_1 \overline{x}_2 x_3$
0	0	0	$\overline{x}_1 \overline{x}_2 \overline{x}_3$

Given a Boolean function in the form of a Boolean table, a Boolean sum of minterms can be formed that has the value 1 when the Boolean function has the value 1, and it has the value 0 when the Boolean function has the value 0. Therefore the minterms in the Boolean sum correspond to those combinations of values for which the Boolean function has the value 1. The sum of minterms that represents the function is called the **sum-of-products expansion** or the **disjunctive normal form** of the Boolean function. Note that this sum is unique except for the order in which the minterms appear in the sum and the order of literals in each minterm.

Example 8.6

Construct the sum-of-products expansion by determining the values of the Boolean function $F(x, y, z) = \overline{x}(\overline{z} + \overline{y}) + \overline{y}\,\overline{z}$ for all possible values of the variables x, y, and z.

Solution

Table 8.6 presents the Boolean function $F(x, y, z)$ for all possible values of the variables x, y, and z. The sum-of-products expansion of $F(x, y, z) = \overline{x}\,\overline{z} + \overline{x}\,\overline{y} + \overline{y}\,\overline{z}$ is the Boolean sum of four minterms corresponding to the four rows of the table that give the value 1 for the function. We therefore have

$$F(x, y, z) = x\overline{y}\,\overline{z} + \overline{x}y\overline{z} + \overline{x}\,\overline{y}z + \overline{x}\,\overline{y}\,\overline{z}.$$

Another way to construct the sum-of-products expansion is to use Boolean identities to expand the product and then simplify. For every product in the sum-of-product expansion that does not involve the variable x_i, multiply the product by $(x_i + \overline{x}_i)$, rewrite the expression for the function so that no parentheses remain, and then delete any repeated products (i.e., remove any duplicate terms). This step is possible due to the fact that $x_i + \overline{x}_i = 1$, and the sum of two identical products P is P.

TABLE 8.6 Boolean table for Examples 8.6 and 8.7.

x	y	z	$\overline{x}$	$\overline{y}$	$\overline{z}$	$\overline{x}\,\overline{y}$	$\overline{x}\,\overline{z}$	$\overline{y}\,\overline{z}$	$F(x,y,z)$	xyz	$xy\overline{z}$	$x\overline{y}\,\overline{z}$	$G(x,y,z)$
1	1	1	0	0	0	0	0	0	0	1	0	0	1
1	1	0	0	0	1	0	0	0	0	0	1	0	1
1	0	1	0	1	0	0	0	0	0	0	0	0	0
1	0	0	0	1	1	0	0	1	1	0	0	1	1
0	1	1	1	0	0	0	0	0	0	0	0	0	0
0	1	0	1	0	1	0	1	0	1	0	0	0	0
0	0	1	1	1	0	1	0	0	1	0	0	0	0
0	0	0	1	1	1	1	1	1	1	0	0	0	0

Example 8.7

Determine the sum-of-products expansion for the function $G(x, y, z) = x\overline{\overline{y}\overline{z}}$. Comment on the result.

Solution

To obtain the sum-of-product expansion, we first use De Morgan's law, then simplify to remove parentheses and form the sum of product terms, next bring in each variable that does not involve a product term, and finally remove duplicate terms. Therefore we have

$$G(x, y, z) = x\overline{\overline{y}\overline{z}} = x(y + \overline{z}) = xy + x\overline{z} = xy(z + \overline{z}) + x\overline{z}(y + \overline{y})$$
$$= xyz + xy\overline{z} + x\overline{z}y + x\overline{z}\,\overline{y} = xyz + xy\overline{z} + x\overline{y}\,\overline{z}.$$

The last column of Table 8.6 presents the Boolean function $G(x, y, z)$ and confirms the above result.

8.5 Functional Completeness

A set of Boolean operators is **functionally complete** if every Boolean function can be defined using them. The triad of Boolean operators $\{+, \cdot, {}^-\}$ forms a functionally complete set. By using De Morgan's law, $x + y = \overline{\overline{x + y}} = \overline{\overline{x} \cdot \overline{y}}$, the Boolean operator "$+$" can be defined using "$\cdot$" and "$^-$." Therefore the pair of operators $\{\cdot, {}^-\}$ is a functionally complete set of two Boolean operators. Also, by using De Morgan's law, $x \cdot y = \overline{\overline{x \cdot y}} = \overline{\overline{x} + \overline{y}}$, the Boolean operator "$\cdot$" can be defined using "$+$" and "$^-$." Therefore the pair of operators $\{+, {}^-\}$ is also a functionally complete set of two Boolean operators. Note that the set $\{+, \cdot\}$ is not functionally complete because it is not possible to express the Boolean function $\overline{x}$ using the set $\{+, \cdot\}$.

It is important to note that there are sets each with only one operator that is functionally complete, namely, **NAND** (not AND) and **NOR** (not OR) operators, as respectively defined below:

$$NAND: x \uparrow y \triangleq \overline{xy} = \begin{cases} 0 & \text{if } x = y = 1 \\ \\ 1 & \text{otherwise} \end{cases}$$

and

$$NOR: x \downarrow y \triangleq \overline{x + y} = \begin{cases} 1 & \text{if } x = y = 0 \\ \\ 0 & \text{otherwise} \end{cases}$$

TABLE 8.7 Functional completeness of NAND and NOR operators.

Boolean operations	NAND operator "↑"	NOR operator "↓"
Boolean product "·" Boolean sum "+" Boolean complement "⁻"	$x.y = (x \uparrow y) \uparrow (x \uparrow y)$ $x + y = (x \uparrow x) \uparrow (y \uparrow y)$ $\overline{x} = x \uparrow x$	$x.y = (x \downarrow x) \downarrow (y \downarrow y)$ $x + y = (x \downarrow y) \downarrow (x \downarrow y)$ $\overline{x} = x \downarrow x$

Both NAND and NOR operators are functionally complete because either operator can express each of the triad of Boolean operators $\{+, \cdot, ^-\}$, as illustrated in Table 8.7.

Example 8.8
Verify that $\{\uparrow\}$ is functionally complete.

Solution
Because $\{\cdot, ^-\}$ is functionally complete, we only need to show that both operators "⁻" and "·" can be expressed in terms of $\{\uparrow\}$. By the definition of NAND operator and the idempotent law, we have

$$\overline{xy} = x \uparrow y \quad \rightarrow \quad \overline{xx} = x \uparrow x \quad \rightarrow \quad \overline{x} = x \uparrow x.$$

By using the above result, the definition of NAND operator, and the involution law, we have

$$xy = \overline{(\overline{xy})} = \overline{xy} \uparrow \overline{xy} = (x \uparrow y) \uparrow (x \uparrow y).$$

Example 8.9
Using Boolean tables, prove the following identities, where ↑ represents a NAND operator:
(a) $x \cdot y = (x \uparrow y) \uparrow (x \uparrow y)$.
(b) $x + y = (x \uparrow x) \uparrow (y \uparrow y)$.
(c) $\overline{x} = x \uparrow x$.

Solution
Noting that ↑ represents a NAND operator, we can build Boolean tables for both sides of each identity:
(a) As shown in Table 8.8(a), the third and the fifth columns are identical.
(b) As shown in Table 8.8(b), the third and the eighth columns are identical.
(c) As shown in Table 8.8(c), the second and the third columns are identical.

TABLE 8.8 Boolean tables for Example 8.9.

(a)

x	y	x·y	$\overline{x \cdot y} \equiv x \uparrow y$	$(x \uparrow y) \uparrow (x \uparrow y)$
1	1	1	0	1
1	0	0	1	0
0	1	0	1	0
0	0	0	1	0

(b)

x	y	x + y	x·x	$\overline{x \cdot x} \equiv x \uparrow x$	y·y	$\overline{y \cdot y} \equiv y \uparrow y$	$(x \uparrow x) \uparrow (y \uparrow y)$
1	1	1	1	0	1	0	1
1	0	1	1	0	0	1	1
0	1	1	0	1	1	0	1
0	0	0	0	1	0	1	0

(c)

x	$\overline{x}$	$\overline{x} \equiv x \uparrow x$
1	0	0
1	0	0
0	1	1
0	1	1

Example 8.10

Using Boolean tables, prove the following identities, where $\downarrow$ represents a NOR operator:

(a) $x \cdot y = (x \downarrow x) \downarrow (y \downarrow y)$.
(b) $x + y = (x \downarrow y) \downarrow (x \downarrow y)$.
(c) $\overline{x} = x \downarrow x$.

Solution

Noting that $\downarrow$ represents a NOR operator, we can build Boolean tables for both sides of each identity:

(a) As shown in Table 8.9(a), the third and the eighth columns are identical.
(b) As shown in Table 8.9(b), the third and the fifth columns are identical.
(c) As shown in Table 8.9(c), the second and the third columns are identical.

In summary, the five sets $\{+, \cdot, {}^-\}$, $\{+, {}^-\}$, $\{\cdot, {}^-\}$, $\{\uparrow\}$, and $\{\downarrow\}$ are all functionally complete, because each set can define any Boolean function.

8.6 Logic Gates

Logic circuits are used in all digital devices, from cell phones and computers to calculators and routers, and are built from basic elements of circuits called **gates.** Each type of gate

TABLE 8.9 Boolean tables for Example 8.10.

(a)

x	y	xy	$\bar{x}$	$\bar{y}$	$\overline{x+x}\equiv x\downarrow x$	$\overline{y+y}\equiv y\downarrow y$	$(x\downarrow x)\downarrow(y\downarrow y)$
1	1	1	0	0	0	0	1
1	0	0	0	1	0	1	0
0	1	0	1	0	1	0	0
0	0	0	1	1	1	1	0

(b)

x	y	$x+y$	$\overline{x+y}\equiv x\downarrow y$	$(x\downarrow y)\downarrow(x\downarrow y)$
1	1	1	0	1
1	0	1	0	1
0	1	1	0	1
0	0	0	1	0

(c)

x	$\bar{x}$	$\bar{x}\equiv x\downarrow x$
1	0	0
1	0	0
0	1	1
0	1	1

implements a Boolean operation. Each gate has one or more inputs but only one output. A gate is a function from B^n to B, where B denotes the Boolean algebra $\{0, 1\}$ and n represents the number of inputs. **Combinational circuits** are made up of different types of logic gates, where the output is a function of only the input, and not on the current state of the circuit, that is, combinational circuits have no memory capabilities. Combinational circuits are designed to perform a variety of tasks. There are three basic logic gates, namely, the AND gate, the OR gate, and inverter (the NOT gate), that can implement the three basic Boolean operations of product, sum, and complement, respectively. We follow the convention that the lines entering the gate symbol from the left are inputs and the single line on the right is the gate output.

An *inverter*, a **NOT gate**, produces an output bit that is complement to the input bit. Fig. 8.1(a) shows an inverter. An inverter resembles the logical operator NOT. For instance, if the input data sequence for an inverter is 1100011, the corresponding output data sequence is then 0011100.

An **OR gate** receives two or more arbitrary Boolean variables and outputs the Boolean sum of the input values. Fig. 8.1(b) shows the basic OR gate consisting of only two inputs. Note that n inputs into an OR gate yield an output of 1 if and only if at least one input is 1. For instance, if the input data for an OR gate are 00110011 and 10101010, the output data sequence is then 10111011. Note that the OR gate yields 0 only when all input bits are 0.

An **AND gate** receives two or more arbitrary Boolean variables and outputs the Boolean product of the input values. Fig. 8.1(c) shows the basic AND gate, which

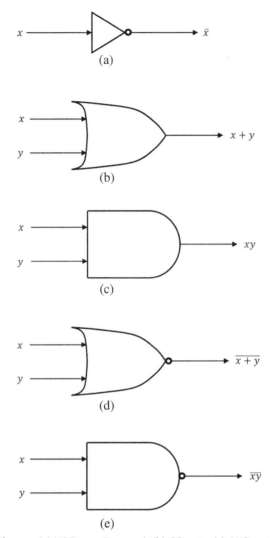

Fig. 8.1 Basic types of gates: (a) NOT gate (inverter), (b) OR gate, (c) AND gate, (d) NOR gate, and (e) NAND gate.

consists of only two inputs. Note that n inputs into an AND gate yields an output of 1 if and only if every input is 1. For instance, if the input data for an AND gate are 00110011 and 10101010, the output data sequence is then 00100010.

Example 8.11
Design a logic circuit for a light fixture controlled by four switches, where flipping any one of the switches turns the light on if it is off and turns it off if it is on.

Solution

We first construct the Boolean table with four variables x, y, z, and w, when one of the four switches changes, the light status changes, as shown in Table 8.10. The Boolean expression reflecting the implementation of the circuit with four switches x, y, z, and w is as follows:

$$wxyz + \overline{w}\,\overline{x}yz + \overline{w}x\overline{y}z + \overline{w}xy\overline{z} + w\overline{x}\,\overline{y}z + w\overline{x}y\overline{z} + wx\overline{y}\,\overline{z} + \overline{w}\,\overline{x}\,\overline{y}\,\overline{z}.$$

The circuit will have 32 inputs, combined by AND gates in groups of four, with inverters where necessary, to produce outputs corresponding to the eight minterms in this expression. These outputs are combined with one big OR gate consisting of eight inputs.

There are two additional gates, namely a NAND gate, equivalent to an AND gate followed by an inverter, and a NOR gate, equivalent to an OR gate followed by an inverter. NAND and NOR gates are shown in Fig. 8.1(d) and (e), respectively, which are basically AND and OR gates each followed by a circle reflecting an inverter. The output of a NAND gate is 0 if and only if all the inputs are 1, and the output of a NOR gate is 1 if and only if all the inputs are 0. NAND and NOR gates each can be combined to perform the basic three Boolean operations of AND, OR, and NOT gates because each is functionally complete. Table 8.11 summarizes the Boolean tables for all these five gates. Many basic logic circuits used in digital devices are built from NAND

TABLE 8.10 Boolean table for Example 8.11.

x	y	z	w	$F(x,y,z,w)$
1	1	1	1	1
1	1	1	0	0
1	1	0	1	0
1	1	0	0	1
1	0	1	1	0
1	0	1	0	1
1	0	0	1	1
1	0	0	0	0
0	1	1	1	0
0	1	1	0	1
0	1	0	1	1
0	1	0	0	0
0	0	1	1	1
0	0	1	0	0
0	0	0	1	0
0	0	0	0	1

TABLE 8.11 Boolean tables for basic logic gates.

Input (x)	Input (y)	NOT gate $(\bar{x})$	OR gate $(x + y)$	AND gate (xy)	NOR gate $\overline{x + y}$	NAND gate $\overline{xy}$
1	1	0	1	1	0	0
1	0	0	1	0	0	1
0	1	1	1	0	0	1
0	0	1	0	0	1	1

gates. It is advantageous to reduce the number of kinds of logic gates, but the price to be paid is an increase in the total number of gates.

Example 8.12

The statement $x\bar{y} + z$ can be implemented by using three basic gates OR, AND, and NOT gates. Assuming only NAND gates should be used to implement such a statement, how many NAND gates are then required?

Solution

Assuming $w = x\bar{y}$, we have

$$w = x\bar{y} = (x \uparrow \bar{y}) \uparrow (x \uparrow \bar{y}) = (x \uparrow (y \uparrow y)) \uparrow (x \uparrow (y \uparrow y)).$$

Therefore we have

$$x\bar{y} + z = w + z = (w \uparrow w) \uparrow (z \uparrow z)$$
$$= ((x \uparrow (y \uparrow y)) \uparrow (x \uparrow (y \uparrow y)) \uparrow (x \uparrow (y \uparrow y)) \uparrow (x \uparrow (y \uparrow y))) \uparrow (z \uparrow z).$$

The final logical equivalence contains only one kind of operator as opposed to three kinds of operators in the original statement. However, there are only three operators in the original statement, whereas there are 13 logical operators in the new statement.

8.7 Minimization of Combinational Circuits

It is important to note that combinational circuits are equivalent if and only if their corresponding Boolean expressions are equal or their Boolean tables are identical. In order to design a combinational circuit, we need to have a table specifying the output for each combination of input values. We then determine the sum-of-product expansion to find a set of logic gates that can implement the combinational circuit. However, the sum-of-product expansion generally contains more terms than necessary. The Boolean identities along with the **binary expression simplification rule**, which states $ef + \bar{e}f = f$, where e and f are binary expressions, can be used iteratively to reduce an expression into a simpler, but equivalent, expression. In order to minimize the number of logic gates, it is important to produce Boolean sums of products that represent a Boolean function with the fewest products of literals such that these products contain the fewest literals possible among all sums of products. This process is called the **minimization of the Boolean function**, by which a circuit with the fewest gates and fewest inputs can be constructed.

Example 8.13

Simplify the following Boolean expression so as to be able to obtain a simpler combinational circuit:

$$F(x, y, z) = xy\overline{z} + x\overline{y}z + x\overline{y}\,\overline{z} + xyz + \overline{x}yz.$$

Solution

Using idempotent law $(x + x = x)$, we add the xyz term to the expression, which is already in the expression. Using the unity property $(x + \overline{x} = 1)$, the identity law $(x \cdot 1 = x)$, and the binary expression simplification rule law, we can then simplify the expression:

$$
\begin{aligned}
F(x, y, z) &= xy\overline{z} + x\overline{y}z + x\overline{y}\,\overline{z} + xyz + \overline{x}yz \\
&= xy\overline{z} + x\overline{y}z + x\overline{y}\,\overline{z} + xyz + (xyz + \overline{x}yz) \\
&= x(y\overline{z} + \overline{y}z + \overline{y}\,\overline{z} + yz) + yz(x + \overline{x}) = x(y + \overline{y})(z + \overline{z}) + yz \\
&= x \cdot 1 \cdot 1 + yz = x \cdot 1 + yz = x + yz.
\end{aligned}
$$

If we do not simplify the original expression, we need three inverters, five AND gates, and four OR gates, whereas after simplification, we need only one AND gate and one OR gate.

Reducing the number of gates on a chip can lead to an increase in circuit reliability, a decrease in cost production, an increase in the number of circuits on a chip, and a reduction in processing time required by a circuit. However, simplification of a Boolean algebra to reduce the number of logic gates may be a very difficult task because grouping various terms and applying the laws of Boolean algebra may not always be quite straightforward.

Minimizing Boolean functions with many variables is a computationally intensive problem, but there are methods that can significantly simplify, but not necessarily minimize, Boolean expressions with a large number of literals. One such method is the Karnaugh map, which is an effective graphical method involving just a few variables, as it becomes significantly more difficult when the number of variables is beyond a few.

The essence of the **Karnaugh map** lies in grouping minterms that differ by exactly one literal. It proceeds by circling pairs of minterms that are candidates for simplification. The Karnaugh map has a rectangular grid of squares. Each square presents a possible minterm in a sum-of-products expansion of the Boolean expression that represents the circuit. We label the squares, so the minterms in any two adjacent squares in each row and in each column differ by exactly one literal. Each square contains a 1 if the corresponding minterm exists in the Boolean expression.

We draw loops around squares that represent minterms that can be combined and then find the corresponding sum of products. It is, however, important to note that we need to identify the largest possible loops of squares first and to cover all the 1s with the fewest loops using the largest loops. Each square representing a minterm must either be used to form a product using fewer literals or be part of the expansion. Two squares are adjacent if and only if the minterms they represent differ in only one literal. Note that if there is a 1 in every square of the Karnaugh map, then the Boolean expression can be combined into the Boolean expression 1 that involves none of the variables.

A 2×2 Karnaugh map corresponds to Boolean expressions $F(x, y)$ with two variables x and y. Accordingly, the four possible minterms with two literals, xy, $x\overline{y}$, $\overline{x}y$, and $\overline{x}\overline{y}$, are represented by the four squares in the map. Noting that two squares are defined as adjacent if they have a side in common, each square is adjacent to two other squares in a 2×2 Karnaugh map. The simplification of a sum-of-products expansion in two variables is carried out by identifying those loops of two and four squares that represent minterms that can be combined. The minterms in two adjacent squares can be combined to involve just one of the two variables, the variable that is common to both squares. For instance, $\overline{x}y + \overline{x}\overline{y}$ can be simplified to $\overline{x}$, as $\overline{x}y$ and $\overline{x}\overline{y}$ are adjacent squares and $\overline{x}$ is common to both squares.

Example 8.14
Find the Karnaugh map and simplify each of the following Boolean expressions:
(a) $xy + x\overline{y}$.
(b) $xy + \overline{x}\overline{y}$.
(c) $xy + x\overline{y} + \overline{x}y$.
(d) $xy + x\overline{y} + \overline{x}y + \overline{x}\overline{y}$.

Solution
Using the Karnaugh map, the minterms for each expression, as shown in Fig. 8.2, are grouped. The simplifications are as follows:
(a) xy and $x\overline{y}$ are adjacent and both involve x. We thus have $xy + x\overline{y} = x$.
(b) xy and $\overline{x}\overline{y}$ are not adjacent. Therefore $xy + \overline{x}\overline{y}$ cannot be simplified.
(c) xy and $x\overline{y}$ are adjacent and both involve x and also xy and $\overline{x}y$ are adjacent and both involve y. We thus have $xy + x\overline{y} + \overline{x}y = x + y$.

(d) xy and $x\bar{y}$ are adjacent and both involve x and also $\bar{x}y$ and $\bar{x}\,\bar{y}$ are adjacent and both involve $\bar{x}$. We thus have $xy + x\bar{y} + \bar{x}y + \bar{x}\,\bar{y} = x + \bar{x} = 1$.

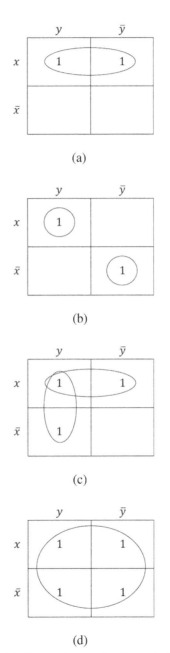

Fig. 8.2 Karnaugh maps for Example 8.14.

A 2×4 Karnaugh map corresponds to Boolean expressions $F(x, y, z)$ with three variables x, y, and z. Accordingly, the eight possible minterms with three literals (xyz, $xy\overline{z}$, $x\overline{y}\,\overline{z}$, $x\overline{y}z$, $\overline{x}yz$, $\overline{x}y\overline{z}$, $\overline{x}\,\overline{y}\,\overline{z}$, $\overline{x}\,\overline{y}z$) are represented by the eight squares in the map. A square can have three adjacent squares. The simplification of a sum-of-products expansion in three variables is carried out by identifying those loops of two, four, and eight squares that represent minterms that can be combined.

Note that in each of the two rows, the first square and the fourth square are considered to be adjacent because they differ by only one literal. The minterms in two adjacent squares can be combined to involve just two variables, the variables that are common to both squares. The minterms in the four squares of a row can be combined to involve just one variable, the variable that is common to all four squares. The minterms in the four squares that share a corner (i.e., two adjacent squares in a row and the two adjacent squares exactly below them) can be combined to involve just one variable, the variable that is common to all four squares.

Example 8.15
Find the Karnaugh map and simplify each of the following Boolean expressions:
(a) $xyz + x\overline{y}z + \overline{x}y\overline{z} + \overline{x}\,\overline{y}\,\overline{z}$.
(b) $xyz + x\overline{y}z + x\overline{y}\,\overline{z} + xy\overline{z}$.
(c) $xyz + \overline{x}yz + \overline{x}\,\overline{y}z + x\overline{y}z$.
(d) $xyz + xy\overline{z} + x\overline{y}z + x\overline{y}\,\overline{z} + \overline{x}y\overline{z}$.

Solution
Using the Karnaugh map, the minterms for each expression, as shown in Fig. 8.3, are grouped. The simplifications are as follows:
(a) $xz + \overline{x}\,\overline{z}$.
(b) x.
(c) z.
(d) $x + y\overline{z}$.

A 4×4 Karnaugh map corresponds to Boolean expressions $F(x, y, z, w)$ with four variables x, y, z, and w. Accordingly, the sixteen possible minterms with four literals ($xyzw$, $xy\overline{z}w$, $x\overline{y}\,\overline{z}w$, $x\overline{y}zw$, $\overline{x}yzw$, $\overline{x}y\overline{z}w$, $\overline{x}\,\overline{y}\,\overline{z}w$, $\overline{x}\,\overline{y}zw$, $xyz\overline{w}$, $xy\overline{z}\,\overline{w}$, $x\overline{y}\,\overline{z}\,\overline{w}$, $x\overline{y}z\overline{w}$, $\overline{x}yz\overline{w}$, $\overline{x}y\overline{z}\,\overline{w}$, $\overline{x}\,\overline{y}\,\overline{z}\,\overline{w}$, $\overline{x}\,\overline{y}z\overline{w}$) are represented by the sixteen squares in the map. Each square is adjacent to four other squares. The simplification of a sum-of-products expansion in four variables is carried out by identifying those loops of 2, 4, 8, and 16 squares that represent minterms that can be combined.

Note that in each of the four rows of squares, the first square and the fourth square are considered to be adjacent, as they differ by only one literal, and in each of the four columns of squares, the first square and the fourth square are considered to be adjacent, as they differ only in one literal. The minterms in two adjacent squares can be combined to involve just three variables, the variables that are common to both squares. The minterms

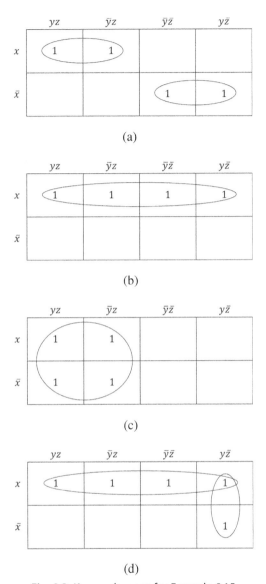

Fig. 8.3 Karnaugh maps for Example 8.15.

in the four squares of a row or in the four squares of a column can be combined to involve just two variables, the variables that are common to all four squares. The minterms in the four squares that share a corner (i.e., two adjacent squares in a row and the two adjacent squares exactly below them) can be combined to involve just two variables, the variables that are common to all four squares. The minterms in the eight squares of two adjacent rows or in the eight squares of two adjacent columns can be combined to involve to just one variable, the variable that is common to all eight squares. We must always begin with the largest loops of squares and use the minimum number of loops.

Example 8.16

Find the Karnaugh map and simplify each of the following Boolean expressions:

(a) $xyzw + \bar{x}ywz + x\bar{y}\,\bar{w}\,\bar{z} + x\bar{y}w\bar{z} + \bar{x}y\bar{w}\,\bar{z}$.

(b) $x\bar{y}wz + x\bar{y}\,\bar{w}z + x\bar{y}\,\bar{w}\,\bar{z} + x\bar{y}w\bar{z} + xywz$.

(c) $xy\bar{w}z + \bar{x}y\bar{w}z + \bar{x}\,\bar{y}\,\bar{w}z + x\bar{y}\,\bar{w}z + \bar{x}y\bar{w}\,\bar{z} + \bar{x}\,\bar{y}\,\bar{w}\,\bar{z}$.

(d) $xywz + x\bar{y}wz + xyw\bar{z} + x\bar{y}w\bar{z}$.

Solution

Using the Karnaugh map, the minterms for each expression, as shown in Fig. 8.4, are grouped. The simplifications are as follows:

(a) $ywz + x\bar{y}\,\bar{z} + \bar{x}y\bar{w}\,\bar{z}$.

(b) $x\bar{y} + xwz$.

(c) $\bar{w}z + \bar{x}\,\bar{w}$.

(d) wx.

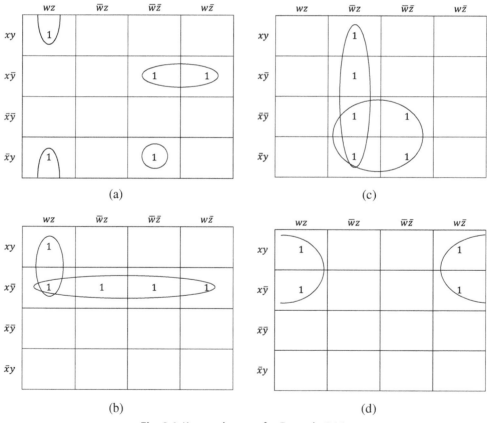

Fig. 8.4 Karnaugh maps for Example 8.16.

In using Karnaugh maps to simplify, cover squares according to the following rules:

- Cover all marked squares at least once.
- Cover the largest possible groups of marked squares.
- Do not cover any unmarked square.
- Use the fewest groups possible.

It is important to highlight that in some combinational circuits, some combination of inputs never occurs. Consequently, we don't care about corresponding output values in such cases. This allows us to construct a simple circuit with the desired output while arbitrarily choosing the output values for those combinations that never occur. To this effect, a d is placed in each square in the Karnaugh map corresponding to a ***don't care condition***, which simply means that the corresponding value of the function can be arbitrarily assigned. In the minimization process, a d can count as a 1 if that leads to the largest blocks of squares in the Karnaugh map, that is, the solution depends on a judicious choice of d's.

Example 8.17
Simplify the Boolean expression presented by the Karnaugh map shown in Fig. 8.5.

Solution
Loops are drawn to take advantage of those d's that lead to the largest blocks of squares. The Boolean expression is thus as follows:

$$\left(x\bar{y}wz + x\bar{y}\,\bar{w}z + \bar{x}\,\bar{y}wz + \bar{x}\,\bar{y}\,\bar{w}z\right) + \left(\bar{x}ywz + \bar{x}yw\bar{z} + xywz + xyw\bar{z}\right)$$

$$= \bar{y}z(xw + x\bar{w} + \bar{x}w + \bar{x}\,\bar{w}) + yw(\bar{x}z + \bar{x}\,\bar{z} + xz + x\bar{z}) = \bar{y}z + yw.$$

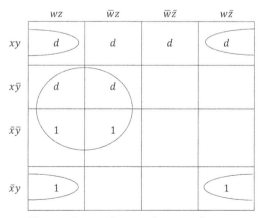

Fig. 8.5 Karnaugh maps for Example 8.17.

Karnaugh maps can be, of course, used to minimize Boolean functions with very few variables, but beyond that, they become extremely complex to analyze. However, the concepts of Karnaugh maps are important because they help understand some of the new algorithms.

There is another procedure, called the **Quine–McCluskey method**, which can simplify sum-of-products expansions and thus minimize the number of circuits. Functionally, the Quine–McCluskey method is identical to the Karnaugh map, but it uses tables. Its tabular form makes it more efficient for use in computer algorithms when a large number of inputs are available, and it also gives a deterministic way to check that the minimal form of a Boolean function has been reached. The method involves two steps: (i) finding those terms that are candidates for inclusion in a minimal expansion as a Boolean sum of Boolean products, and (ii) determining which of these terms are essential to use by successively combining terms into terms with one fewer literal.

Exercises

(8.1)

Determine the values of the following Boolean expressions:

(a) $1 \cdot a + (0 + b) \cdot c$, for $a = 0$, $b = 1$, and $c = 1$.

(b) $\overline{(0 + 1)} + 0 \cdot 1$.

(8.2)

Using Boolean tables, show for each of the following cases both Boolean functions are equivalent:

(a) $x\bar{y} + y\bar{z} + \bar{x}z = \bar{x}y + \bar{y}z + x\bar{z}$.

(b) $x + yz = (x + y)(x + z)$.

(8.3)

Express each Boolean expression $E(x, y, z)$ as a sum of products and then in its complete sum-of-products form:

(a) $E = z(\bar{x} + y) + \bar{y}$.

(b) $E = \overline{y(x + yz)}$.

(c) $E = \overline{(\bar{x} + y)} + \bar{x}y$.

(8.4)

Let $E = x\bar{y} + xy\bar{z} + \bar{x}y\bar{z}$. Prove the following statements:

(a) $x\bar{z} + E = E$.

(b) $x + E \neq E$.

(8.5)

Construct a digital circuit using AND and OR gates that implements majority voting for five individuals, where each individual can vote either yes or no for each proposal, and a proposal is passed if it receives at least three yes votes.

(8.6)

(a) Consider the expression $f(x, y, z) = x\bar{z} + x\bar{y}z + \bar{x}$. Write a Boolean function in sum-of-products expansion.

(b) Consider the expression $f(x, y) = x\bar{y} + \bar{x}y$. Write a Boolean function in product-of-sums expansion.

(8.7)

Determine the Boolean expression for the following cases:

(a) Half adder: Add two bits without considering a carry from a previous addition; compute the sum bit and the carry bit. Note that the circuit has two inputs x and y and two output bits s and c.

(b) Full adder: Add two bits with considering a carry from a previous addition; compute the sum bit and the carry bit. Note that the circuit has three inputs $(x, y,$ and $c_i)$ and two output bits $(s$ and $c_{i+1})$. Expand on how half adders can be used to implement a full adder.

(8.8)

Construct logic circuits for the following Boolean expressions, where w is the output and x, y, z are the inputs:

(a) $w = \overline{(x + yz)} + y$.

(b) $w = \overline{(\bar{x}y)} + \overline{(x + z)}$.

(c) $w = x\bar{y} + \overline{(x + y)} + \overline{(\bar{x}y)}$.

(8.9)

Using the Karnaugh map, simplify the following Boolean expressions:

(a) $F = xyzw + xy\bar{z}w + x\bar{y}z\bar{w} + x\bar{y}\,\bar{z}\,\bar{w} + \bar{x}\,\bar{y}\,\bar{z}\,\bar{w} + \bar{x}yzw + \bar{x}yz\bar{w} + \bar{x}y\bar{z}\,\bar{w} + \bar{x}y\bar{z}w$.

(b) $G = xy\bar{z}\,\bar{w} + xy\bar{z}w + x\bar{y}zw + x\bar{y}z\bar{w} + \bar{x}\,\bar{y}zw + \bar{x}\,\bar{y}z\bar{w} + \bar{x}y\bar{z}\,\bar{w}$.

(8.10)

Simplify the Boolean expression presented by the Karnaugh map shown in Fig. 8.6.

	yz	$y\bar{z}$	$\bar{y}\bar{z}$	$\bar{y}z$
wx	d	d	d	d
$w\bar{x}$	d	d	1	1
$\bar{w}\bar{x}$	1	1	1	
$\bar{w}x$	1			1

Fig. 8.6 Karnaugh map for Exercise 8.10.

CHAPTER 9

Relations

Contents

There are many forms of relations. Relationships exist between people, such as parent-child, student-teacher, and employer-employee. Relationships exist also in mathematics, such as less than, parallel to, a subset of, logarithm of, and factorial of. In fact, all mathematical functions are a special type of relation. In a way, a relation considers the existence of a certain connection often between pairs of objects in a definite order. In this chapter, we briefly present the mathematics of relations defined on sets, focus on ways to represent relations, and explore various properties they may have.

9.1 Relations on Sets

In the context of mathematics of relations, relationships between two sets are often based on ordered pairs made up of two related elements, each belonging to a set. An ***ordered pair*** of elements a and b is denoted by (a, b), while noting that $(a, b) \neq (b, a)$ unless $a = b$. The sets of ordered pairs are called ***binary relations***. The binary relations are in contrast to ***n-ary relations***, which express relationships among elements of n sets with $n > 2$ being an integer and thus involve ordered n-tuples. Such a relation is the fundamental structure used in relational databases. The term *relation* by itself generally refers to a binary relation unless otherwise stated or implied.

A ***relation*** between the sets A and B is a subset R of the Cartesian product $A \times B$, where the Cartesian product is defined as $A \times B = \{(a, b) | a \in A \text{ and } b \in B\}$. If $(a, b) \in R$, it is then read as a is related to b. The set A is called the ***domain of the relation***, and the set B is called the ***range of the relation***. If $(a, b) \notin R$, it is then read as a is not related to b. If $A = B$, the relation is said to be a ***relation on*** A.

Discrete Mathematics
ISBN 978-0-12-820656-0, https://doi.org/10.1016/B978-0-12-820656-0.00009-5
155

The relation R is a ***one-to-one relation*** if no element of B appears as a second coordinate in more than one ordered pair in R, and the relation R is an ***onto relation*** if every element of B appears as a second coordinate in at least one ordered pair in R.

Unlike functions, every relation has an inverse. If R is a relation between the sets A and B, then the inverse relation of R, denoted by R^{-1}, is a subset of the Cartesian product $B \times A$. In other words, the ***inverse relation*** is defined as follows: $R^{-1} = \{(b, a)|(a, b) \in R\}$. The domain and range of R^{-1} are equal, respectively, to the range and domain of R. Moreover, if R is a relation on A, then R^{-1} is also a relation on A. The ***complementary relation*** $\overline{R}$ is the set of ordered pairs, which is defined as follows: $\overline{R} = \{(a, b)|(a, b) \notin R\}$.

Example 9.1

(a) Let $A = \{1, 2, 3\}$ and $B = \{7, 8, 9\}$. Determine $A \times B$ and $B \times A$.

(b) Let $A = \{12, 14, 18, 20\}$ and $B = \{6, 8, 10\}$, and define a relation R from A to B as follows: $R = \{(a, b)|a$ is a multiple of $b\}$. Determine the domain and range of R.

(c) Let $A = \{2, 3, 4\}$ and $B = \{8, 9, 10\}$. Let R be the divisibility relation from A to B (i.e., for all $(a, b) \in A \times B$, a divides b). Determine R^{-1} and describe it in words.

Solution

(a) $A \times B = \{(1,7), (1,8), (1,9), (2,7), (2,8), (2,9), (3,7), (3,8), (3,9)\}$

$B \times A = \{(7,1), (8,1), (9,1), (7,2), (8,2), (9,2), (7,3), (8,3), (9,3)\}$.

(b) Because we have $R = \{(12,6), (18,6), (20,10)\}$, the domain of R is $\{12, 18, 20\}$ and the range of R is $\{6, 10\}$.

(c) Because we have the relation $R = \{(2,8), (2,10), (3,9), (4,8)\}$, we have the inverse relation $R^{-1} = \{(8,2), (10,2), (9,3), (8,4)\}$. Therefore for all $(b, a) \in R^{-1}$, b is a multiple of a.

9.2 Properties of Relations

A binary relation on a set A is a binary relation from A to A that is a subset of $A \times A$. There are various ways to classify relations on a set. In this section, we focus on the most important properties that a relation R on a set A can have, namely, reflexive, symmetric, and transitive. In order to prove a relation has one of these properties, the method of exhaustion or the method of generalization needs to be employed.

A relation R on a set A is ***reflexive*** if and only if $(a, a) \in R$ for every element $a \in A$. In informal terms, in a reflexive relation, each element is related to itself. An example of a reflexive relation is the equality relation on the set of real numbers, as every real number is equal to itself, another example of a reflexive relation is the divides relation on the set of positive integers, as every positive integer divides itself. Using quantifiers, the relation R on the set A is reflexive if $\forall a((a,\ a) \in R)$.

A relation R on a set A is **antireflexive**, also known as **irreflexive**, if and only if $(a,\ a) \notin R$ for every element $a \in A$. In informal terms, no element in A is related to itself. An example of an antireflexive relation is the greater-than relation on the real numbers. Note that antireflexive does not mean not reflexive, as it is possible to define a relation where some elements are related to themselves, but others are not (i.e., neither all nor none is).

Example 9.2
(a) Provide examples for reflexive and antireflexive relations and for a relation that is neither reflexive nor antireflexive.

(b) Assuming the following three relations are defined on the set $A = \{1,\ 2,\ 3\}$, determine which of the relations are reflexive, which are not reflexive, and which are antireflexive:

$$R_1 = \{(1,1), (1,2), (2,2), (3,1), (3,3)\}$$

$$R_2 = \{(1,1), (1,2), (2,3), (3,1), (3,3)\}$$

$$R_3 = \{(1,3), (1,2), (2,3), (3,1), (3,1)\}$$

Solution
(a) Consider the relation that the product of two positive integers is even on the set A. If A is the set of even numbers, the relation is then reflexive because the square of an even number is even. If A is the set of odd numbers, the relation is then antireflexive because the square of an odd number is not even. If A is the set of natural numbers (i.e., positive integers), the relation is then neither reflexive nor antireflexive.

(b) R_1 is reflexive as it contains all the three pairs $(1,\ 1)$, $(2,\ 2)$, and $(3,\ 3)$. R_2 is not reflexive, as it does not contain all the three pairs $(1,\ 1)$, $(2,\ 2)$, and $(3,\ 3)$; more specifically, $(2,\ 2)$ does not belong to R_2. R_3 is antireflexive, as it does not contain even one of the three pairs $(1,\ 1)$, $(2,\ 2)$, and $(3,\ 3)$.

A relation R on a set A is **symmetric** if and only if $(a,\ b) \in R$, then $(b,\ a) \in R$ for all $a,\ b \in A$. Thus R is not symmetric if there exists $a \in A$ and $b \in A$ such that $(a, b) \in R$ but $(b,\ a) \notin R$. In informal terms, in a symmetric relation, if any one element is related to any other element, then the second element is related to the first as well. An example of a symmetric relation is the equality relation on the set of real numbers because if $a = b$ is true, then $b = a$ is also true. Using quantifiers, the relation R on the set A is symmetric if $\forall a \forall b((a,\ b) \in R \rightarrow (b,\ a) \in R)$.

A relation R on a set A is ***antisymmetric*** if and only if $(a, b) \in R$, then $(b, a) \notin R$ for all $a, b \in A$ and $a \neq b$. In informal terms, in an antisymmetric relation, if any one element is related to any other element, then the second element cannot be related to the first. An example of an antisymmetric relation is the divisibility relation on the natural numbers. Using quantifiers, the relation R on the set A is antisymmetric if $\forall a \forall b(((a, b) \in R \wedge (b, a) \in R) \rightarrow (b = a))$.

A relation R is called ***asymmetric*** if $(a, b) \in R$ implies that $(b, a) \notin R$. Note that anti-symmetric does not mean not symmetric, as it is possible to define a relation that may lack both these properties. A relation cannot be both symmetric and antisymmetric if it contains some pair of the form (a, b), where $a \neq b$.

Example 9.3
Assuming the following four relations are defined on the set $A = \{1, 2, 3, 4\}$, determine which of the relations are symmetric or antisymmetric:
(a) $R_1 = \{(1, 1), (1, 2), (2, 3), (1, 3), (4, 4)\}$
(b) $R_2 = \{(1, 1), (1, 2), (2, 1), (2, 2), (3, 3), (4, 4)\}$
(c) $R_3 = \{(1, 3), (3, 1), (2, 3)\}$
(d) $R_4 = \{(1, 1), (2, 2)\}$

Solution
(a) R_1 is not symmetric because $(1, 2) \in R_1$, but $(2, 1) \notin R_1$. However, R_1 is antisymmetric.
(b) R_2 is not antisymmetric, as $(1, 2)$ and $(2, 1)$ belong to R_2, but $2 \neq 1$. However, R_2 is symmetric.
(c) R_3 is neither symmetric nor antisymmetric.
(d) R_4 is both symmetric and antisymmetric.

A relation R on a set A is called ***transitive*** if and only if $(a, b) \in R$ and $(b, c) \in R$, then $(a, c) \in R$, for all $a, b, c \in A$. In informal terms, if any one element is related to a second element and that second element is related to a third element, then the first element is related to the third element. Examples of transitive relation may include the less than $(<)$ relation or the subset $(\subseteq)$ relation. Using quantifiers, the relation R on the set A is transitive if $\forall a \forall b \forall c((a, b) \in R \wedge (b, c) \in R \rightarrow (a, c) \in R)$.

The relation R is not transitive, also called ***intransitive***, if for all $a, b, c \in A$, $(a, b) \in R$, $(b, c) \in R$, but $(a, c) \notin R$. An example of intransitive relation is if in plane geometry the straight lines a and b are perpendicular to one another and the straight lines b and c are perpendicular to one another, then the straight lines a and c are not perpendicular to one another. Using quantifiers, the relation R on the set A is intransitive if $\forall a \forall b \forall c((a, b) \in R \wedge (b, c) \in R \rightarrow (a, c) \notin R)$.

Table 9.1 summarizes the properties of a binary relation R on a set A and their requirements, where $a, b, c \in A$.

TABLE 9.1 Requirements for various properties of a binary relation.

Property	Requirement
Reflexive	$(a, a) \in R$ for every element $a \in A$
Irreflexive	$(a, a) \notin R$ for every element $a \in A$
Symmetric	$(a, b) \in R \rightarrow (b, a) \in R$, for all $a, b \in A$
Antisymmetric	$(a, b) \in R \rightarrow (b, a) \notin R$, for all $a, b \in A$ and $a \neq b$
Transitive	$(a, b) \in R$ and $(b, c) \in R \rightarrow (a, c) \in R$, for all $a, b, c \in A$
Intransitive	$(a, b) \in R$ and $(b, c) \in R \rightarrow (a, c) \notin R$, for all $a, b, c \in A$

Example 9.4

Consider the "likes" relation on the set of people. Highlight various properties of this relation.

Solution

- Reflexivity: Everyone likes themselves.
- Irreflexivity: No one likes themselves.
- Symmetry: If a likes b, then b likes a.
- Antisymmetry: No pair of distinct people like each other.
- Transitivity: If a likes b and b likes c, then a likes c too.
- Intransitivity: If a likes b and b likes c, then a does not like c.

Example 9.5

Determine which of the following relations on the set of real numbers are reflexive and/or symmetric and/or transitive:

(a) The equality relation.
(b) The inequality relation.
(c) The greater-than-or-equal-to or less-than-or-equal-to relation.
(d) The greater-than or less-than relation.

Solution

Using Table 9.1, Table 9.2 presents the properties of the relations.

TABLE 9.2 Properties of relations for Example 9.5.

Set	Relation	Reflexive?	Symmetric?	Transitive?
Any nonempty set	$=$	Yes	Yes	Yes
Any nonempty set	$\neq$	No	Yes	No
Real numbers	$\leq$ or $\geq$	Yes	No	Yes
Real numbers	$<$ or $>$	No	No	Yes

Example 9.6

Determine whether the following relations R on the set of all people living in the world is reflexive and/or symmetric and/or transitive:

(a) $(a, b) \in R$ if and only if a has more money than b.
(b) $(a, b) \in R$ if and only if a and b were born on the same day.
(c) $(a, b) \in R$ if and only if a and b have a common grandparent.

Solution

(a) The relation is neither reflexive nor symmetric, but it is transitive.
(b) The relation is reflexive, symmetric, and transitive.
(c) The relation is both reflexive and symmetric, but it is not transitive.

The total number of binary relations on a set with n elements is 2^{n^2}. Table 9.3 presents formulas for the number of binary relations with various properties while noting that they can all be derived using mathematical induction.

Example 9.7

Determine all relations on the set $A = \{1, 2\}$, and identify the relations that are reflexive and/or symmetric.

Solution

In the set A, we have two elements, therefore $n = 2$. As shown in Table 9.4, the total number of relations is 16, out of which 4 are reflexive and 8 are symmetric.

9.3 Representations of Relations

There are various ways to represent a binary relation between two finite sets. Suppose that the relation is from the set A to the set B, where the elements of A and B have been listed in some arbitrary order. A set of ordered pairs reflecting a binary relation from A to B can be represented by tables, arrow diagrams, digraphs, and matrices.

TABLE 9.3 Formulas for various properties of a binary relation.

Type of relation	Number of relations
Reflexive	$2^{n(n-1)}$
Irreflexive	$2^{n(n-1)}$
Symmetric	$2^{\frac{n(n+1)}{2}}$
Antisymmetric	$2^n \times 3^{\frac{n(n-1)}{2}}$
Asymmetric	$3^{\frac{n(n-1)}{2}}$

TABLE 9.4 Properties of relations for Example 9.7.

Relations	Reflexive?	Symmetric?
$R_0 = \{\varnothing\}$	No	Yes
$R_1 = \{(1,1)\}$	No	Yes
$R_2 = \{(1,2)\}$	No	No
$R_3 = \{(2,1)\}$	No	No
$R_4 = \{(2,2)\}$	No	Yes
$R_5 = \{(1,1),(1,2)\}$	No	No
$R_6 = \{(1,1),(2,1\}$	No	No
$R_7 = \{(1,1),(2,2)\}$	Yes	Yes
$R_8 = \{(1,2),(2,1)\}$	No	Yes
$R_9 = \{(1,2),(2,2)\}$	No	No
$R_{10} = \{(2,1),(2,2)\}$	No	No
$R_{11} = \{(1,1),(1,2),(2,1)\}$	No	Yes
$R_{12} = \{(1,1),(1,2),(2,2)\}$	Yes	No
$R_{13} = \{(1,1),(2,1),(2,2)\}$	Yes	No
$R_{14} = \{(1,2),(2,1),(2,2)\}$	No	Yes
$R_{15} = \{(1,1),(1,2),(2,1),(2,2)\}$	Yes	Yes

Tables can be used to represent binary relations on the same set or on two different sets. In a *table*, columns are labeled by the elements of the finite set A, and rows are labeled by the elements of the finite set B. Only the entries of the table that show the set of the ordered pairs are marked. In other words, if a certain entry in the table highlights an ordered pair that is not in the set of ordered pairs reflecting the relation of interest, it is then left unmarked.

Arrow diagrams can show binary relations on the same set or on two different sets using two disjoint disks. In an *arrow diagram*, the elements of the finite set A (the domain of the relation) are shown in the left-hand disk and the elements of the finite set B (the range of the relation) are shown in the right-hand disk; then arrows from the elements in the left-hand disk to the elements in the right-hand disk are drawn to represent all ordered pairs reflecting the relation of interest.

Example 9.8

Noting $A = \{a_1, a_2, a_3\}$ and $B = \{b_1, b_2, b_3, b_4\}$, the relation R from the set A to the set B is defined by the set of ordered pairs $\{(a_1, b_1), (a_1, b_2), (a_2, b_3), (a_3, b_3), (a_3, b_4)\}$. Represent this relation in a table and an arrow diagram.

Solution

The relation R can be represented graphically using a table and an arrow diagram, as shown in Fig. 9.1.

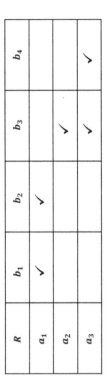

R	b_1	b_2	b_3	b_4
a_1	✓	✓		
a_2			✓	
a_3			✓	✓

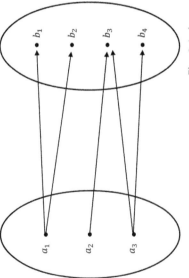

Fig. 9.1 Arrow diagram and table for Example 9.8.

Digraphs, also known as directed graphs, will be extensively discussed in the chapter on graphs. However, we briefly mention it here in the context of representation of relations. To draw the digraph of a binary relation on a set A, **points**, **vertices**, or **nodes**, representing the elements of A are drawn. Each ordered pair is represented using an **arc**, a **link**, or an **edge**, with its direction indicated by an arrow. A **directed graph**, also known as **digraph**, consists of a set V of vertices together with a set E of edges. In the edge (a_1, a_2), a_1 is called the **initial vertex**, and a_2 is called the **terminal vertex**. Note that when the initial vertex is the same as the terminal vertex, the edge is called a **loop**.

Note that the digraph representing a relation can be used to determine the relation properties in an insightful way. The digraph of a reflexive relation has a loop at every vertex of the digraph. The digraph of a symmetric relation has the property that whenever there is a direct edge from one vertex to another, there is also a direct edge in the opposite direction. The digraph of an antisymmetric relation has the property that between any two distinct vertices, there is at most one direct edge. The digraph of a transitive relation has the property that whenever there are directed edges from, say, the first node to the second node and from the second node to the third node, there is also a directed edge from the first node to the third node.

Example 9.9

Highlight the features of the digraphs of the following relations defined on the set of real numbers:
(a) The equality relation.
(b) The inequality relation.

Solution

(a) Every node has a loop. If there is a direct edge from one node to another, then there is a direct edge in the opposite direction. If there are directed edges from node a to node b and from node b to node c, then there is also a directed edge from node a to node c.

(b) There are no loops. If there is a direct edge from one node to another, then there is no direct edge in the opposite direction. If there are directed edges from node a to node b and from node b to node c, then there is also a directed edge from node a to node c.

A zero-one matrix is an effective way to represent a relation, as it allows a computer to easily analyze a relation. In a zero-one matrix, columns are labeled by the elements of the finite set A and rows are labeled by the elements of the finite set B. Note that in the zero-one matrix of a binary relation on a set, the same ordering for the rows as for the columns is used. In the **zero-one matrix** of the relation R, denoted by M_R, each entry that belongs to the set of the ordered pairs in the relation is set to 1; otherwise, it is set to 0.

Note that the zero-one matrix representing a relation on a set can be used to determine the relation properties. Whenever the zero-one matrix of a relation on a set, which is a square matrix, has 1s on the main diagonal, the relation is reflexive, while noting that the elements off the main diagonal can be either 0 or 1. Whenever the zero-one matrix of a relation on a set and its transpose are the same (i.e., it is a symmetric matrix), the relation is symmetric. The relation is antisymmetric if an element in row i and column j is 1 with $i \neq j$, then the element in row j and column i is 0. The relation is transitive if and only if whenever an entry in row i and column j in the square of the zero-one matrix is nonzero, the entry in row i and column j in the zero-one matrix is also nonzero.

Example 9.10

Determine if each of the following relations is reflexive, symmetric, antisymmetric, and/or transitive:

(a) Using digraph representation of the following relation on $\{a_1, a_2, a_3\}$:

$$R = \{(a_1, a_2), (a_1, a_3), (a_2, a_3), (a_3, a_2)\}.$$

(b) Using zero-one matrix representation of the following relation on $\{a_1, a_2, a_3\}$:

$$R = \{(a_1, a_1), (a_2, a_2), (a_3, a_3), (a_1, a_3), (a_3, a_1)\}.$$

Solution

(a) As Fig. 9.2 shows the digraph, the relation R is not reflexive but is irreflexive because there are no loops. It is not symmetric, as the edge (a_1, a_2) is present but not the edge (a_2, a_1). It is not antisymmetric because both edges (a_2, a_3) and (a_3, a_2) are present. It is not transitive because the edges (a_2, a_3) and (a_3, a_2) are not accompanied by the edge (a_2, a_2).

(b) We have

$$M_R = \begin{pmatrix} 1 & 0 & 1 \\ 0 & 1 & 0 \\ 1 & 0 & 1 \end{pmatrix} \rightarrow M_R^T = \begin{pmatrix} 1 & 0 & 1 \\ 0 & 1 & 0 \\ 1 & 0 & 1 \end{pmatrix} \ \& \ M_R^2 = \begin{pmatrix} 2 & 0 & 2 \\ 0 & 1 & 0 \\ 2 & 0 & 2 \end{pmatrix}.$$

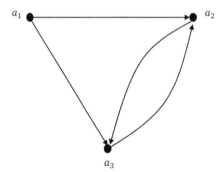

Fig. 9.2 Digraph for Example 9.10.

The relation R is reflexive, as all the diagonal elements of the matrix M_R are equal to 1. Because we have $M_R = M_R^T$, the relation R is symmetric. The relation R is not antisymmetric, for there are two 1s symmetrically placed about the main diagonal (i.e., at positions row 1, column 3 and row 3, column 1). Whenever entry row i, column j in M_R^2 is nonzero, and entry row i, column j in M_R is also nonzero, the relation R is therefore transitive.

9.4 Operations on Relations

Relations can be combined to produce new relations. Operations on relations may include union, intersection, difference, and composition.

Let R and S be any two relations from A to B. The **union of two relations** R and S is defined as $R \cup S = \{(a, b) | (a, b) \in R$ and/or $(a, b) \in S\}$; the **intersection of two relations** R and S is defined as $R \cap S = \{(a, b) | (a, b) \in R$ and $(a, b) \in S\}$; and the **difference of two relations** R and S is defined as $R - S = \{(a, b) | (a, b) \in R$ and $(a, b) \notin S\}$. Graphically (i.e., in terms of digraphs), $R \cup S$ consists of all edges in R together with those in S, $R \cap S$ consists of all common edges in R and S, and $R - S$ consists of all edges in R that are not in S.

Suppose the zero-one matrices for the relations R and S are represented by M_R and M_S, respectively. The zero-one matrix representing the union of these relations, denoted by $M_{R \cup S}$, has a 1 in the position where either M_R or M_S has a 1 or both of them have a 1. The zero-one matrix representing the intersection of these relations, denoted by $M_{R \cap S}$, has a 1 in the position where both M_R and M_S have a 1. The zero-one matrix representing the difference between the relations R and S, denoted by M_{R-S}, has a 1 in the position where M_R has a 1 but M_S does not have a 1.

Example 9.11

Consider the following relations on $\{a_1, a_2, a_3\}$:

$$R = \{(a_1, a_1), (a_1, a_2), (a_2, a_3)\}$$

and

$$S = \{(a_1, a_1), (a_1, a_3), (a_2, a_2), (a_2, a_3), (a_3, a_3)\}.$$

Identify the zero-one matrices for the relations reflecting the union and intersection of these two relations, and determine the corresponding relations.

Solution

The zero-one matrices of R, S, $R \cup S$, and $R \cap S$ are as follows:

$$M_R = \begin{pmatrix} 1 & 1 & 0 \\ 0 & 0 & 1 \\ 0 & 0 & 0 \end{pmatrix}, M_S = \begin{pmatrix} 1 & 0 & 1 \\ 0 & 1 & 1 \\ 0 & 0 & 1 \end{pmatrix}, M_{R \cup S} = \begin{pmatrix} 1 & 1 & 1 \\ 0 & 1 & 1 \\ 0 & 0 & 1 \end{pmatrix} \& M_{R \cap S} = \begin{pmatrix} 1 & 0 & 0 \\ 0 & 0 & 1 \\ 0 & 0 & 0 \end{pmatrix}.$$

We thus have

$$R \cup S = \{(a_1, a_1), (a_1, a_2), (a_1, a_3), (a_2, a_2), (a_2, a_3), (a_3, a_3)\}$$

and

$$R \cap S = \{(a_1, a_1), (a_2, a_3)\}.$$

Let A, B, and C be sets, R be a relation from A to B, S be a relation from B to C, with $a \in A$, $b \in B$, and $c \in C$, while noting that A, B, and C have m, n, and p elements, respectively. Then R and S give rise to a relation from A to C, denoted by $S \circ R$, called the **composition of two relations** R and S, and defined by $(a, c) \in (S \circ R)$ if there exists an element b in B such that $(a, b) \in R$ and $(b, c) \in S$. Note that the composition of relations R and S is denoted by $S \circ R$ rather than $R \circ S$. This is done in order to conform with the usual use of $g \circ f$ to denote the composition of f and g, where f and g are functions. However, when a relation R is composed with itself, then the meaning of $R \circ R$ is unambiguous.

Suppose R is a relation on a set A, that is R is a relation from a set A to itself. The **powers of a relation** R can be recursively defined from the composition of two relations. Therefore $R \circ R = R^2$ is always defined, and similarly, $R^n = R^{n-1} \circ R$ is defined for all integers $n \geq 2$.

Example 9.12

Let $R = \{(3, 2), (7, 7), (2, 7), (5, 3)\}$. Find the powers R^n, $n = 2, 3, \ldots$.

Solution

$$R = \{(3, 2), (7, 7), (2, 7), (5, 3)\} \rightarrow R^2 = R \circ R = \{(7, 7), (2, 7), (3, 7), (5, 2)\} \rightarrow$$

$$R^3 = R^2 \circ R = \{(7, 7), (2, 7), (3, 7), (5, 7)\} \rightarrow$$
$$R^4 = R^3 \circ R = \{(7, 7), (2, 7), (3, 7), (5, 7)\}.$$

We can thus conclude that all powers R^n for $n \geq 3$ are the same.

The zero-one matrices for the relations R, S, and $S \circ R$, denoted by M_R, M_S, and $M_{S \circ R}$, respectively, display an interesting connection, as the matrix resulting from the Boolean product of M_R and M_S (i.e., $M_R \odot M_S$) and the matrix $M_{S \circ R}$ have the same nonzero entries, while noting that M_R, M_S, and $M_{S \circ R}$ have sizes $m \times n$, $n \times p$, and $m \times p$, respectively. In other words, we have the following Boolean product:

$$M_{S \circ R} = M_R \odot M_S.$$

Example 9.13

Let $A = \{a_1, a_2, a_3\}$, $B = \{b_1, b_2, b_3, b_4\}$, and $C = \{c_1, c_2, c_3, c_4\}$. Determine $S \circ R$ when the relation R is defined as $R = \{(a_1, b_1), (a_1, b_3), (a_2, b_2)\}$ from A to B, and the relation S is defined as $S = \{(b_1, c_2), (b_1, c_3), (b_2, c_1), (b_2, c_4), (b_4, c_3)\}$ from B to C.

Solution

The corresponding zero-one matrices are as follows:

$$M_R = \begin{pmatrix} 1 & 0 & 1 & 0 \\ 0 & 1 & 0 & 0 \\ 0 & 0 & 0 & 0 \end{pmatrix} \quad \& \quad M_S = \begin{pmatrix} 0 & 1 & 1 & 0 \\ 1 & 0 & 0 & 1 \\ 0 & 0 & 0 & 0 \\ 0 & 0 & 1 & 0 \end{pmatrix} \rightarrow$$

$$M_{S \circ R} = M_R \odot M_S = \begin{pmatrix} 0 & 1 & 1 & 0 \\ 1 & 0 & 0 & 1 \\ 0 & 0 & 0 & 0 \end{pmatrix}.$$

The nonzero entries of $M_{S \circ R}$ indicate that $S \circ R = \{(a_1, c_2), (a_1, c_3), (a_2, c_1), (a_2, c_4)\}$.

An interesting application of the composition operation lies in databases. For instance, the file R is considered to be a relation from the set of names of people to the set of their annual incomes, and the file S is considered to be a relation from the set of people's annual incomes to the set of their annual income taxes. Therefore the composition $S \circ R$ is a relation from the set of names of people to the set of their annual income taxes.

9.5 Closure Properties

A relation R may not have a desired property, such as reflexivity, symmetry, or transitivity. If there is a relation containing R and having the desired property, then the smallest such relation is the **closure of the relation** R with respect to the property. Assuming R is a relation on a set A with n elements, the reflexive closure of R, the symmetric closure of R, and the transitive closure of R exist. Moreover, these closures are also unique, in that there cannot be even two distinct reflexive closures, symmetric closures, or transitive closures of some relation.

The **reflexive closure** of the relation R is the smallest relation R_r, such that $R \subset R_r$ and R_r is reflexive on the set A. The relation R_r is obtained by simply adding to R all pairs of the form (a, a) with $a \in A$ that do not already belong to R. In other words, the reflexive closure of R is $R \cup \Delta_A$, where $\Delta_A = \{(a, a) | a \in A\}$ is known as the **diagonal relation** on A.

The **symmetric closure** of the relation R is the smallest relation R_s, such that $R \subset R_s$ and R_s is symmetric on the set A. The relation R_s is obtained by simply adding to R all pairs in the form (b, a) whenever (a, b) belongs to R. In other words, the symmetric closure of R is $R \cup R^{-1}$, where $R^{-1} = \{(b, a) | (a, b) \in R\}$.

The **transitive closure** of the relation R is the smallest relation R_t, such that $R \subset R_t$ and R_t is transitive on the set A with n elements. Every possible matched pair of the form $(a, b) \leftrightarrow (b, c)$ is examined, and then make sure that the ordered pair (a, c) is either in the relation or is added to the relation. Obviously, obtaining the transitive closure is more complicated than obtaining either the reflexive closure or the symmetric closure. The relation R_t is obtained by simply including all pairs that belong to the relations R, $R^2 = R \circ R$, ..., and $R^n = R^{n-1} \circ R$. In other words, the transitive closure of R is $R \cup R^2 \cup ... \cup R^n$.

Example 9.14
Consider the relation $R = \{(1, 2), (2, 3), (3, 3)\}$ on the set $A = \{1, 2, 3\}$. Determine the reflexive, symmetric, and transitive closures of the relation R.

Solution
We have

$$R_r = R \cup \{(1,1), (2,2), (3,3)\} = \{(1,1), (1,2), (2,2), (2,3), (3,3)\}$$

and

$$R_s = R \cup \{(2,1), (3,2)\} = \{(1,2), (2,1), (2,3), (3,2), (3,3)\}.$$

As $n = 3$, we then obtain

$$\begin{cases} R = \{(1, 2), (2, 3), (3, 3)\} \\ R^2 = R \circ R = \{(1, 3), (2, 3), (3, 3)\} \quad \rightarrow \\ R^3 = R^2 \circ R = \{(1, 3), (2, 3), (3, 3)\} \end{cases}$$

$$R_t = R \cup R^2 \cup R^3 = \{(1,2), (1,3), (2,3), (3,3)\}.$$

Note also that the zero-one matrix for the transitive closure is the join of the zero-one matrices of the first three powers of the zero-one matrix of R. We therefore have

$$M_R = \begin{pmatrix} 0 & 1 & 0 \\ 0 & 0 & 1 \\ 0 & 0 & 1 \end{pmatrix} \rightarrow M_{R^2} = \begin{pmatrix} 0 & 0 & 1 \\ 0 & 0 & 1 \\ 0 & 0 & 1 \end{pmatrix} \rightarrow M_{R^3} = \begin{pmatrix} 0 & 0 & 1 \\ 0 & 0 & 1 \\ 0 & 0 & 1 \end{pmatrix} \rightarrow$$

$$M_t = M_R \vee M_{R^2} \vee M_{R^3} = \begin{pmatrix} 0 & 1 & 1 \\ 0 & 0 & 1 \\ 0 & 0 & 1 \end{pmatrix} \rightarrow R_t = \{(1,2), (1,3), (2,3), (3,3)\},$$

which confirms the earlier result.

It can be shown that the number of bit operations required for the transitive closure of a relation on a set with n elements using the join of the zero-one matrices of the first n powers of the zero-one matrix is $2n^3(n-1)$. However, there are more efficient algorithms. For instance, there is an algorithm that requires only $2n^3$ operations, that is, a reduction in computation by a factor of $(n-1)$. Table 9.5 shows how to obtain various closures of a relation R with matrix M_R on a set A whose cardinality is n. Note that M_{R^i} is the ith Boolean power of the matrix M_R for the relation R.

9.6 Equivalence Relations

The central idea of equivalence relations is the idea of grouping things that look different but are in some way alike. Equivalence relations matter whenever it is important to show that an element of a set is in a certain class of elements, instead of finding out about its particular identity.

Let A be a set and R a relation on A; R is an **equivalence relation** if and only if R is reflexive, symmetric, and transitive. If we let A be a set with partition $\{A_i\}$ (i.e., $A_i \cap A_j = \varnothing$, whenever $i \neq j$), and let R be the relation induced by the partition, then the relation R is an equivalence relation. For instance, if we have the set $A = \{1, 2, 3\}$ and let

TABLE 9.5 Various closures of a binary relation.

Relation	Set	Matrix
Reflexive closure	$R \cup \{(a,\ a) \vert a \in A\}$	$M_R \vee I_n$
Symmetric closure	$R \cup R^{-1}$	$M_R \vee M_{R^{-1}}$
Transitive closure	$\overset{n}{\underset{i=1}{\cup}} R^i$	$M_R \vee M_{R^2} \vee \ldots \vee M_{R^n}$

$\{\{1,\ 2\},\ \{3\}\}$ be a partition of A, then the relation R caused by the partition of A, that is, $R = \{(1,\ 1),\ (1,\ 2),\ (2,\ 1),\ (2,\ 2),\ (3,\ 3)\}$, is reflexive, symmetric, and transitive.

Let R be an equivalence relation on a set A. The set of all elements that are related to an element a of A is called the ***equivalence class*** of a and is denoted by $[a]$. If $x \in [a]$, then x is a ***representative*** of the class $[a]$. Moreover, if we let R be an equivalence relation on a set A, then the distinct equivalence classes of R form a partition of A, and every partition of A induces an equivalence relation on A. For instance, let $R = \{(1,\ 1),\ (1,\ 2),\ (2,\ 1),\ (2,\ 2),\ (3,\ 3)\}$ be an equivalence relation on the set $A = \{1,\ 2,\ 3\}$. We thus have the equivalent class $[1] = \{1,\ 2\}$ where 1 and 2 are its representatives, and the equivalent class $[3] = \{3\}$ where 3 is its representative. Accordingly, the partition of A induced by R is $[\{1,\ 2\},\ \{3\}]$.

Example 9.15

(a) Let $A = \{0,\ 1,\ 2,\ 3,\ 4\}$ whose partition is as follows: $\{0,\ 3,\ 4\}$, $\{1\}$, and $\{2\}$. Determine the relation R induced by this partition. Is the relation R an equivalence relation?

(b) Let $R = \{(1,\ 1),\ (1,\ 5),\ (2,\ 2),\ (2,\ 3),\ (3,\ 2),\ (3,\ 3),\ (4,\ 4),\ (5,\ 1),\ (5,\ 5)\}$ be the equivalence relation on the set $A = \{1,\ 2,\ 3,\ 4,\ 5\}$. Determine the partition of A induced by R (i.e., find the equivalence classes of R).

Solution

(a) We can have the following ordered pairs:

$$\begin{cases} \{0,3,4\} \rightarrow \{(0,0),(0,3),(0,4),(3,0),(3,3),(3,4),(4,0),(4,3),(4,4)\} \\ \{1,1\} \rightarrow \{(1,1)\} \\ \{2,2\} \rightarrow \{(2,2)\} \end{cases} \rightarrow$$

$$R = \{(0,0),(0,3),(0,4),(1,1),(2,2),(3,0),(3,3),(3,4),(4,0),(4,3),(4,4)\}.$$

The relation R is an equivalence relation on $\{0,\ 1,\ 2,\ 3,\ 4\}$, simply because it is clearly reflexive, symmetric, and transitive.

(b) The elements related to 1 are 1 and 5, hence $[1] = \{1,\ 5\}$. We then select an element not belonging to $[1]$, say 2. The elements related to 2 are 2 and 3, hence $[2] = \{2,\ 3\}$. The only element that does not belong to $[1]$ or $[2]$ is 4. The only element related to 4 is 4, hence $[4] = \{4\}$. Accordingly, the partition of A induced by the relation R is $[\{1,\ 5\},\ \{2,\ 3\},\ \{4\}]$.

The approach to find the smallest equivalence relation containing a given relation constitutes first taking the transitive closure of the relation, then taking the reflexive closure of that relation, and finally taking its symmetric closure.

9.7 Partial Orderings

Having relations to order some or all of the elements of sets are often very much needed. A relation on the set of tasks required to build a house and a relation on the words listed in a dictionary are some examples of partial orderings. Let R be a relation defined on a set A. R is a **partial order relation** or **partial ordering relation** if and only if R is reflexive, antisymmetric, and transitive, as these properties characterize relations that can be employed to order the elements of sets. For instance, the less-than-or-equal-to ($\leq$) relation on a set of real numbers, the subset ($\subseteq$) relation on the power set of sets, and the divisibility (|) relation on a set of positive integers are all partial orderings.

A set A together with a partial ordering R is called a **partially ordered set**, or **poset**, and is denoted by the pair (A, R). Members of the set A are called elements of the poset. Note that if R is a partial order on a set A, the notation $a \preccurlyeq b$ is sometimes used to indicate that $(a, b) \in R$ in an arbitrary poset (A, R). Suppose that R is a partial order relation on a set A. If a, $b \in A$ and either $a \preccurlyeq b$ or $b \preccurlyeq a$, then the elements a and b are called **comparable**; otherwise, they are incomparable. In other words, if a, $b \in A$ and neither $a \preccurlyeq b$ nor $b \preccurlyeq a$, then the elements a and b are called **incomparable** or **noncomparable**.

If R is a partial order relation on a set A, and every pair of elements in A is comparable, then R is called a **total order relation** or **linear order relation** on A, and A is called **totally ordered set** or **linearly ordered set**. For instance, the less-than-or-equal-to relation R on the positive integers A is a total order, for if a and b are integers, either $a \leq b$ or $b \leq a$, whereas the divisibility relation R on the set of positive integers A is a partial order, as it has both comparable elements, such as 4 and 8, and incomparable elements, such as 5 and 9.

Example 9.16

Consider the set of integers $\mathbf{Z}$ and define the relation $R = \{(a, b)|b = a^r\}$ for some positive integer r.

(a) Show that R is a partial order relation on $\mathbf{Z}$.

(b) Show that R is not a total order relation on $\mathbf{Z}$.

Solution

(a) In order to prove that the relation R on the set $\mathbf{Z}$ is a partial ordering, we need to prove it is (i) reflexive, (ii) antisymmetric, and (iii) transitive.

    (i) R is reflexive, as we have $a = a^1$.

    (ii) Suppose $(a, b) \in R$ and $(b, a) \in R$, that is, we have $b = a^s$ and $a = b^t$ for some positive integers s and t. We then have $a = a^{st}$, which in turn leads to the following three possibilities:

$$\begin{cases} st = 1 \ \rightarrow s = 1 \text{ and } t = 1 \ \rightarrow a = b \\ a = 1 \ \rightarrow b = 1^s = 1 \quad\quad\quad \rightarrow a = b \ \rightarrow R \text{ is antisymmetric, as } a = b. \\ a = -1 \rightarrow b = -1 \text{ (as } b \neq 1) \ \rightarrow a = b \end{cases}$$

(iii) Suppose $(a, b) \in R$ and $(b, c) \in R$, that is, we have $b = a^s$ and $c = b^t$ for some positive integers s and t. We then have $c = a^{st}$, that is $(a, c) \in R$. Hence R is transitive.

(b) In order to prove that the relation R on the set Z is a total ordering, we need to prove that every pair of elements in Z is comparable. Because the integers 3 and 5 are incomparable, R is not a total ordering.

The directed graph for a finite poset can be simplified quite significantly. For instance, because a partial order is reflexive, each vertex has a loop, which can be deleted. In addition, all edges whose existence is implied by transitivity can be dropped. Moreover, if the remaining edges are drawn upward and all arrows are removed, the resulting diagram is called the ***Hasse diagram*** of a poset.

Example 9.17

Construct the Hasse diagram for the partial ordering $(A, |)$, where $A = \{1, 2, 5, 25, 50\}$, that is, for all $a, b \in A$, we have $a|b$ if and only if we have $b = ka$ for some integer k.

Solution

The steps to obtain the Hasse diagram are as follows:

(i) Fig. 9.3a shows the digraph of the poset.

(ii) Fig. 9.3b shows when loops in Fig. 9.3a are dropped.

(iii) Fig. 9.3c shows when all edges implied by transitivity in Fig. 9.3b are deleted.

(iv) Fig. 9.3d shows when all arrows are omitted in Fig. 9.3c and edges are drawn upward.

To recover the directed graph of a relation from the Hasse diagram, these steps are required: first, reinsert the direction markers on the arrows making all arrows point upward, next add loops at each vertex, and finally for each sequence of arrows from one node to a second and that second node to a third, add an arrow from the first node to the third.

9.8 Relational Databases

A ***database*** is an organized collection of structured information or data. A database is usually controlled by a ***database management system***. The data can be easily accessed, managed, modified, updated, controlled, and organized. A database management system responds to queries, where a ***query*** is a request for information from the database. Most databases use ***structured query language***, widely known as SQL, for writing and querying data.

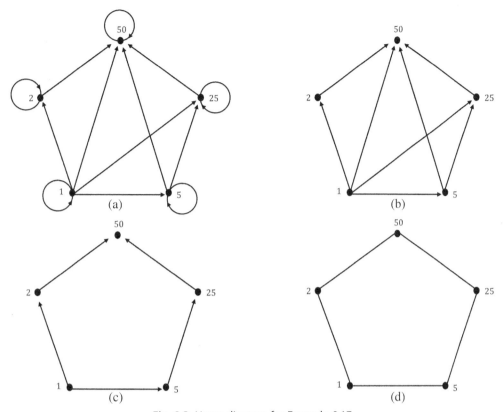

Fig. 9.3 Hasse diagram for Example 9.17.

A **relational database** consists of a collection of n-ary relations involving ordered n-tuples. An n-ary relation is a subset of a Cartesian product of n sets. An **n-ary relation** on the sets $A_1, A_2, \ldots, A_n$, called the **domains of the relation**, is a subset of $A_1 \times A_2 \times \ldots \times A_n$, where n is called its **degree**. The special cases of 2-ary, 3-ary, and 4-ary relations are called **binary**, **ternary**, and **quaternary**, respectively. For instance, the equation $x^2 + y^2 + z^2 = 1$ determines a ternary relation R on the set of real numbers, where a triple (x, y, z) is the coordinates of a point on the sphere with radius 1 whose center is at the origin $(0, 0, 0)$. There are basically three systems of designations that are commonly used in describing relational databases, and it is not uncommon to mix the three systems of terminology.

From a mathematical standpoint, a relational database consists of a number of n-ary relations, where each **relation** has n-tuples and the coordinated positions in each tuple are called **attributes**. An attribute must be single valued and not a set. Each attribute has an **attribute name** and all attribute names form the **attribute set**.

From the standpoint of storing a database on a computer, each relation is stored as a separate *file*, the tuples are considered to be **records** and the attributes are called *fields*.

From a two-dimensional structure standpoint, a relation is visualized as a **table**, the *rows* are ordered tuples of attributes that represent a unique entity, and the **columns** in a table represent attributes. The order of the rows is unimportant and there are no duplicate rows. The entry in a given row and column is single valued.

A domain, that is an attribute, of an *n*-ary relation is called a **primary key** when the value of the *n*-tuple from this domain uniquely determines the *n*-tuple (i.e., when no two *n*-tuples in relation have the same value from this domain). Each table has a primary key, consisting of one or more attributes. When the values of a set of domains determine an *n*-tuple in a relation, the Cartesian product of these domains is called a **composite key**. It is important that primary and composite keys remain valid when new records (rows, tuples, records) are added to the database.

Example 9.18

Table 9.6 provides relevant information regarding the employees in a company. Identify the relevant terms in the context of relational databases.

Solution

- Table ↔ Relation ↔ File.
- Rows ↔ Tuples ↔ Records (e.g., Avicenna, 11256566, March 2, 1960, $200,000, 30 years).
- Columns ↔ Attributes ↔ Fields (e.g., Salary).
- Degree $n = 5$.
- Primary key: ID number.

TABLE 9.6 Relevant information for Example 9.18.

Name	ID number	Date of birth	Salary	Years of service
G. Carlin	10997722	June 13, 1955	$500,000	40
A. Avicenna	11256566	March 2, 1960	$200,000	30
J. Rumi	12042579	June 22, 1965	$180,000	28
F. Nietzsche	12345678	April 6, 1970	$170,000	27
C. Guevara	12444490	June 5, 1974	$160,000	26
M. Gandhi	12537487	July 22, 1975	$150,000	25
J. Maxwell	15588892	May 1, 1980	$120,000	20
N. Mandela	16392630	March 8, 1985	$100,000	12
M. X	18172530	August 5, 1990	$80,000	8
A. Einstein	18253377	May 31, 1995	$60,000	5
M. Angelou	19119123	March 6, 1997	$50,000	2

Exercises

(9.1)

Suppose a, b, and $n > 0$ are integers such that n divides $b - a$ (i.e., a is congruent to b modulo n), that is, $a \equiv b \pmod{n}$. Determine if congruence modulo n is an equivalence relation on the integers.

(9.2)

Let $R = \{(1, y), (1, z), (3, y), (4, x), (4, z)\}$ be a relation from the set $A = \{1, 2, 3, 4\}$ to the set $B = \{x, y, z, w, t\}$.

(a) Determine the domain and range of the relation R.

(b) Using the zero-one matrix of the relation R, determine the inverse relation R.

(9.3)

Let $A = \{2, 3, 4, 6, 12\}$. Let R be the divisibility relation from A to A, that is, for all (a, b), $a \in A$ divides $b \in A$.

(a) Determine the set that represents the relation R.

(b) Determine the zero-one matrix representing the relation R.

(c) Determine if the relation R is reflexive and/or symmetric and/or transitive.

(9.4)

Represent the relation $R = \{(a, b) | a$ is a factor of $b\}$ defined on the set $A = \{2, 3, 4, 6, 8, 12\}$. Is the relation R reflexive and/or symmetric and/or transitive?

(9.5)

The relations R and S have been defined on the set $A = \{0, 1, 2, 3\}$. Determine if each of the following relations is reflexive and/or symmetric and/or transitive:

(a) $R = \{(0,0), (0, 1), (0, 3), (1, 0), (1, 1), (2, 2), (3, 0), (3, 3)\}$.

(b) $S = \{(0,0), (0, 2), (0, 3), (2, 3)\}$.

(9.6)

(a) Which of the relations $R = \{(a, b), (b, c), (a, c)\}$, $S = \{(a, a), (a, b), (a, c), (b, a), (b, c)\}$, $T = \{(a, a), (b, b), (c, c)\}$, and $U = \{(a, b)\}$ on $\{a, b, c\}$ are transitive?

(b) Determine if any one of the relations represented by the following zero-one matrices are equivalence relations:

$$M_{R_1} = \begin{pmatrix} 1 & 1 & 1 \\ 0 & 1 & 1 \\ 1 & 1 & 1 \end{pmatrix}, \quad M_{R_2} = \begin{pmatrix} 1 & 0 & 1 & 0 \\ 0 & 1 & 0 & 1 \\ 1 & 0 & 1 & 0 \\ 0 & 1 & 0 & 1 \end{pmatrix}, \quad \& \; M_{R_3} = \begin{pmatrix} 1 & 1 & 1 & 0 \\ 1 & 1 & 1 & 0 \\ 1 & 1 & 1 & 0 \\ 0 & 0 & 0 & 1 \end{pmatrix}$$

(9.7)

Find the transitive closures of the following relations defined on the set $\{a,\ b,\ c\}$:

(a) $R = \{(a, b), (b, a), (b, c)\}$.

(b) $S = \{(a, a), (b, b), (c, c)\}$.

(c) $T = \varnothing$.

(9.8)

Determine the transitive closure of the relation $R = \{(a, b), (b, a), (b, c), (c, d), (d, a)\}$ on the set $\{a, b, c, d\}$.

(9.9)

(a) Let R be the relation on the set A, where A is the set of real numbers and $R = \{(a, b) | a - b$ is an integer$\}$. Show that R is an equivalence relation.

(b) Let R be the relation on the set A, where A is the set of integers and $R = \{(a, b) | a \equiv b \pmod{m}$, integer $m > 1\}$. Show that R is an equivalence relation.

(9.10)

(a) Let R be the divisibility relation on the set A, where A is the set of positive integers. Show that R is not an equivalence relation.

(b) Let R be the relation on the set A, where A is the set of real numbers and $R = \{(a, b) | |a - b| < 1\}$. Show that R is not an equivalence relation.

CHAPTER 10

Number Theory

Contents

Number theory, known as a pure branch of mathematics, is about the properties of integers. Although integers are familiar and their properties seem simple, number theory is a challenging subject. For instance, until just recently, ***Fermat's last theorem***, which states the equation $x^n + y^n = z^n$, where x, y, and z are integers and $xyz \neq 0$, has no solutions for an integer $n > 2$, had remained unsolved for more than 300 years. Other examples, including ***Goldbach's conjecture***, which states that every even integer greater than two is the sum of two primes, and the ***twin prime conjecture***, which asserts that there are infinitely many twin primes (pairs of primes that differ by 2), are yet to be proven. Number theory has become increasingly important because of its applications to modern cryptography. In this chapter, the fundamental yet basic concepts of number theory are briefly discussed.

10.1 Numeral Systems

A ***numeral*** is any symbol used to represent a number. Before embarking on a brief discussion of integers and their properties, it may be important to briefly describe the widely known numeral systems representing integers, namely the Roman numerals and the Hindu-Arabic numerals. The Roman numeral system was used by most Europeans until the fourteenth century, when they were replaced throughout most of Europe with the much more effective Hindu-Arabic numerals still used today.

In the ***Roman numeral system***, numerals are represented by seven distinct letters. The basic numerals used by the Romans are as follows: I = 1, V = 5, X = 10, L = 50, C = 100, D = 500, and M = 1000. These seven numerals can be combined together to represent larger integers, based on some basic rules. For example, the integer $1173 (= 1000 + 100 + 50 + 20 + 3)$ would be represented as MCLXXIII. Although the

Discrete Mathematics
ISBN 978-0-12-820656-0, https://doi.org/10.1016/B978-0-12-820656-0.00010-1

Roman numeral system allowed easy addition and subtraction, multiplication and division proved to be much more difficult. The lack of an effective system for utilizing fractions and irrational numbers, combined with the imperative absence of the important concept of zero, hindered mathematical advances.

The **Hindu–Arabic numeral system**, which is based on 10 distinct symbols, reflects human anatomy with its 10 fingers. Note that there are various symbol sets representing the 10 distinct symbols in the Hindu-Arabic numeral system that are used in different parts of the world. The most powerful aspect of the Hindu-Arabic system is the existence of a separate numeral for zero that can serve both as a placeholder and as a symbol for "none." The modern system of notation, using 10 different numerals including a zero, was invented in India and reached its present form by the seventh century. This system was then spread to Europe by the Arabs, hence the name the Hindu-Arabic numeral system. It is now universally used to represent numbers.

10.2 Divisibility

Let a and b be integers with $a \neq 0$. If there is an integer c such that $b = ac$, then we say a **divides** b, a is a **factor** of b, a is a **divisor** of b, b is **divisible** by a, or b is a **multiple** of a.

The notation $a|b$ denotes a divides b, that is, $\frac{b}{a}$ is an integer. For instance, $16|48$ implies 16 divides 48, as $\frac{48}{16} = 3$ is an integer. In contrast, the notation $a \nmid b$ denotes a does not divide b, which in turn means $\frac{b}{a}$ is not an integer. For instance, $20 \nmid 48$ implies 20 does not divide 48, as $\frac{48}{20} = 2.4$ is not an integer. Note that if a is a nonzero integer, then $a|0$, as $0 = 0 \times a$.

With $b = ac$, every integer b is divisible by $a = \pm 1$ (i.e., $c = \pm b$) and by $a = \pm b$ (i.e., $c = \pm 1$), where they are called the **trivial divisors** of b.

Assuming $a \neq 0$, b, c, m, and n are integers, some of the properties of divisibility of integers are as follows:

- If $a|b$ and $b|c$, where $b \neq 0$, then $a|c$.
- If $a|b$, then $a|nb$.
- If $a|b$ and $a|c$, then $a|(mb + nc)$.
- If $a|b$ and $b|a$, where $b \neq 0$, then $a = \pm b$.
- If $a|1$, then $a = \pm 1$.
- If $a|b$, then $ma|mb$.

The process of long division is known as the **division algorithm** or the **quotient-remainder theorem**. Assuming a is an integer and d is a positive integer, then there exist unique integers q and r such that $a = dq + r$. Note that d is called the **divisor**, a is called the **dividend**, q is called the **quotient**, and $0 \leq r < d$ is called the **remainder**.

It is important to highlight that the notations $q = a \operatorname{div} d$ and $r = a \bmod d$ are often used to express the quotient and remainder, respectively. For $0 \le r < d$, we thus have the following:

$$a = dq + r \quad \leftrightarrow \quad \begin{cases} q = a \operatorname{div} d \\ r = a \bmod d \end{cases}$$

Note that we also have the following:

$$\begin{cases} a \operatorname{div} d = \left\lfloor \dfrac{a}{d} \right\rfloor \\ a \bmod d = a - d \left\lfloor \dfrac{a}{d} \right\rfloor \end{cases}$$

Example 10.1

For each of the parts (a) and (b), determine integers q and r such that $a = dq + r$ and $0 \le r < d$.

(a) $a = 54$ and $d = 4$.

(b) $a = -51$ and $d = 5$.

(c) Assuming a is an integer, determine $4a \bmod 11$ if we have $a \bmod 11 = 6$.

Solution

(a) $54 = 4 \times 13 + 2 \to q = 13$ and $r = 2$. We thus have $2 = 54 \bmod 4$ and $13 = 54 \operatorname{div} 4$.

(b) $-51 = 5 \times (-11) + 4 \to q = -11$ and $r = 4$ (as the remainder must be positive). We thus have $4 = -51 \bmod 5$ and $-11 = -51 \operatorname{div} 5$.

(c) $a \bmod 11 = 6 \to a = 11q + 6 \to 4a = 4(11q + 6) = 44q + 24 \to 4a = 11(4q + 2) + 2 \to 4a \bmod 11 = 2$.

10.3 Prime Numbers

Primes are the building blocks of positive integers, and as **Euclid's theorem** states, there are infinitely many primes. Prime numbers, once of only theoretical interest, now are important in many applications, especially in modern cryptography, where large primes play a pivotal role.

All integers greater than 1 are grouped into two mutually exclusive sets of integers: one set consists of prime numbers (or simply primes), and the other consists of composite numbers. An integer $p \ge 2$ is **prime** if it is divisible only by 1 and itself (i.e., p). If an integer greater than 1 is not prime, it is then **composite**. In other words, an integer $n \geq 2$ is composite if and only if there exists an integer a such that $a|n$ with $1 < a < n$, that is, $\frac{n}{a}$ is an integer. Note that 1 is neither prime nor composite.

The **fundamental theorem of arithmetic** states that every integer greater than 1 is either prime or the product of two or more primes. In other words, if an integer n is greater than 1, then there is prime $p \leq n$ such that $p|n$. For instance, 101 is a prime number as there are no positive integers, but 1 and 101 that divide 101, and 102 is a composite number that can be expressed as the product of the prime numbers 2, 3, and 17 (i.e., $102 = 2 \times 3 \times 17$).

Every integer $n > 1$ can be expressed uniquely as $n = p_1 p_2 ... p_k$, with $p_1 \leq p_2 \leq ... \leq p_k$ as primes, where k is a positive integer. For instance, we have $10,800 = 2 \times 2 \times 2 \times 2 \times 3 \times 3 \times 3 \times 5 \times 5$. The unique factorization of an integer $n > 1$ formed by grouping together equal prime factors produces the **unique prime-power factorization** $n = p_1^{m_1} p_2^{m_2} ... p_j^{m_j}$, where $p_1 < p_2 < ... < p_j$ are distinct primes, and $m_1, m_2, ..., m_j$ are positive integers. For instance, with $n = 10,800$, we have $p_1 = 2$, $p_2 = 3$, $p_3 = 5$ and $m_1 = 4$, $m_2 = 3$, $m_3 = 2$, as $10,800 = 2^4 \times 3^3 \times 5^2$.

If n is composite, then n has a prime factor less than or equal to $\lfloor \sqrt{n} \rfloor$, that is, n is composite if and only if n has a divisor d satisfying $2 \leq d \leq \lfloor \sqrt{n} \rfloor$. If n is not divisible by any prime, from 2, which is the smallest prime, up to the largest prime that is not exceeding $\lfloor \sqrt{n} \rfloor$, then n is prime. However, if n is divisible by a prime factor p, then the procedure is continued by prime factorization of $\frac{n}{p}$, while noting that $\frac{n}{p}$ has no prime factors less than p. Again, if $\frac{n}{p}$ is not divisible by any prime, from p up to the largest prime that is not exceeding $\left\lfloor \sqrt{\frac{n}{p}} \right\rfloor$, then $\frac{n}{p}$ is prime. If $\frac{n}{p}$ has a prime factor q, then continue by factoring $\frac{n}{pq}$. This process continues until the factorization has been reduced to a prime.

Example 10.2
Determine the prime factorization of each of the following integers:
(a) 1547
(b) 1601

Solution
(a) To find the prime factorization of 1547, first perform divisions of 1547 by successive primes, beginning with 2 and no greater than $\sqrt{1547}$. None of the primes 2, 3, and 5 divides 1547. However, 7 divides 1547, as we have $\frac{1547}{7} = 221$. We then divide 221 by successive primes, beginning with 7 itself. Neither of the primes 7 and 11 divides 221. However, 13 divides 221, as we

have $\frac{221}{13} = 17$. As 17 is prime, the procedure is completed. We thus have $1547 = 7 \times 13 \times 17$.

(b) We first list all primes less than or equal to $\sqrt{1601}$, namely 2, 3, 5, 7, 11, 13, 17, 19, 23, 29, 31, and 37. None of them is a factor of 1601, so 1601 is prime.

There is an important quantity in number theory, referred to as **Euler's totient function** and denoted by $\varphi(n)$, defined as the number of positive integers less than n and relatively prime to n. By convention, $\varphi(1) = 1$. For instance, if $n = 9$, then $\varphi(9) = 6$, namely, the set of relatively primes is $\{1, 2, 4, 5, 7, 8\}$. It should be noted that for prime p, we have $\varphi(p) = p - 1$. For instance, if $p = 11$, then $\varphi(11) = 10$, namely, the set of relatively primes is $\{1, 2, 3, 4, 5, 6, 7, 8, 9, 10\}$. If p and q are two prime numbers, with $p \neq q$, we then have

$$\varphi(n) = \varphi(pq) = \varphi(p) \times \varphi(q) = (p-1)(q-1).$$

For instance, if $p = 7$ and $q = 3$, we then have $\varphi(21) = \varphi(7 \times 3) = \varphi(7) \times \varphi(3) = (7-1)(3-1) = 12$, namely, the set of relatively primes is $\{1, 2, 4, 5, 8, 10, 11, 13, 16, 17, 19, 20\}$.

10.4 Greatest Common Divisors and Least Common Multiples

The **greatest common divisor (gcd)** of two nonzero integers a and b, denoted by $\gcd(a, b)$, is the largest integer d such that $d|a$ and $d|b$, except that $\gcd(0, 0) = 0$. Note that if $\gcd(a, b) = d$, then $\gcd\left(\frac{a}{d}, \frac{b}{d}\right) = 1$. For instance, the set of divisors of 24 is $\{1, 2, 3, 4, 6, 8, 12, 24\}$ and the set of divisors of 42 is $\{1, 2, 3, 6, 7, 14, 21\}$. Because the set of common divisors is $\{1, 2, 3, 6\}$, we have $\gcd(24, 42) = 6$, also $\gcd\left(\frac{24}{6}, \frac{42}{6}\right) = \gcd(4, 7) = 1$.

The integers a and b are **relatively prime** if their gcd is 1. For instance, neither 15 nor 16 is prime; however, 15 and 16 are relatively prime, as their gcd is 1. In addition, integers are called **pairwise relatively prime** if the gcd of any two integers is 1. For instance, none of the integers 25, 26, and 27 is prime, yet they are pairwise relatively prime.

The gcd of two nonzero integers exists if the set of their common divisors is nonempty and finite. The methods to determine the gcd of two integers a and b are as follows:

Brute-force method: First, find all the positive divisors of each integer, then determine the set of all common divisors of both integers, and finally select the largest common divisor in the set.

Prime factorization: The prime factorizations of integers a, b, and $\gcd(a, b)$ are as follows:

$$\begin{cases} a = p_1^{a_1} p_2^{a_2} \cdots p_n^{a_n} \\ b = p_1^{b_1} p_2^{b_2} \cdots p_n^{b_n} \end{cases} \rightarrow \gcd(a, b) = p_1^{\min(a_1,\, b_1)} p_2^{\min(a_2,\, b_2)} \cdots p_n^{\min(a_n,\, b_n)}$$

where $p_1 < p_2 < \ldots < p_n$ are distinct primes, each exponent is a nonnegative integer with $\min(x, y)$ representing the minimum of the two nonnegative integers x and y, all primes occurring in the prime factorization of either a or b (i.e., $p_1, p_2, \ldots, p_n$) are included in both factorizations, and an exponent may be zero if necessary.

The Euclidean algorithm: Assuming $a \geq b$, $r_0 = a$, and $r_1 = b$, successive application of the division algorithm yields the following sequence of equations:

$$\begin{aligned} r_0 &= r_1 q_1 + r_2 & 0 &\leq r_2 < r_1 \\ r_1 &= r_2 q_2 + r_3 & 0 &\leq r_3 < r_2 \\ &\vdots & &\vdots \\ r_{n-2} &= r_{n-1} q_{n-1} + r_n & 0 &\leq r_n < r_{n-1} \\ r_{n-1} &= r_n q_n + r_{n+1} & r_{n+1} &= 0, \end{aligned}$$

where $n \geq 1$ is an integer. Because the remainders are nonnegative and getting smaller, the sequence of remainders $r_2 > r_3 > \ldots \geq 0$ must eventually terminate with a remainder of zero. Using mathematical induction, we can show

$$\gcd(r_0, r_1) = \gcd(r_1, r_2) = \ldots = \gcd(r_{n-1}, r_n) = \gcd(r_n, r_{n+1}) = \gcd(r_n, 0) = r_n.$$

In summary, by applying the division algorithm successively, the gcd is the last (i.e., the smallest) nonzero remainder in the sequence of divisions.

Example 10.3
Determine the gcd of 72 and 108 using the three above-mentioned methods.

Solution
(i) The divisors of 72 include 2, 3, 4, 6, 8, 9, 12, 18, 24, 36, and 72, whereas the divisors of 108 include 2, 3, 4, 6, 9, 12, 18, 27, 36, 54, and 108. The common divisors are then 2, 3, 4, 6, 9, 12, 18, 36. The gcd is thus 36.

(ii) The prime factorizations of 72 and 108 are $2^3 \times 3^2$ and $2^2 \times 3^3$, respectively. The gcd is thus $2^{\min(3,\, 2)} \times 3^{\min(2,\, 3)} = 2^2 \times 3^2 = 36$.

(iii) Using the Euclidean algorithm, we have $108 = 72 \times 1 + 36$. As we have $72 = 36 \times 2 + 0$, 36, which is the last (i.e., the smallest) nonzero remainder in the sequence of divisions, is the gcd.

Finding the gcd of two large integers using the brute-force method or prime-factorization method often proves to be time-consuming. However, the Euclidean algorithm is more computationally efficient, and it is thus the preferred method to find the gcd.

Example 10.4
Determine the gcd of 2766 and 9960.

Solution
Successive application of the division algorithm yields the following sequence of equations:

$$9960 = 2766 \times 3 + 1662$$
$$2766 = 1662 \times 1 + 1104$$
$$1662 = 1104 \times 1 + 558$$
$$1104 = 558 \times 1 + 546$$
$$558 = 546 \times 1 + 12$$
$$546 = 12 \times 45 + 6$$
$$12 = 6 \times 2 + 0$$

The gcd is 6, as it is the last (i.e., the smallest) nonzero remainder in the sequence of divisions.

If a and b are positive integers, then there exist some integers s and t such that $\gcd(a, b) = sa + tb$. This equation is called **Bezout's identity**, and s and t are referred to as **Bezout coefficients** of a and b. Note that if a and b are relatively prime (i.e., $\gcd(a, b) = 1$), we then have $sa + tb = 1$. In order to express $\gcd(a, b)$ as a linear combination of integers a and b (i.e., to determine the integers s and t), a method based on working backward through the divisions of the Euclidean algorithm, known as the **extended Euclidean algorithm**, can be employed.

Example 10.5
Express the gcd of 210 and 54 as a linear combination of 210 and 54, that is, determine the integers s and t in $\gcd(210, 54) = 210s + 54t$.

Solution
Using the Euclidean algorithm, we first find $\gcd(210, 54)$. Successive application of the division algorithm yields the following sequence of equations:

$$210 = 54 \times 3 + 48$$

$$54 = 48 \times 1 + 6$$

$$48 = 6 \times 9 + 0$$

Note that $\gcd(210, 54) = 6$, as 6 is the last nonzero remainder in the sequence of divisions. We now employ the Euclidean algorithm in the reverse order as follows:

$$\gcd(210, 54) = 6 = 54 - 1 \times 48 = 54 - 1 \times (210 - 54 \times 3)$$
$$= -1 \times 210 + 4 \times 54 = 210s + 54t \rightarrow s = -1 \ \& \ t = 4.$$

The **least common multiple (lcm)** of the positive integers a and b, denoted by $\mathrm{lcm}(a, b)$, is the smallest positive integer that is divisible by both a and b. The prime factorizations of integers a, b, and $\mathrm{lcm}(a, b)$ are as follows:

$$\begin{cases} a = p_1^{a_1} p_2^{a_2} \cdots p_n^{a_n} \\ b = p_1^{b_1} p_2^{b_2} \cdots p_n^{b_n} \end{cases} \rightarrow \mathrm{lcm}(a, b) = p_1^{\max(a_1, \ b_1)} p_2^{\max(a_2, \ b_2)} \cdots p_n^{\max(a_n, \ b_n)}$$

where $p_1 < p_2 < \ldots < p_n$ are distinct primes, each exponent is a nonnegative integer with $\max(x, y)$ representing the maximum of the two numbers x and y, and all primes occurring in the prime factorization of either a or b, that is, $p_1, p_2, \ldots, p_n$, are included in both factorizations.

It is important to note that for the positive integers a and b, we have the following identity:

$$a \times b \equiv \gcd(a, \ b) \times \mathrm{lcm}(a, b).$$

Example 10.6
Determine the lcm of 72 and 108 using the prime factorization. Using $\mathrm{lcm}(72, 108)$, determine $\gcd(72, 108)$.

Solution
The prime factorizations of 72 and 108 are $2^3 \times 3^2$ and $2^2 \times 3^3$, respectively. Their lcm is thus $2^{\max(3, \ 2)} \times 3^{\max(2, \ 3)} = 2^3 \times 3^3 = 216$. As we have $a \times b = \gcd(a, \ b) \times \mathrm{lcm}(a, b)$, we have $72 \times 108 = \gcd(72, 108) \times 216$. We therefore have $\gcd(72, 108) = 36$.

10.5 Divisibility Test

A **divisibility test** is a quick way to determine whether an integer, called dividend, is divisible by a smaller integer, called divisor, without performing the division. The test is usually based on the examination of the digits of the dividend in a way that solely depends on what the divisor is. Consider an integer a with n digits $\{a_{n-1}, a_{n-2}, \ldots, a_1, a_0\}$ whose decimal representation is then as follows:

$$a = a_{n-1}(10^{n-1}) + a_{n-2}(10^{n-2}) + \ldots + a_1(10^1) + a_0(10^0).$$

Note that a_0 is the least significant digit and a_{n-1} is the most significant digit. As an example, we have $71524 = 7 \times (10^4) + 1 \times (10^3) + 5 \times (10^2) + 2 \times (10^1) + 4 \times (10^0)$, where the least significant digit is 4 and the most significant digit is 7.

The divisibility rules for dividing an integer a by the integers 1, 2, 3, 4, 5, 6, 7, 8, 9, or 10 are as follows:

Divisibility by 1: $1|a \rightarrow$ No special condition on the coefficients $a_{n-1}, a_{n-2}, \ldots, a_0$ (i.e., every integer is divisible by 1).

Divisibility by 2: $2|a \rightarrow a \bmod 2 = a_0 \bmod 2 = 0 \rightarrow a_0 \in \{0, 2, 4, 6, 8\}$ (i.e., the least significant digit must be even).

Divisibility by 3: $3|a \rightarrow a \bmod 3 = (a_{n-1} + a_{n-2}, \ldots + a_0) \bmod 3 = 0$ (i.e., the sum of all digits must be divisible by 3).

Divisibility by 4: $4|a \rightarrow a \bmod 4 = (a_1(10^1) + a_0(10^0)) \bmod 4 = 0$ (i.e., the number representing the last two digits must be divisible by 4).

Divisibility by 5: $5|a \rightarrow a \bmod 5 = a_0 \bmod 5 = 0$ (i.e., the last digit must be a 0 or a 5).

Divisibility by 6: $6|a \rightarrow a \bmod 6 = 0 \rightarrow a \bmod 2 = 0$ and $a \bmod 3 = 0$ (i.e., the integer must be divisible by both 2 and 3).

Divisibility by 7: $7|a \rightarrow a \bmod 7 = ((a_{n-1}(10^{n-2}) + a_{n-2}(10^{n-3}) + \ldots + a_1(10^0)) - 2(a_0)) \bmod 7 = 0$. Note that the process may need to be repeated.

Divisibility by 8: $8|a \rightarrow a \bmod 8 = (a_2(10^2) + a_1(10^1) + a_0(10^0)) \bmod 8 = 0$ (i.e., the number representing the last three digits must be divisible by 8).

Divisibility by 9: $9|a \rightarrow a \bmod 9 = (a_{n-1} + a_{n-2}, \ldots + a_0) \bmod 9 = 0$ (i.e., the sum of all digits must be divisible by 9).

Divisibility by 10: $10|a \rightarrow a \bmod 10 = a_0 \bmod 10 = 0$ (i.e., the last digit must be a 0).

Note that for some divisors, such as 7, there are multiple rules of divisibility, and only one of them is given here. Moreover, applying the divisibility by 7 to a large dividend may require several iterations (i.e., the process needs to be repeated until the divisibility becomes obvious). In addition, there are divisibility rules for notable prime divisors greater than 10, such as 11, 13, and beyond.

In order to test divisibility by any number expressed as the product of prime factors, we must separately test for divisibility by the highest power of each of its prime factors. For example, testing divisibility by $72 = 2^3 \times 3^2$ is equivalent to testing divisibility by both $2^3(= 8)$ and $3^2(= 9)$, which are relatively prime. In other words, checking the divisibility by both 4 and 18 (as we have $72 = 4 \times 18$) or by both 3 and 24 (as we have $72 = 3 \times 24$) would not be sufficient, simply because 4 and 18 are not relatively prime, nor are 3 and 24.

Example 10.7
Determine which one of the integers from 1 to 10 inclusive divides 2520.

Solution
- As any integer is divisible by 1, 2520 is divisible by 1.
- As its last digit (i.e., 0) is even, 2520 is divisible by 2.
- As the sum of all its digits (i.e., 9) is divisible by 3, 2520 is divisible by 3.
- As its last two digits (i.e., 20) is divisible by 4, 2520 is divisible by 4.
- As its last digit is a 0, 2520 is divisible by 5.
- As it is divisible by 2 and 3 both, 2520 is divisible by 6.
- As it is not clear if $252 - 2(0) = 252$ is divisible by 7, we need to continue the process. As $25 - 2(2) = 21$ is divisible by 7, 2520 is divisible by 7.
- As its last three digits (i.e., 520) is divisible by 8, 2520 is divisible by 8.
- As the sum of all its digits (i.e., 9) is divisible by 9, 2520 is divisible by 9.
- As its last digit is a 0, 2520 is divisible by 10.

   Note that the integers from 1 to 10 inclusive all divide 2520. As an alternative method, we could show that 2520 is divisible by 5, 7, 8, and 9, which are pairwise relatively prime, and we have $2520 = 5 \times 7 \times 8 \times 9$. The integer 2520 is divisible by 8, it is thus divisible by 2 and 4; it is divisible by 9, thus it is divisible by 3; it is divisible by 3 and 2, therefore it is divisible by 6; and it is divisible by 2 and 5, hence it is divisible by 10. Moreover, 2520 is the smallest integer that is divisible by all integers from 1 to 10 inclusive.

10.6 Congruences

Modular arithmetic is an important aspect of divisibility. Some trivial examples of modular arithmetic may include that 2 hours and 20 minutes after 7:45 is 10:05 and that 17 days after a Tuesday is a Friday. Congruences have many applications, such as generating pseudorandom numbers for computer simulations, generating parity check bits to detect and correct errors in digital transmission and storage, storing huge records in a rather small table and retrieving them quickly, and above all modern cryptography.

   Assuming a and b are integers and m is a positive integer, **a is congruent to b modulo m**, denoted by $a \equiv b \pmod{m}$, if m divides $a - b$, that is, $m|(a-b)$ or equivalently $a - b = km$ for some integer k. We say that $a \equiv b \pmod{m}$ is a **congruence** and that m is its **modulus**.

   Congruence is an equivalence relation, as for integers a, b, c, and $m > 0$, we have
- Reflexivity property: $a \equiv a \pmod{m}$.
- Symmetry property: $a \equiv b \pmod{m} \rightarrow b \equiv a \pmod{m}$.

- Transitivity property: $a \equiv b \pmod{m}$ and $b \equiv c \pmod{m}$ $\rightarrow$ $a \equiv c \pmod{m}$.

While noting that the notation $a \equiv b \pmod{m}$ represents a relation on the set of integers and the notation $a \bmod m = b$ represents a function, we have

$$a \equiv b \pmod{m} \text{ if and only if } a \bmod m = b \bmod m$$

where a, b, and $m > 0$ are all integers. In addition, for integers a, b, c, d, n, and $m > 0$, we have

$$\begin{cases} a \equiv b \pmod{m} \\ c \equiv d \pmod{m} \end{cases} \rightarrow \begin{cases} a + c \equiv b + d \pmod{m} \\ a - c \equiv b - d \pmod{m} \\ ac \equiv bd \pmod{m} \\ a^n \equiv b^n \pmod{m} \end{cases}$$

Note that $ac \equiv bc \pmod{m}$ does not imply $a \equiv b \pmod{m}$ unless $\gcd(c, m) = 1$. For instance, $85 \equiv 55 \pmod{10}$ does not yield $17 \equiv 11 \pmod{10}$ as $\gcd(5, 10) \neq 1$, but $85 \equiv 55 \pmod{6}$ does yield $17 \equiv 11 \pmod{6}$ as $\gcd(5, 6) = 1$.

Fermat's little theorem, which is very useful in computing the remainder modulo prime of large powers of integers, can be expressed in terms of congruences. **Fermat's little theorem** states that if p is prime and a is an integer not divisible by p, then

$$a^{p-1} \equiv 1 \pmod{p} \leftrightarrow a^{p-1} \bmod p = 1,$$

or equivalently,

$$a^p \equiv a \pmod{p} \leftrightarrow a^p \bmod p = a.$$

Example 10.8

Find $3^9 \bmod 5$, using Fermat's little theorem.

Solution

Noting 5 is prime and 3 is not divisible by 5, we employ Fermat's little theorem as follows:

$$3^4 \equiv 1 \pmod{5} \rightarrow 3^4 = 5k_1 + 1,$$

where k_1 is an integer. Using this result, we proceed as follows:

$$3^9 = 3 \times 3^8 = 3 \times \left(3^4\right)^2 = 3 \times (5k_1 + 1)^2 = 3 \times \left(25k_1^2 + 10k_1 + 1\right)$$
$$= 3 \times \left(5 \times \left(5k_1^2 + 2k_1\right) + 1\right) = 3 \times (5k_2 + 1) = 5 \times (3k_2) + 3$$
$$= 5k_3 + 3 \rightarrow 3^9 \equiv 3 \pmod{5},$$

where $k_2 = 5k_1^2 + 2k_1$ and $k_3 = 3k_2$ are integers. We thus have $3^9 \bmod 5 = 3$.

Example 10.9

Find $3^{201} \bmod 11$, using Fermat's little theorem.

Solution

Note that 11 is prime and 3 is not divisible by 11, we can thus employ Fermat's little theorem as follows:

$$3^{10} \equiv 1 \ (\bmod \ 11) \rightarrow 3^{10} = 11k_1 + 1 \rightarrow \left(3^{10}\right)^{20} = (11k_1 + 1)^{20}$$
$$= 11k_2 + 1.$$

While noting $201 = 1 + 20 \times 10$, we then have

$$3^{201} = 3 \times \left(3^{10}\right)^{20} = 3(11k_2 + 1) = 11(3k_2) + 3$$
$$= 11k_3 + 3 \rightarrow 3^{201} \equiv 3 \ (\bmod \ 11),$$

where k_1, k_2, and k_3 are integers. We thus have $3^{201} \bmod 11 = 3$.

Euler's theorem states that for every a and n that are relatively prime, we have

$$a^{\varphi(n)} \equiv 1 \ (\bmod \ n),$$

where $\varphi(n)$ is Euler's totient function. As an example, $a = 3$ and $n = 10$ are relatively prime, and $\varphi(10) = 4$. We thus have $3^{\varphi(10)} = 3^4 = 81 \equiv 1 \ (\bmod \ 10)$. As another example, for $a = 2$, if we have $n = 11$ and consequently $\varphi(11) = 10$, we then have $2^{\varphi(11)} = 2^{10} = 1024 \equiv 1 \ (\bmod \ 11)$.

A congruence of the form $ax \equiv b \ (\bmod \ m)$ is known as a ***linear congruence in one variable***, where $m > 0$, a, and b are all integers and x is an unknown variable. A major application of linear congruences lies in cryptography.

For all integers a and $m > 1$, if $\gcd(a, m) = 1$ (i.e., a and m are relatively prime), then there exists a unique integer $\bar{a}$ such that $\bar{a}a \equiv 1 \ (\bmod \ m)$, where the integer $\bar{a}$ is called an ***inverse of a modulo m*** and $0 < \bar{a} < m$. As an example, knowing that 5 and 11 are relatively prime, we can find an inverse of 5 modulo 11, that is, we need to find a multiple of 5 that is one more than a multiple of 11. For instance, we have $(-13) \times 5 \equiv 1 \ (\bmod \ 11)$, but -13 is not a positive integer less than 11. We can add a multiple of 11 to -13, so the result will be a positive integer less than 11. More specifically, we can have $(-13 + 2 \times 11) \times 5 \equiv 1 \ (\bmod \ 11)$, that is, we have $9 \times 5 \equiv 1 \ (\bmod \ 11)$. Therefore 9 is an inverse of 5 modulo 11, as we have $0 < 9 < 11$. To find an inverse of a modulo m when a and m are relatively prime, the method of inspection can be helpful if m is small. However, a more efficient algorithm, known as the extended Euclidean algorithm, can be employed, which is based on the reverse steps in the Euclidean algorithm.

Example 10.10

Noting 43 and 660 are relatively prime numbers, find an inverse of 43 modulo 660 using the extended Euclidean algorithm.

Solution

We use the Euclidean algorithm not to find $\gcd(43, 660)$, as we already know $\gcd(43, 660) = 1$, but to employ the extended Euclidean algorithm in order to find an inverse of 43 modulo 660. The steps are as follows:

$$660 = 43 \times 15 + 15$$
$$43 = 15 \times 2 + 13$$
$$15 = 13 \times 1 + 2$$
$$13 = 2 \times 6 + 1.$$

We now use the above results in the reverse order to find 1 in terms of 43 and 660:

$$1 = 13 - 2 \times 6 = 13 - (15 - 13) \times 6 = 7 \times 13 - 6 \times 15$$
$$= 7 \times (43 - 15 \times 2) - 6 \times 15 = 7 \times 43 - 20 \times 15$$
$$= 7 \times 43 - 20 \times (660 - 43 \times 15) = 307 \times 43 - 20 \times 660.$$

Therefore 307 is an inverse of 43 modulo 660, as $43 \times 307 \equiv 1 \pmod{660}$.

There are applications that require solutions to systems of linear congruences. The Chinese remainder theorem provides a unique solution when the modulo of a system of linear congruences are pairwise relatively prime. The **Chinese remainder theorem** states that if m_1, m_2, ..., m_n are pairwise relatively prime positive integers and $a_1, a_2, ..., a_n$ are arbitrary integers, then the following system of linear congruences has a unique solution x modulo $m = m_1 m_2 ... m_n$, with $0 \leq x < m$:

$$\begin{cases} x \equiv a_1 \pmod{m_1} \\ \\ x \equiv a_2 \pmod{m_2} \\ \vdots \\ x \equiv a_n \pmod{m_n} \end{cases}$$

where the simultaneous solution is as follows:

$$x \equiv \left(\sum_{i=1}^{n} a_i \, \beta_i \gamma_i \right) \pmod{m}$$

with $\beta_i = \frac{m}{m_i}$ and an integer γ_i is an inverse of β_i modulo m_i, that is, $\beta_i \gamma_i \equiv 1 \pmod{m_i}$.

Example 10.11

Determine the number when it is divided by 3 the remainder is 2, when it is divided by 5 the remainder is 4, and when it is divided by 7 the remainder is 6.

Solution

We must find a unique solution x for the following system of linear congruences:

$$\begin{cases} x \equiv 2 \ (\text{mod } 3) \\ x \equiv 4 \ (\text{mod } 5) \\ x \equiv 6 \ (\text{mod } 7). \end{cases}$$

As 3, 5, and 7 are pairwise relatively prime, we can use the Chinese remainder theorem. Noting y_1, y_2, and y_3 can be found by using the extended Euclidean algorithm or the method of inspection, we can then have

$$\begin{cases} m_1 = 3 \\ m_2 = 5 \\ m_3 = 7 \end{cases} \rightarrow m = m_1 m_2 m_3 = 105 \rightarrow \begin{cases} \beta_1 = \dfrac{m}{m_1} = 35 \\ \beta_2 = \dfrac{m}{m_2} = 21 \\ \beta_3 = \dfrac{m}{m_3} = 15 \end{cases} \rightarrow \begin{cases} y_1 = 2 \\ y_2 = 1 \\ y_3 = 1 \end{cases}$$

with $a_1 = 2$, $a_2 = 4$, $a_3 = 6$, the solution to the simultaneous congruences is thus as follows:

$$x \equiv (a_1 \beta_1 y_1 + a_2 \beta_2 y_2 + a_3 \beta_3 y_3) \ (\text{mod } m) \equiv (2 \times 35 \times 2 + 4 \times 21 \times 1 + 6$$
$$\times 15 \times 1) \ (\text{mod } 105) \equiv 314 \ (\text{mod } 105) \equiv 104.$$

It is important to note that the **sum and product of two integers in the modular arithmetic** using the same divisor are as follows:

$$\begin{cases} (a + b) \bmod m = ((a \bmod m) + (b \bmod m)) \bmod m \\ (a \times b) \bmod m = ((a \bmod m) \times (b \bmod m)) \bmod m \end{cases}$$

In modern cryptography, **exponentiation in modular arithmetic**, also known as **fast modular exponentiation,** is often much needed. It is important to calculate c^d mod m, where c, d, and m are very large integers. Computing c^d and then dividing it by m to determine its remainder is totally impractical. To this effect, a two-step approach is generally taken. As the first step, the exponent d can be written in the binary form, that is, d is written as the sum of terms, each in the form of 2^k, where k is a nonnegative integer. As the second step, the product property of modular arithmetic is used to reduce the number of calculations. To this end, the algorithm successively determines c^{2^0} mod m, c^{2^1} mod m, c^{2^2} mod m, c^{2^3} mod m, and so on, multiplies together only those terms of interest as defined by d, and then finds the remainder of the product when it is divided by m.

Example 10.12
Find 3^{644} mod 645.

Solution
Because $3^{644} \cong 1.84 \times 10^{307}$ is an extremely large number, it is practically impossible to find the remainder without using fast modular exponentiation. To this effect, we first represent the exponent in the binary form, we thus have

$$644 = 2^9 + 2^7 + 2^2 = 512 + 128 + 4 \;\rightarrow\; 3^{644} \text{ mod } 645$$
$$= \left(3^{512} \times 3^{128} \times 3^4\right) \text{ mod } 645.$$

We now successively determine 3^{2^1} mod 645, 3^{2^2} mod 645, and all other terms up to and including 3^{2^9} mod 645, as follows:

Exponent $= 2^0 = 1 \rightarrow 3^1$ mod 645 $= 3$ mod 645 $= 3$.

Exponent $= 2^1 = 2 \rightarrow 3^2$ mod 645 $= \left(\left(3^1 \text{ mod } 645\right) \times \left(3^1 \text{ mod } 645\right)\right)$ mod 645 $= (3 \times 3)$ mod 645 $= 9$.

Exponent $= 2^2 = 4 \rightarrow 3^4$ mod 645 $= \left(\left(3^2 \text{ mod } 645\right) \times \left(3^2 \text{ mod } 645\right)\right)$ mod 645 $= (9 \times 9)$ mod 645 $= 81$.

Exponent $= 2^3 = 8 \rightarrow 3^8$ mod 645 $= \left(\left(3^4 \text{ mod } 645\right) \times \left(3^4 \text{ mod } 645\right)\right)$ mod 645 $= (81 \times 81)$ mod 645 $= 111$.

Exponent $= 2^4 = 16 \rightarrow 3^{16}$ mod 645 $= \left(\left(3^8 \text{ mod } 645\right) \times \left(3^8 \text{ mod } 645\right)\right)$ mod 645 $= (111 \times 111)$ mod 645 $= 66$.

Exponent $= 2^5 = 32 \rightarrow 3^{32}$ mod 645 $= \left(\left(3^{16} \text{ mod } 645\right) \times \left(3^{16} \text{ mod } 645\right)\right)$ mod 645 $= (66 \times 66)$ mod 645 $= 486$.

Exponent $= 2^6 = 64 \rightarrow 3^{64}$ mod 645 $= \left(\left(3^{32} \text{ mod } 645\right) \times \left(3^{32} \text{ mod } 645\right)\right)$ mod 645 $= (486 \times 486)$ mod 645 $= 126$.

Exponent $= 2^7 = 128 \rightarrow 3^{128}$ mod 645 $= \left(\left(3^{64} \text{ mod } 645\right) \times \left(3^{64} \text{ mod } 645\right)\right)$ mod 645 $= (126 \times 126)$ mod 645 $= 396$.

Exponent $= 2^8 = 256 \rightarrow 3^{256}$ mod 645 $= \left(\left(3^{128} \text{ mod } 645\right) \times \left(3^{128} \text{ mod } 645\right)\right)$ mod 645 $= (396 \times 396)$ mod 645 $= 81$.

Exponent $= 2^9 = 512 \rightarrow 3^{512}$ mod 645 $= \left(\left(3^{256} \text{ mod } 645\right) \times \left(3^{256} \text{ mod } 645\right)\right)$ mod 645 $= (81 \times 81)$ mod 645 $= 111$.

Based on the product of integers in the modular arithmetic and the modular exponentiation, we have

$$3^{644} \bmod 645 = \left(3^{512} \times 3^{128} \times 3^4\right) \bmod 645$$
$$= \left(\left(3^{512} \bmod 645\right) \times \left(3^{128} \bmod 645\right) \times \left(3^4 \bmod 645\right)\right) \bmod 645$$
$$= (111 \times 396 \times 81) \bmod 645 = 36.$$

Data (record) that is stored in a computer memory (table) typically has two parts: a **key** that uniquely identifies that piece of data and a **value** that is the information of interest. Often, a key is a large number k consisting of many digits, say a 13-digit ISBN to identify books. One way to store the records is to place the record with key k into location k of the table, thus theoretically requiring a huge table (e.g., 10^{13} locations in the case of a 13-digit ISBN). This is very wasteful of computer memory space, as the number of records to be stored is relatively small. A solution to this problem is to use hashing functions defined from larger to smaller sets of integers.

In order to map data of arbitrarily large size to small fixed size, a **hashing function** h is used to assign memory location $h(k)$ to the record that has k as its key. Frequently, mod functions are used as hashing functions to convert keys into memory (list) locations. The location numbers are formally called **indices**. The index produced by h determines the spot in the list (i.e., the memory location) to store the value of the record.

The most common hashing function is $h(k) = k \bmod m$, where m is the number of available memory locations. Therefore to find $h(k)$, we need only to find the remainder when k is divided by m. This hashing function can be easily evaluated so that files can be quickly retrieved. In addition, this hashing function is onto, so all memory locations are possible. For instance, the memory location assigned by the hashing function $h(k) = k \bmod 65536$ to the records of a book with ISBN 9780124076822 is 19,222.

Because a hashing function is not one to one, simply because there are much more possible keys than memory locations, more than one record may be assigned to a memory location. In such a case, we say a **collision** has occurred. In other words, $h(k_1) = h(k_2)$, but $k_1 \neq k_2$. To handle collisions, a **collision resolution policy** is required. One simple way to resolve a collision is to find the next unoccupied memory location following the occupied memory location assigned by the hashing function. If we come to the end of the list without finding a memory location, then we would continue the search back at the beginning of the list, as if the array were circular. This method of collision resolution is called **linear probing**. Note that there are other methods to resolve collisions.

10.7 Representations of Integers

Any positive integer a can be uniquely represented in the following form:

$$a = a_k\left(b^k\right) + a_{k-1}\left(b^{k-1}\right) + \ldots + a_1(b) + a_0,$$

where b, known as the **base**, is an integer greater than 1, k is a nonnegative integer, a_0, $a_1, \ldots, a_k$ are nonnegative integers less than b, and $a_k \neq 0$. This representation of the integer a is called the **base$-b$ expansion** of a, which can be denoted by $(a_k a_{k-1} \ldots a_1 a_0)_b$. Note that b represents the number of different symbols that can be used in a numeral system. There are various representations of integers, including the following expansions:

- **Decimal expansions**: $b = 10 \rightarrow 10$ symbols $\{0, 1, 2, 3, 4, 5, 6, 7, 8, 9\}$.
- **Binary expansions**: $b = 2 \rightarrow 2$ symbols $\{0, 1\}$.
- **Octal expansions**: $b = 8 \rightarrow 8$ symbols $\{0, 1, 2, 3, 4, 5, 6, 7\}$.
- **Hexadecimal expansions**: $b = 16 \rightarrow 16$ symbols $\{0, 1, 2, 3, 4, 5, 6, 7, 8, 9,$ A, B, C, D, E, F$\}$.

Note that the subscript 10 representing the base in the decimal expansion is commonly omitted, and the binary expansion of an integer is just a bit string. The binary expansion is widely used in digital devices and networks to represent and carry out arithmetic with integers. The hexadecimal system is commonly used to describe locations in memory because it can represent every byte (8 bits) as two consecutive hexadecimal digits instead of the eight digits that would be required by the binary expansion. While noting that in hexadecimal expansions, A $=$ 10, B $=$ 11, C $=$ 12, D $=$ 13, E $=$ 14, and F $=$ 15, it is much easier to read hexadecimal numbers than binary numbers.

Example 10.13
Express the following expansions in base 10:
(a) $(11001101)_2$.
(b) $(CD)_{16}$.

Solution
(a) $(11001101)_2 = 1(2^7) + 1(2^6) + 0(2^5) + 0(2^4) + 1(2^3) + 1(2^2) + 0(2^1) + 1(2^0) = 205$.
(b) $(CD)_{16} = 12(16^1) + 13(16^0) = 205$.

Base conversion of an integer a in the decimal expansion into any nondecimal base b is as follows: divide a and its successive quotients by b until a zero quotient is reached, then pick the remainders in the reverse order.

Example 10.14
Express 3489, which is in the decimal expansion, in the following representations:
(a) The binary expansion, that is, the base is 2.
(b) The hexadecimal expansion, that is, the base is 16.

Solution

(a) $3489 = 2 \times 1744 + 1 \rightarrow 1744 = 2 \times 872 + 0 \rightarrow 872 = 2 \times 436 + 0 \rightarrow 436 = 2 \times 218 + 0 \rightarrow 218 = 2 \times 109 + 0 \rightarrow 109 = 2 \times 54 + 1 \rightarrow 54 = 2 \times 27 + 0 \rightarrow 27 = 2 \times 13 + 1 \rightarrow 13 = 2 \times 6 + 1 \rightarrow 6 = 2 \times 3 + 0 \rightarrow 3 = 2 \times 1 + 1 \rightarrow 1 = 2 \times 0 + 1.$

The successive remainders that we have found (i.e., 1, 0, 0, 0, 0, 1, 0, 1, 1, 0, 1, 1) are digits from the right to the left of 3489 in base 2. Hence $(3489)_{10} = (110110100001)_2$.

(b) $3489 = 16 \times 218 + 1 \rightarrow 218 = 16 \times 13 + 10 \rightarrow 13 = 16 \times 0 + 13.$

The successive remainders that we have found, 1, 10($=$ A), and 13($=$ D), are digits from the right to the left of 3489 in base 16. Hence $(3489)_{10} = (DA1)_{16}$. Note that for the conversion of the hexadecimal expansion to the binary expansion, each hexadecimal digit corresponds to a block of four binary digits.

10.8 Binary Operations

The computational methods of binary arithmetic are analogous to those of decimal arithmetic. In binary arithmetic, the number $2 = (10)_2$ in binary notation plays a role similar to that of the number 10 in decimal arithmetic.

In **binary addition**, carryovers of binary addition are performed in the same manner as in decimal addition. We thus have $0 + 0 = 0$, $0 + 1 = 1$, $1 + 0 = 1$, and $1 + 1 = 10$, all in base 2.

The **binary multiplication** is carried out by multiplying the multiplicand by one bit of the multiplier at a time and the result of the partial product for each bit is placed in such a manner that the least significant bit is under the corresponding multiplier bit. Finally, the partial products are added to get the complete product.

In **binary subtraction**, like decimal subtraction, it may be necessary to borrow. However, the method introduced here is based on not using borrow, as one's complement and two's complement of binary numbers are used to perform subtraction.

The **one's complement** of a binary number can be obtained by inverting each bit from 1 to 0 or from 0 to 1. For example, one's complement of the binary number 110010 is 001101. To get **two's complement** of a binary number is to first obtain the one's complement of the number and then add 1 to the least significant bit. For example, two's complement of the binary number 10010 is $01101 + 00001 = 01110$.

The **binary subtraction** can be carried out using the following steps:

(i) Find the two's complement of the subtrahend, and then add it to the minuend.

(ii) If the final carryover of the sum is 1, it is dropped and the result is positive, and if there is no carryover, the two's complement of the sum will be the result and it is negative.

The **binary division** is similar to that employed in the decimal system. However, in the case of binary numbers, the operation is simpler because the quotient can have either 1 or 0 depending upon the divisor.

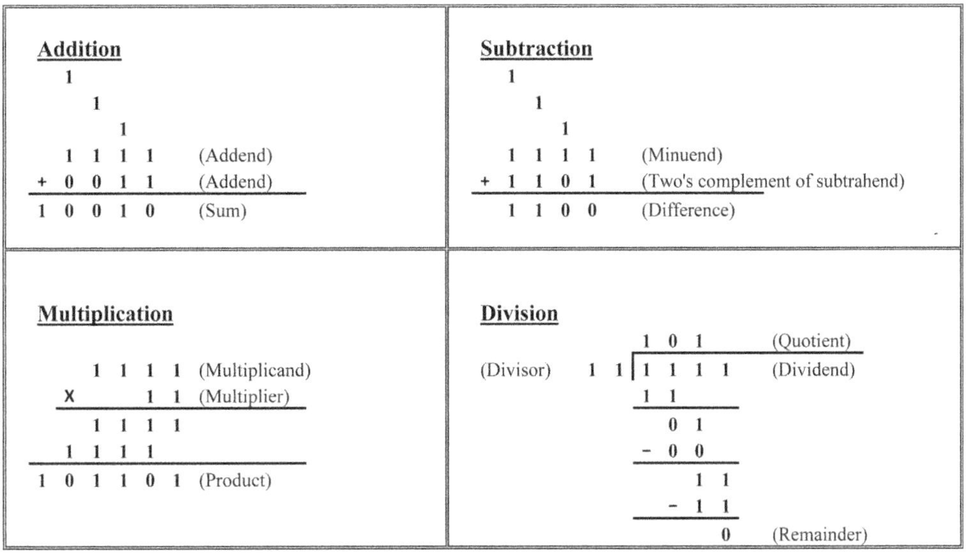

Addition	Subtraction

1
   1
     1
   1 1 1 1   (Addend)
 + 0 0 1 1   (Addend)
 1 0 0 1 0   (Sum)
	1
	   1
	     1
	   1 1 1 1 (Minuend)
	 + 1 1 0 1 (Two's complement of subtrahend)
	 1 1 0 0 (Difference)

Addition

```
 1
   1
     1
 1 1 1 1   (Addend)
+0 0 1 1   (Addend)
 1 0 0 1 0 (Sum)
```

Subtraction

```
 1
   1
     1
 1 1 1 1   (Minuend)
+1 1 0 1   (Two's complement of subtrahend)
 1 1 0 0   (Difference)
```

Multiplication

```
    1 1 1 1  (Multiplicand)
×     1 1   (Multiplier)
    1 1 1 1
  1 1 1 1
1 0 1 1 0 1 (Product)
```

Division

```
              1 0 1      (Quotient)
(Divisor)  1 1│1 1 1 1   (Dividend)
              1 1
               0 1
             - 0 0
                1 1
              - 1 1
                  0      (Remainder)
```

Fig. 10.1 Binary operations for Example 10.15.

Example 10.15

Consider the two binary numbers $m = (1111)_2$ and $n = (11)_2$. Perform the following binary operations: (a) $m + n$, (b) $m \times n$, (c) $m - n$, and (d) $m \div n$.

Solution

The details of the binary operations are presented in Fig. 10.1.
(a) Addition: $\quad\quad m + n = (1111)_2 + (11)_2 = (10010)_2$.
(b) Multiplication: $\quad m \times n = (1111)_2 \times (11)_2 = (101101)_2$.
(c) Subtraction: $\quad\quad m - n = (1111)_2 - (11)_2 = (1100)_2$.
(d) Division: $\quad\quad\quad m \div n = (1111)_2 \div (11)_2 = (101)_2$.

Exercises

(10.1)

(a) Assuming we have $\gcd(a, 105) = 15$ and $\text{lcm}(a, 105) = 210$, determine the integer a.

(b) Assuming we have $\gcd(a, b) = 6$ and $\text{lcm}(a, b) = 72$, determine the possible values for the positive integers a and b.

(10.2)

Determine the gcd and lcm of 82,320 and 950,796 using prime factorization.

(10.3)

(a) Convert 130, which is in the decimal expansion, to the binary expansion.

(b) Convert 20,385, which is in the decimal expansion, to the hexadecimal expansion.

(c) Convert 10110101, which is in the binary expansion, to the decimal expansion.

(10.4)

Using the division algorithm, determine q and r for each of the following cases:

(a) $a = 4461$ and $b = 16$.

(b) $a = -262$ and $b = 3$.

(10.5)

(a) Determine the gcd of 2310 & 2431 using the Euclidean algorithm.

(b) Show that 209 and 221 are relatively prime, using the Euclidean algorithm.

(10.6)

Consider the two binary numbers $m = (10110)_2$ and $n = (1011)_2$. Perform the following binary operations: (a) $m + n$, (b) $m \times n$, (c) $m - n$, and (d) $m \div n$.

(10.7)

Evaluate the following functions to obtain the corresponding remainder or quotient:

(a) $-101 \bmod 13$.

(b) $199 \bmod 19$.

(c) $228 \operatorname{div} 119$.

(d) $-111 \operatorname{div} 99$.

(10.8)

Use Fermat's little theorem to compute the following expressions:

(a) $3^{302} \bmod 11$.

(b) $5^{2003} \bmod 13$.

(10.9)

A sequence of pseudorandom numbers $\{x_n\}$, with $0 \le x_n < m$ for all positive integers n, can be successively generated by using the recursively defined function $x_{n+1} = (ax_n + c) \bmod m$, for a given modulus m, multiplier $2 \le a < m$, increment $0 \le c < m$, and seed $0 \le x_0 < m$. Assuming $m = 9$, $a = 7$, $c = 5$, and $x_0 = 3$, determine the first 10 pseudorandom numbers.

(10.10)

Determine the smallest integer that is divisible by 1, 2, 3, 4, 5, 6, 7, 8, 9, 10, and 11.

CHAPTER 11

Cryptography

Contents

Governments for military intelligence and diplomatic purposes over the past couple of millennia have been protecting information through secret messaging. In view of the advent of the Internet, there is an indispensable need to protect the information, in its transmission as well as its storage from unauthorized access and malicious actions. Cryptography is the field to keep information secret and offer protection of data from unwanted access and from any manipulation, impersonation, and forgery. An overview of cryptography is briefly presented in this chapter.

11.1 Classical Cryptography

Cryptography is about making secret communication to make messages secure in the presence of adversaries. By *encryption*, an original message, called *plaintext*, is transformed into a coded message, called *ciphertext*. This transformation is performed before the plaintext is transmitted or stored. The reverse process is called *decryption* and is performed after the ciphertext is received or retrieved. The algorithm used for encryption and decryption is often called a *cipher*, and the process of encryption and decryption requires a *secret key*. A key is a number (value) that the cipher operates on, without which the unauthorized parties must not be able to recover the original message.

In *classical cryptography*, symbols, characters, letters, and digits were directly manipulated with the sole goal to provide secrecy through obscurity. It appears that encrypted messages were first developed in ancient Egypt using *disordered hieroglyphics*, which consisted of visual symbols and characters. There were then other forms of message concealment, which were developed by the Greeks, namely *stenography*, through which a secret message was hidden within an ordinary nonsecret message, and by the Spartans, namely *scytale*, by which a narrow strip of parchment was wound on a rod and the message written across the adjoining edges.

Discrete Mathematics
ISBN 978-0-12-820656-0, https://doi.org/10.1016/B978-0-12-820656-0.00011-3

A well-known classical cryptography technique, which was developed by Romans, was the **Caesar cipher**, a simple encryption method based on substitution. The Caesar cipher shifts each letter in the alphabet by three letters forward, for instance, the letter G becomes J. It thus requires a letter three places further along, while wrapping the letters at the end of the alphabet around to the letters at the beginning of the alphabet that is X wraps around to A, Y to B, and Z to C.

Mathematically described, in the Caesar cipher, each letter is coded by its position relative to others. To this effect, an integer $i \in \{0, 1, \ldots, 24, 25\}$ replaces a letter whose position in the alphabet $\{A, B, \ldots, Y, Z\}$ is the ith; for instance, D is the fourth letter in the alphabet, that is $i = 3$, D is thus replaced by 3. Assuming the nonnegative integer $p \leq 25$, the functions providing the encrypted message and the decrypted message are $f(p) = (p+3) \bmod 26$ and $f^{-1}(p) = (p-3) \bmod 26$, respectively.

A slight generalization of the Caesar, cipher called the **shift cipher** or the **additive cipher**, is when 3 is replaced by the integer b, called a **key**. In other words, the numerical equivalent of each letter is shifted by b, thus yielding the following functions:

$$\begin{cases} \text{Encryption} \to f(p) = (p + b) \bmod 26 \\ \text{Decryption} \to f^{-1}(p) = (p - b) \bmod 26 \end{cases}$$

Example 11.1
Using the shift cipher with key $b = 7$, encrypt the message DESTROY ALL THE EVIDENCE.

Solution
The following steps, shown in Table 11.1, must be taken to encrypt the message:
- **(i)** Break the message DESTROY ALL THE EVIDENCE into a set of individual letters.
- **(ii)** Translate each letter to the corresponding number.
- **(iii)** Apply $f(p) = (p+7) \bmod 26$ to each number in part (ii).
- **(iv)** Translate the new set of numbers to get a set of encrypted letters.
    The encrypted text is thus KLZAYVFHSSAOLLCPKLUJL.

Example 11.2
Using the shift cipher with key $b = 11$, decrypt the message LEELNVHTESH SLEJZFSLGP.

Solution
The following steps, shown in Table 11.2, must be taken to decrypt the message:
- **(i)** Break the message "LEELNVHTESHSLEJZFSLGP" into a set of individual letters.

Table 11.1 Steps in shift cipher for Example 11.1.

(i)	D	E	S	T	R	O	Y	A	L	L	T	H	E	E	V	I	D	E	N	C	E
(ii)	3	4	18	19	17	14	24	0	11	11	19	7	4	4	21	8	3	4	13	2	4
(iii)	10	11	25	0	24	21	5	7	18	18	0	14	11	11	2	15	10	11	20	9	11
(iv)	K	L	Z	A	Y	V	F	H	S	S	A	O	L	L	C	P	K	L	U	J	L

Table 11.2 Steps in shift cipher for Example 11.2.

(i)	L	E	E	L	N	V	H	T	E	S	H	S	L	E	J	Z	F	S	L	G	P
(ii)	11	4	4	11	13	21	7	19	4	18	7	18	11	4	9	25	5	18	11	6	15
(iii)	0	19	19	0	2	10	22	8	19	7	22	7	0	19	24	14	20	7	0	21	4
(iv)	A	T	T	A	C	K	W	I	T	H	W	H	A	T	Y	O	U	H	A	V	E

(ii) Translate each letter to the corresponding number.

(iii) Apply $f^{-1}(p) = (p-11) \bmod 26$ to each number in part (ii).

(iv) Translate the new set of numbers to get a set of decrypted letters.

The decrypted text is thus ATTACK WITH WHAT YOU HAVE.

A generalization of the shift cipher, called the **affine cipher**, is when for the given alphabet size n, the encryption function is $f(p) = (mp+b) \bmod n$, where $0 \le p \le n-1$, b, m, and n are all integers, and $\gcd(m, n) = 1$ to ensure $f(p)$ is a one-to-one correspondence. To decrypt a message using an affine cipher, we need to find p using $p \equiv (\overline{m}(f(p) - b))$ $(\bmod\ n)$, where $\overline{m}$ is an inverse of m modulo n, that is, $\overline{m}m \equiv 1\ (\bmod\ n)$.

Example 11.3

Consider the affine cipher $f(p) = (11p+4) \bmod 26$.

(i) Encrypt the letter N.

(ii) Decrypt the letter S.

Solution

(a) We first translate the letter N to the corresponding number; we therefore have $p = 13$. We then apply $f(p) = (11p+4) \bmod 26$ for $p = 13$ to obtain $f(13) = (11 \times 13 + 4) \bmod 26 = 17$. We then translate 17 to get the letter R, the encrypted letter of N.

(b) We first translate the letter S to the corresponding number, and therefore obtain $f(p) = 18$. With $m = 11$, we find $\overline{m} = 19$ as $\overline{m}m \equiv 1\ (\bmod\ 26)$. We thus have $p \equiv (19(18-4))(\bmod\ 26) \equiv (19)(14)(\bmod\ 26) \equiv 266\ (\bmod\ 26) \equiv 6$. Having $p = 6$, we translate 6 to get back the letter G, the decrypted letter of S.

Both shift ciphers and affine ciphers replace a letter by another letter; hence they are called **monoalphabetic ciphers**. In a **polyalphabetic cipher**, each occurrence of a character may have a different substitute; for instance, T could be enciphered as P at the beginning or the end of the text but as M in the middle. Therefore each character in the ciphered message depends on both the corresponding original text character and its position in the message.

Monoalphabetic ciphers and even polyalphabetic ciphers are simple methods of encryption and easy to break, thus being very vulnerable to attacks. This vulnerability led the way to the introduction of block ciphers, which are significantly more effective methods of encryption. A **block cipher** breaks up the original text into fixed blocks of letters and then replaces blocks of letters by other blocks of letters using a one-to-one correspondence transformation. A type of block ciphers is the **transposition cipher**, in which the order of letters in a block of letters are rearranged (reordered) according to a fixed permutation rather than being substituted with other letters of the alphabet. In other words, the letters of a block do not change, but instead they are just shuffled in a deterministic way to encrypt the block.

Example 11.4
Using the transposition cipher based on the permutation of the set $\{1, \ 2, \ 3\}$ with $1 \rightarrow 3$, $2 \rightarrow 1$, and $3 \rightarrow 2$, encrypt the message HELP IS COMING.

Solution
As the given set for the permutation consists of three integers, the plaintext message is then split into blocks of three letters, we thus need to encrypt HEL, PIS, COM, and ING. In order to encrypt each block based on the given mapping, the first letter replaces the third letter, the second letter replaces the first letter, and the third letter replaces the second letter. The encrypted message is thus ELH ISP OMC NGI.

11.2 Modern Cryptography

To encrypt a message, an encryption algorithm, an encryption key, and the plaintext are needed, and to decrypt a message, a decryption algorithm, a decryption key, and the ciphertext are required. The encryption and decryption algorithms are public (i.e., anyone can access them), but the keys are secret and thus need to be protected.

Number theory uniquely plays a pivotal role in modern cryptography. Cryptography has become increasingly complex and its applications more varied. The major requirements for a system employing cryptography are as follows:
- To provide an easy and inexpensive means of encryption and decryption to all authorized users in possession of the appropriate key.
- To ensure that the task of producing the plaintext without the key is made extremely difficult and time-consuming.

Relying on the processing power and speed of modern computers, original messages are no longer encoded in characters in a specified language, nor are they encoded one at a time. *Modern cryptography* operates on binary bit sequences and relies on publicly known algorithms for encoding the message. Secrecy is obtained through a secret key, which is used as the seed for the algorithms. In modern cryptography, encryption and decryption can be carried out rapidly using complicated functions that are designed to be resistant to attack. The computational difficulty of algorithms in conjunction with the fact that only the parties interested in secure communication possess the secret key makes it extremely difficult for anyone else to obtain the original information.

The underlying need for modern cryptography stems from the fact there are essential applications in today's world that require sensitive information to be fully protected. Some of the widely popular applications requiring cryptography are electronic and mobile commerce transactions, email privacy, secure remote surveillance, file transfers of confidential data, secure e-voting, banking data, secure cloud computing, medical records, and secure remote access.

There are numerous threats that can arise in transmission and storage of data, the array of attacks is constantly widening, and network security is continually becoming more challenging. The notion of security is tied to computing power, as a coded message is only as safe as the amount of computing power needed to break it. In short, the goal is to make undecipherability by an adversary as difficult as possible. Protecting information in its storage and retrieval as well as in transactional and messaging services is always of paramount importance.

The primary reasons to make messages secure through cryptographic mechanisms are as follows:

- **Confidentiality**: ensuring the transmitted message containing confidential data is hidden from unauthorized parties.
- **Authentication**: verifying the communicating parties are those they claim to be.
- **Integrity**: confirming that the message content has not been tampered with.
- **Nonrepudiation**: not being able to deny the transmission between the two parties has taken place.

There are fundamentally two types of adversaries. **Passive adversaries** are a threat to confidentiality, as they do not interrupt, alter, or insert any data. **Active adversaries** additionally threaten integrity and authentication. In any event, potential adversaries may have powers and resources ranging from minimal to unlimited.

There are two broad categories of cryptography: **private-key cryptography**, also known as **secret-key cryptography** or **symmetric-key cryptography**, and **public-key cryptography**, also referred to as **asymmetric-key cryptography**. Public-key cryptography is growing, but private-key cryptography is more common. They often complement each other, as there are applications in which both public-key cryptography and private-key cryptography are used.

11.3 Private-Key Cryptography

Private-key cryptography is based on sharing secrecy by permuting or substituting characters in the plaintext. All classical ciphers with no exception fall into private-key cryptography. The private-key cryptography is often used for long messages, for they require less time to encrypt. In private-key cryptography, an n-bit block of plaintext is encrypted and an n-bit block of ciphertext is decrypted. If a message block has fewer than n bits, padding must be added to make it an n-bit block. The common values of n are 64, 128, 256, and 512 bits. Private-key cryptography uses the same k-bit key for both the encryption at the transmitter and the decryption at the receiver, as shown in Fig. 11.1.

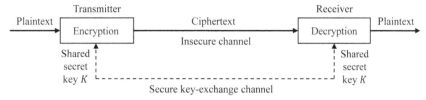

Fig. 11.1 Private-key Cryptography.

In private-key cryptography, once the key is known, both encryption and decryption can be carried out.

In private-key cryptography, the decryption algorithm is the inverse of the encryption algorithm, and the number of private keys for N users to communicate is $\frac{N(N-1)}{2}$, as each pair must have a unique private key. The number of keys grows quadratically, thus making these systems infeasible for larger-scale use. In private-key distribution, a trusted third party, referred to as a **key distribution center** (KDC), is used. In private-key cryptography, each user establishes a shared secret key with KDC. The secret keys, created by KDC, are used exclusively between KDC and the users and not among the users themselves. When a user wants to transmit secretly with another one, the transmitter then asks KDC for a session (temporary) secret key to be used between the two users. A session private key between two parties is used only once.

A widely known example of private-key cryptography is the **Data Encryption Standard** (DES), which has a block cipher structure. DES was developed in the 1970s, and for over 25 years, DES was used by the US government to protect binary data during transmission and storage in computer systems and by the banking industry and businesses to protect financial transactions for commercial data security. At the transmitter, a 64-bit plaintext is created into a 64-bit ciphertext, and at the receiver, a 64-bit cipher text is converted back to a 64-bit plaintext, noting that in DES, the same 56-bit key is used for both encryption and decryption. The key in fact consists of 64 bits. However, only 56 of these are actually used by the algorithm. Eight bits are used solely for error detection and are thereafter discarded. Despite its deprecation as an official standard, DES remains popular; it is used across a wide range of applications, including ATM encryption, email privacy, and secure remote access.

Another widely known example of private-key cryptography is the **Advanced Encryption Standard** (AES), which was established in 2001 as a replacement for DES. The highly complex structure of AES yields a high degree of protection against cryptanalysis. AES has a block size of 128 bits but with three different key lengths of 128, 192, and 256 bits. AES has been adopted by the US government and supersedes DES. AES is based on block ciphers and is efficient in both software and hardware.

11.4 Public-Key Cryptography

Public-key cryptography is based on personal secrecy rather than sharing secrecy. In public-key cryptography, the plaintext and ciphertext are numbers that are manipulated by mathematical functions. In public-key cryptography, a public key and a private key are used (see Fig. 11.2). It is a salient requirement that it must not be possible to determine the private key from the public key. In general, the public key is small, and the private key is large. A pair of keys can be used many times. The number of keys for N users to communicate is $2N$. The algorithm is complex and more efficient for short messages.

In public-key distribution, an organization, known as a **certification authority** (CA), is used. In public-key cryptography, public keys, like secret keys, need to be distributed securely; otherwise, the process can be subject to forgery. CA first checks the identification of a user, asks the user's public key, and writes it on the certificate. To prevent the certificate itself from being forged, CA signs the certificate with its own private key, which is difficult to be forged. The user uploads the signed certificate. Anyone who wants a user's public key downloads the user's signed certificate and then uses the CA's public key to obtain the user's public key.

Message authentication protects two communicating parties from any third party; it does not, however, protect the two parties against each other. In situations where something more than authentication is needed, the digital signature is the most attractive solution; in addition, it can also prevent denial and forgery. Public-key cryptography can be used to provide nonrepudiation by producing a digital signature. A **digital signature** uses a pair of private-public keys belonging to the sender to provide message integrity and message authentication. In short, a digital signature is reminiscent of an ordinary signature, as they both have the benefit of being easy to produce but difficult enough for anyone else to forge. A digital signature can be permanently tied to the content of the message being signed, and it cannot be moved then from one digital file to another, for any attempt will be detected.

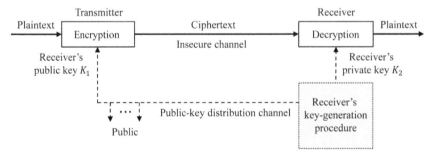

Fig. 11.2 Public-key Cryptography.

In a digital signature system, on the one hand, the messages are normally long, but on the other hand, we must use public-key cryptography systems that are very inefficient when dealing with long messages. The solution is to create a ***digest*** of a message through a hashing function, for instance, a checksum is produced that is much smaller than the message. Therefore the transmitter first produces a digest, then encrypts it using its own private key to produce the signature, and then sends the signature along with the message to the receiver. At the receiver, the signer's public key is applied to the signature to get the transmitted digest, and the digest is also directly determined from the message. If both digests are the same, then only the given transmitter could have issued the message and that the message has maintained its integrity.

It is important to highlight the distinction between how private and public keys are used in digital signature, vis-à-vis in cryptography for confidentiality. In digital signature, the signer (transmitter) signs the message digest with the signer's private key and the verifier (receiver) verifies with the signer's public key. In contrast, in cryptography, the public and private keys of the receiver are used in the process.

Another important application of public-key cryptography is the digital envelope. A ***digital envelope*** is a framework for data encryption, in which the data is encrypted under a secret key using a private-key cryptography, such as DES or AES, while this secret key is encrypted using a public-key cryptography such as RSA, and sent to the other party. In short, a public key cryptography can be used to distribute private keys to pairs of individuals wishing to communicate.

11.5 The RSA Cryptosystem

The RSA, named after the initials of its inventors, is the most widely used public-key cryptography for secure data transmission. Its effectiveness is based on the fact that very large prime numbers are fairly easy to produce on a computer, but it is enormously difficult (i.e., extremely time-consuming) to factor a product of two large unknown prime numbers. Digital envelopes and digital signatures are two important applications of RSA. However, these two security functions are mutually independent (i.e., neither, either, or both can be applied to a message). A digital envelope is the electronic equivalent of putting the message into a sealed envelope to provide privacy and prevent unauthorized alterations. A digital signature is the electronic equivalent of signing and sealing the letter. The digital envelope protects the digital signature.

The following steps show how the RSA keys can be generated:

(i) Choose two very large prime numbers p and q at random, on the order of a couple of hundred digits each, often using probabilistic computer algorithms, and calculate $n = pq$. As the number of digits in n is approximately equal to the sum of digits in p and q, no computer currently can factor it in a reasonable length of time. Note that due to the nature of factoring algorithms, p and q need to be of similar size to keep the RSA system secure. In addition, the large numbers are stored in binary form and generally 2048 to 4096 bits are needed for n to ensure a reasonably high degree of security.

(ii) Compute $\varphi(n) = (p-1)(q-1)$, and choose an arbitrary integer e satisfying $1 < e < \varphi(n)$, which is relatively prime to $\varphi(n)$. In other words, $\gcd(e, \varphi(n)) = 1$. Although for some applications, such as making encryption faster on small devices like smart cards, it is desirable to have small values of e, it is best not to choose a small value for e, as the secrecy of the cipher may then be compromised.

(iii) Find the positive integer d, an inverse of e modulo $\varphi(n)$, that is, find an integer d such that $ed \equiv 1 \pmod{\varphi(n)}$. Note that as $\gcd(e, \varphi(n)) = 1$, we have $de + c\varphi(n) = 1$, where the integers c and d can be found using the extended Euclidean algorithm, with the condition that $0 < d < \varphi(n)$.

The **public key** is $\{n, e\}$, which is widely distributed, and thus anyone with the public key can encrypt a message to send. The **private key** is $\{n, d\}$, where the security of the system depends on d being very difficult to calculate if only n and e are known. Because d is not publicly available, only someone in possession of that value can correctly decrypt the message.

The correctness of the RSA cryptosystem follows from the elementary number theory, in particular from Euler's theorem. The RSA cryptosystem is a block cipher and the following steps reflect how to encrypt using the public key and decrypt using the private key:

(i) Translate each letter in the plaintext message M into a two-digit number because the RSA cipher works only on numbers while noting that A is translated into 00, B is translated into 01, ..., and Z is translated into 25.

(ii) Concatenate the two-digit numbers into a sequence of digits representing the plaintext message, and then divide the sequence of digits into the largest possible equally sized blocks of even number of digits. Note that a block of digits is represented by m, where $0 < m < n$, and $\gcd(m, n) = 1$.

(iii) Pad the plaintext message with a number of random characters at the beginning and the end of the message to foil some potential code breaking attacks (if needed) or if a block is not full, additional characters need to be filled in to make the last block the same size as all other blocks or to ensure $\gcd(m, n) = 1$.

(iv) Determine the cyphertext C block by block using $c = m^e \bmod n$, which is the encryption of m in the RSA cryptosystem.

(v) Determine the plaintext M block by block using $m = c^d \bmod n$, which is the decryption of c in the RSA cryptosystem.

Example 11.5
Assuming $p = 101$, $q = 103$, and the message is PASSION. Determine the blocks of integers that are encrypted using the RSA cryptography.

Solution

We have $n = pq = 101 \times 103 = 10,403$. The translation of the seven letters in PASSION into their numerical equivalents is 15001818081413. We now need to divide this sequence of fourteen digits into the largest possible equally sized blocks of even number of digits, where each block is less than 10,403. This results in blocks of four digits, as a block of six digits is greater than 10,403. Therefore the sequence of 14 digits are groups into 4 blocks of 4 digits, where we have padded the end of the sequence, say, with 24, so as to make the sequence into 16 digits, a multiple of 4. The four blocks of integers to be encrypted block by block are thus as follows: 1500 1818 0814 1324.

Example 11.6

Show how the message 19 can be encrypted using the RSA cryptography by the sender and how accordingly the encrypted message can be decrypted by the receiver.

Solution

Suppose the sender arbitrarily chooses the two prime numbers $p = 83$ and $q = 53$ and thus has $n = pq = 83 \times 53 = 4399$ and $\varphi(n) = (p-1)(q-1) = 82 \times 52 = 4264$. The sender then chooses an arbitrary $e = 23$, which is relatively prime to 4264, because we have $\gcd(4264, 23) = 1$. Therefore $(4399, 23)$ is a valid public key. The encrypted message using fast modular multiplication is then as follows: $c = m^e \bmod n = 19^{23} \bmod 4399 = 2556$.

The valid private key is thus $(4399, 927)$, which the receiver possesses. Note that $d = 927$, as it is an inverse of e modulo $\varphi(n) = 23$ modulo 4264 while noting that $de + c\varphi(n) = 23d + 4264c = 1$. The decrypted message using fast modular multiplication is then as follows: $m = c^d \bmod n = 2556^{927} \bmod 4399 = 19$.

Example 11.7

Suppose in an RSA cryptosystem, we have $p = 23$ and $q = 31$. Encrypt $M = 572$ to get C and then decrypt C to get M back.

Solution

$$\text{Given} \begin{cases} p = 23 \\ \\ q = 31 \end{cases} \rightarrow n = pq = 713 \rightarrow \varphi(n) = (p-1)(q-1) = 660.$$

$$\text{Having} \begin{cases} e < \varphi(n) \\ \\ \gcd(e, \varphi(n)) = 1 \end{cases} \rightarrow \begin{array}{l} \text{Choosing } e = 29 \\ \text{as an arbitrary choice,} \\ \text{while meeting both requirements.} \end{array}$$

$$\text{Having} \begin{cases} d < \varphi(n) \\ \\ de \equiv 1 \ (\text{mod } \varphi(n)) \end{cases} \rightarrow \text{Using the Euclidean algorithm} \rightarrow d = 569.$$

$$\text{Having} \begin{cases} \text{Public key}: \{713, 29\} \\ \\ \text{Private key}: \{713, 569\} \end{cases} \rightarrow \text{Obtaining} \begin{cases} C = 572^{29} \text{ mod } 713 = 113 \\ \\ M = 113^{569} \text{ mod } 713 = 572 \end{cases}$$

Note that by replacing each exponent by its binary expansion and then using the modular exponentiation algorithm, C and M were calculated.

Example 11.8

Suppose in an RSA cryptosystem, we have $p = 53$ and $q = 61$. Encrypt $M = 1717$ to get C and then decrypt C to get M back.

Solution

$$\text{Given} \begin{cases} p = 53 \\ \\ q = 61 \end{cases} \rightarrow n = pq = 3233 \rightarrow \varphi(n) = (p-1)(q-1) = 3120.$$

$$\text{Having} \begin{cases} e < \varphi(n) \\ \\ \gcd(e, \varphi(n)) = 1 \end{cases} \rightarrow \begin{array}{l} \text{Choosing } e = 17 \\ \text{as an arbitrary choice,} \\ \text{while meeting both requirements.} \end{array}$$

$$\text{Having} \begin{cases} d < \varphi(n) \\ \\ de \equiv 1 \ (\text{mod } \varphi(n)) \end{cases} \rightarrow \text{Using the Euclidean algorithm} \rightarrow d = 2753.$$

$$\text{Having} \begin{cases} \text{Public key: } \{3233, 17\} \\ \\ \text{Private key: } \{3233, 2753\} \end{cases} \rightarrow \text{Obtaining} \begin{cases} C = 1717^{17} \bmod 3233 = 2460 \\ \\ M = 2460^{2753} \bmod 3233 = 1717 \end{cases}$$

Note that by replacing each exponent by its binary expansion and then using the modular exponentiation algorithm, C and M were calculated.

Exercises

(11.1)
Encrypt HELLO using the shift cipher with $b = 15$.

(11.2)
Encrypt the message BEWARE OF MARTIANS using the transposition cipher with blocks of four letters, and the permutation $1 \rightarrow 3$, $2 \rightarrow 1$, $3 \rightarrow 4$, and $4 \rightarrow 2$.

(11.3)
Propose a method to decrypt a ciphertext message if a transposition cipher was used.

(11.4)
Suppose the alphabet in a language consists of n letters. Assuming the shift cipher is employed, determine a key b for which the enciphering function and deciphering function are the same.

(11.5)
Determine an inverse of 43 modulo 660, that is, find an integer s such that $43s \equiv 1 \pmod{660}$.

(11.6)
Show why the RSA cryptography works, that is, for the cyphertext C and plaintext M we have

$$C = M^e \bmod n \quad \rightarrow \quad M = C^d \bmod n.$$

(11.7)
In a trivial, certainly unrealistic, example of the RSA cryptosystem, we have $p = 3$ and $q = 13$. Encrypt 20.

(11.8)
Suppose in an RSA cryptosystem, we select two prime numbers $p = 17$ and $q = 11$. After determining a private key and a public key, determine the encryption of 88 and show how 88 can be obtained back from the encrypted number.

(11.9)

Suppose in an RSA cryptosystem, we select two prime numbers $p = 59$, $q = 43$, and $e = 13$. Assuming the encrypted message is 0667 1947 0671, determine the original message.

(11.10)

Show that in the RSA cryptography, when $n = pq$, with p and q being primes, and $\varphi(n) = (p-1)(q-1)$ are both known, we can easily factor n to find p and q.

CHAPTER 12

Algorithms

Contents

The term *algorithm* is the corrupted version of the last name of the famous mathematician al-Khowarizmi, whose book on Hindu numerals was principally responsible for spreading the Hindu-Arabic numeral system throughout Europe and the development of mathematics during the European Renaissance. Algorithms are very widely used in everyday life in one way or another. For instance, there are several search engines on the Internet, where each has its own proprietary algorithm that ranks websites for each keyword or combination of keywords. From seeking the shortest route and online shopping to biometric authentication and video compression, algorithms are everywhere in modern life. This chapter briefly discusses some aspects of algorithms and describes several sorting and search methods.

12.1 Algorithm Requirements

An **algorithm** is a finite unambiguous sequence of instructions, set of rules, or number of steps that involves repetition of an operation or reiteration of a procedure for performing a computation, solving a mathematical problem, or accomplishing some end in a finite amount of time. Some of the important requirements for an algorithm are as follows:
- An algorithm has input values from a specified set.
- An algorithm produces the output values from the input values, which is the solution to a problem.
- An algorithm possesses finiteness, that is, it produces the output after a finite number of steps.
- An algorithm possesses definiteness, that is, all steps of the algorithm are precisely defined using unambiguous, well-defined operations.
- An algorithm possesses correctness, that is, it produces correct output values for any set of input values.

Discrete Mathematics
ISBN 978-0-12-820656-0, https://doi.org/10.1016/B978-0-12-820656-0.00012-5

- An algorithm possesses effectiveness, that is, it performs each step precisely and in a finite amount of time, and no step can be impossible to do, such as division by zero.
- An algorithm is well-ordered, as a computer can only execute an algorithm if it knows the exact order of steps to perform.
- An algorithm possesses generality, that is, it should accept any general set of input values.

Example 12.1

Noting a chessboard (a square board divided into 64 alternating dark and light squares) is referred to as an 8×8 board, the total number of squares of all sizes (from the largest square to the smallest squares) in an $n \times n$ chessboard is as follows:

$$S = \sum_{i=1}^{n} i^2$$

Describe the steps of an algorithm that finds the sum S for an integer $n \geq 1$.

Solution

Noting that this algorithm meets all requirements, the following sequence of steps is required:

(1) Input n.
(2) Set $S = 0$.
(3) Set $i = 1$.
(4) Compute $i^2 = i \times i$.
(5) Add i^2 to S.
(6) Add 1 to i.
(7) Compare i to n; if $i > n$, then output S and stop.
(8) Go back to step (4).

An algorithm is called optimal for the solution of a problem with respect to a specific operation (e.g., the number of comparisons) if there is no algorithm for solving this problem using a fewer number of operations.

In order to implement an algorithm using a computer, a flowchart or a pseudocode may be required. A *flowchart* is a graphical method of presenting an algorithm to illustrate its flow of execution. A flowchart shows how the flow of execution proceeds from one statement to the next until the end of the algorithm is reached. A *pseudocode* is a semi-formal language, an intermediate step between a text description of an algorithm and an implementation of the algorithm in a programming language. There are no required syntax rules for pseudocode, as it is intended for human reading rather than machine reading. However, there are many useful notational conventions to eliminate ambiguity and aid in clarity. The flowchart or pseudocode of an algorithm can be converted into a programming language of choice, such as Java or C++.

There are some problems for which it can be shown that no algorithm exists for solving them, one such problem is the halting problem. The **halting problem** is a procedure that takes as input a computer program and the input to the program and determines whether the program will eventually finish running (i.e., will ultimately stop) or continue to run forever. It is obviously very important to have such a procedure to test whether a program may enter into an infinite loop when writing and debugging programs. Alan Turing proved that a general algorithm to solve the halting problem for all possible program-input pairs cannot exist. Based on proof by contradiction, it can be shown that the halting problem on Turing machines (an abstract computational model that performs computations by reading and writing to an unlimited amount of memory) is undecidable, hence unsolvable, as the halting problem seeks an algorithm that works for all programs and all input data.

12.2 Algorithmic Paradigms

An *algorithmic paradigm* is an abstraction higher than the notion of an algorithm. An algorithmic paradigm is a generic model, based on a particular approach, which underlies the design of a class of algorithms for solving a multitude of problems. Some of the important algorithmic paradigms are as follows:

- *Brute-force algorithms*: The brute-force algorithm is a simple algorithm in concept that blindly iterates all possible solutions to search for one or more than one solution that may solve a problem without any regard to the heavy computational requirements. It takes an inefficient approach for solving problems, as it does not take advantage of the special structure of the problem. Examples include using all possible permutations of numbers to open a safe, finding the largest number in a list of numbers, sorting problems, such as the bubble, insertion, and selection sorts, and polling a multitude of communication devices to determine those having messages to transmit.

- *Divide-and-conquer algorithms*: The divide-and-conquer algorithm is an effective algorithm that works by recursively breaking down a problem into two or more subproblems of the same or related type until these become simple enough to be solved directly and rather easily. The solutions to the subproblems are then combined to give a solution to the original problem. Examples include sorting problems, such as quicksort and merge sort, binary search, solving the closest pair problem, routing mail by sorting letters into separate bags for different geographical areas, and applying the law of total probability.

- **Backtracking algorithms**: The backtracking algorithm incrementally builds candidates to the solution and abandons a candidate (i.e., backtracks) as soon as it determines that the candidate cannot possibly be a part of a valid solution. It is generally applied to find solutions to some constrained optimization problems. Examples include the eight queens puzzle, which asks for all arrangements of eight chess queens so that no queen attacks any other, crosswords, Sudoku, the knapsack problem, and finding spanning trees.
- **Dynamic programming**: The dynamic programming algorithm can be effectively used for solving a complex problem by recursively breaking down the problem. It requires that overlapping subproblems exist, and the optimal solution of the problem can be obtained using optimal solutions of its subproblems stored in memory with the help of a recurrence relation. Examples include the scheduling problem, Fibonacci numbers, matrix chain multiplication, and the traveling salesman problem.
- **Probabilistic algorithms**: The probabilistic algorithm can solve problems that cannot be easily solved by deterministic algorithms. In contrast to a deterministic algorithm, which always follows the same steps for a given input and has to go through a very large number of possible cases, the probabilistic algorithm makes some random choices at some steps, which may lead to different outputs in much fewer steps, but with a tiny probability that the final answer may not be correct. Examples include Monte Carlo algorithm in quality control, collisions in hashing functions, and Bayesian spam filters.
- **Greedy algorithms**: The greedy algorithm is one of the simplest and most intuitive algorithms that is used in optimization problems, and it often leads to an optimal solution. The algorithm makes the optimal choice at each step as it attempts to find the minimum or maximum value of some parameter. Greedy algorithms sometimes fail to produce optimal solutions because they do not consider all the data, as the choice made by a greedy algorithm may depend on choices it has made so far, but it is not aware of future choices it could make. A greedy algorithm in a shortsighted manner identifies an optimal subproblem in the problem. With the goal of minimizing or maximizing the parameter of interest, create an iterative way to go through all of the subproblems so as to build a solution. Examples may include finding the shortest path through a graph and a Huffman code used in lossless digital compression of data.

Example 12.2
Use the greedy algorithm to find an optimal solution in each of the following two cases:

(a) Use the least number of coins to make 92 cents change with quarters, dimes, nickels, and pennies.

(b) Determine the path with the largest sum in Fig. 12.1.

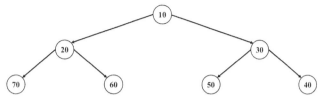

Fig. 12.1 Possible paths for Example 12.2.

Solution

(a) In this case, the greedy algorithm produces an optimal solution. To make change for 92 cents for the least number of coins, the focus at any step is on the largest possible coin, that is, on quarters, dimes, nickels, and pennies, respectively. First, three quarters are selected, leaving 17 cents; next, one dime is selected, leaving 7 cents; next a nickel is selected, leaving 2 cents; and finally 2 pennies. The total number of coins is thus $7(= 3 + 1 + 1 + 2)$, as $92 = (3 \times 25) + (1 \times 10) + (1 \times 5) + (2 \times 1)$.

(b) In this case, the greedy algorithm does not produce an optimal solution. The greedy algorithm fails to find the largest sum simply because it makes decisions based only on the information it has at any one step, without regard to the overall problem. In order to reach the largest sum, at each step, the greedy algorithm chooses what appears to be the optimal immediate choice. It therefore chooses 30 instead of 20 at the second step and thus does not reach the best solution, which is $100 (= 10 + 20 + 70)$, and it mistakenly finds $90 (= 10 + 30 + 50)$ as the largest sum.

12.3 Complexity of Algorithms

The two main measures for the computational complexity of an algorithm are as follows:

- *Space complexity*: It is measured by the maximum amount of computer memory needed in the execution of the algorithm, and the requirement is frequently a multiple of the data size.
- *Time complexity*: It is measured by counting the number of key operations using the size of the input as its argument.

The term *complexity* generally refers to the running time of the algorithm. The function $f(n)$, representing the time complexity of an algorithm, is measured by the following two factors:

(i) The size n of the input data and the characteristics of the particular input data.

(ii) The number of basic key operations that must be performed when the algorithm is executed, while noting they generally include addition, subtraction, multiplication, division, and comparison.

Example 12.3

Discuss the algorithms required to perform the following mathematical operations:

(a) Matrix multiplication using direct method.

(b) Polynomial evaluation using Horner's method.

Solution

(a) In the multiplication of two matrices, order matters (i.e., matrix multiplication is not commutative). In order to multiply two matrices, we must first make sure that the number of columns in the first matrix equals the number of rows in the second matrix, as this is the prerequisite for multiplication. The widely known direct method is as follows: if the first matrix is an $m \times n$ matrix and the second matrix is an $n \times p$ matrix, where m, n, and p are all positive integers, then their matrix product is the $m \times p$ matrix, whose entries are given by the dot product of the corresponding row of the first matrix and the corresponding column of the second matrix. In other words, we multiply the elements of row i of the first matrix by the elements of column j in the second matrix to obtain the element of the ith row and jth column of the product. The algorithm to perform this matrix multiplication thus requires mpn multiplications and $mp(n-1)$ additions. Note that there are other algorithms that require a smaller number of multiplications and additions to multiply two matrices than the direct method does.

(b) Consider the following function $f(x)$, which is a polynomial of degree n:

$$f(x) = a_n x^n + a_{n-1} x^{n-1} + \dots + a_2 x^2 + a_1 x + a_0.$$

Suppose we want to evaluate $f(x)$ for $x = c$, that is, $f(c)$. We can consider two different algorithms for a polynomial evaluation. The first is known as the direct method, which requires $n + (n-1) + (n-2) + \dots + 1 = \frac{n(n+1)}{2}$ multiplications and n additions. The second is known as **Horner's method**, which is based on rewriting the polynomial by successively factoring out x, and is as follows:

$$f(x) = a_0 + a_1 x + a_2 x^2 + \dots + a_{n-1} x^{n-1} + a_n x^n$$
$$= a_0 + x(a_1 + x(a_2 + \dots + x(a_{n-1} + x a_n))).$$

Horner's method is significantly more efficient than the direct method, as it requires only n multiplications and n additions.

Noting the function $f(n)$ represents the time complexity of an algorithm for the input data of size n, there are typically two types of time complexity to analyze:

(i) The ***average-case time complexity***, that is, the expected value of $f(n)$, is usually difficult to analyze, and it is generally assumed that the distribution of the possible inputs is uniform when the actual distribution is unknown, and the uniform distribution may not actually apply to real situations.

(ii) The ***worst-case time complexity***, that is, the maximum value of $f(n)$, is easier than the average-case complexity to analyze, as this complexity is based on the largest number of operations required, as a solution for any possible input is guaranteed.

12.4 Measuring Algorithm Efficiency

Algorithmic efficiency is a property of an algorithm that relates to the amount of computational resources used by the algorithm. An algorithm must be analyzed to determine its resource usage, and the efficiency of an algorithm can be measured based on the usage of different resources. The asymptotic growth of functions is commonly used in the analysis of algorithms to estimate the run time and the amount of memory they require. It is very important to provide approximations that make it easy to estimate the large-scale differences in algorithmic efficiency while ignoring differences of a constant factor and differences that occur only for small sets of input data.

The estimates of resources required by an algorithm are represented by the big-Oh (O), big-Omega (Ω), and big-Theta (Θ) notations without being concerned about constant multipliers or smaller order terms. Note that the three notations $f(x) = O(g(x))$, $f(x) = \Omega(g(x))$, and $f(x) = \Theta(g(x))$ all stand for collections of functions. Hence the equality sign does not mean equality of functions. Moreover, as we are dealing with functions representing complexity, these functions take on only positive values. Therefore all references to absolute values can be dropped for such functions.

The growth of a function representing the complexity of an algorithm can be estimated using the big-O notation as its input grows. Let $f(x)$ and $g(x)$ be real-valued functions defined on the same set of nonnegative real numbers. Then $f(x)$ is $O(g(x))$, which is read as $f(x)$ is ***big-Oh*** of $g(x)$, if there are real constants C and k such that $|f(x)| \leq C|g(x)|$ whenever $x > k$. This definition indicates that $f(x)$ grows slower than some fixed multiple of $g(x)$ as x grows slowly without bound. However, the rate of growth of multiple of $g(x)$ should be preferably close to the rate of growth of $f(x)$, where $g(x)$ provides an upper bound for the size of $f(x)$ for large values of x. The constants C and k are called ***witnesses*** to the relationship $f(x)$ is $O(g(x))$ and are not unique. Although there are infinitely many pairs of witnesses, we need only one pair of witnesses. The approach to find a pair of witnesses is to first find a value of k for which the size of $|f(x)|$ can be easily estimated when $x > k$, and then find a value of C for which $|f(x)| \le C|g(x)|$ when $x > k$. In short, if for a sufficiently large value of x, the values of $f(x)$ are less than those of a multiple of $g(x)$, then $f(x)$ is $O(g(x))$.

The following properties of the big-O estimate are of importance:

- When we have $f(x) = O(g(x))$ and $|h(x)| > |g(x)|$, we then have $f(x) = O(h(x))$.
- When $f(x)$ is a polynomial of degree n, we then have $f(x) = O(x^n)$.
- If we have $f_1(x) = O(g_1(x))$ and $f_2(x) = O(g_2(x))$, then we have $f_1(x) + f_2(x) = O(\max(|g_1(x)|, |g_2(x)|))$. Therefore if we have $f_1(x) = O(g(x))$ and $f_2(x) = O(g(x))$, then we have $f_1(x) + f_2(x) = O(g(x))$.
- If we have $f_1(x) = O(g(x))$ and $f_2(x) = O(f_1(x))$, then we have $f_2(x) = O(g(x))$.
- If we have $f_1(x) = O(g_1(x))$ and $f_2(x) = O(g_2(x))$, then we have $f_1(x)f_2(x) = O(g_1(x)g_2(x))$.

Note that if $f(x)$ is a sum of several terms and there is one with the largest growth rate, then that term can be kept and all others omitted, and also if $f(x)$ is a product of several factors, then constants (terms in the product that do not depend on x) can be omitted.

Example 12.4

Show that $f(x) = x^3 + 3x^2 + 3x + 1$ is $O(x^3)$ for three pairs of witnesses.

Solution

(a) If we choose $k = 1$ (i.e., $x > 1$), we then have $3x^3 > 3x^2 > 3x$ and $x^3 > 1$, where these inequalities are obtained by multiplying both sides of $x > 1$ by appropriate terms. An upper bound on $f(x)$ is thus as follows:

$$0 < x^3 + 3x^2 + 3x + 1 < x^3 + 3x^3 + 3x^3 + x^3 = 8x^3 \to C = 8.$$

(b) If we choose $k = 2$ (i.e., $x > 2$), we have $\frac{3}{2}x^3 > 3x^2$, $\frac{3}{4}x^3 > \frac{3}{2}x^2 > 3x$, and $\frac{x^3}{8} > 1$, where these inequalities are obtained by multiplying both sides of $x > 2$ by appropriate terms. An upper bound on $f(x)$ is thus as follows:

$$0 < x^3 + 3x^2 + 3x + 1 < x^3 + \frac{3}{2}x^3 + \frac{3}{4}x^3 + \frac{1}{8}x^3 = \frac{27}{8}x^3 \to C = \frac{27}{8}.$$

(c) If we choose $k = 3$ (i.e., $x > 3$), we have $x^3 > 3x^2$, $\frac{x^3}{3} > x^2 > 3x$, and $\frac{x^3}{27} > 1$, where these inequalities are obtained by multiplying both sides of $x > 3$ by appropriate terms. An upper bound on $f(x)$ is thus as follows:

$$0 < x^3 + 3x^2 + 3x + 1 < x^3 + x^3 + \frac{1}{3}x^3 + \frac{1}{27}x^3 = \frac{64}{27}x^3 \to C = \frac{64}{27}.$$

Fig. 12.2 shows $f(x) = O(x^3)$ for all these three pairs of witnesses.

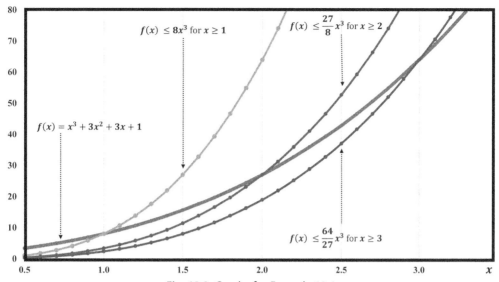

Fig. 12.2 Graphs for Example 12.4.

Example 12.5

Suppose we have $f(x) = x^n$, where n is a positive integer.

(a) Show that x^n is $O(x^{n+1})$.

(b) Show that x^{n+1} is not $O(x^n)$.

Solution

(a) Note that when $x > 1$, as we multiply both sides of $x > 1$ by x^n we have $x^n < x^{n+1}$. Consequently, we can take $k = 1$ and $C = 1$ as a pair of witnesses.

(b) We use a proof by a contradiction to show that no pair of witnesses C and k exists such that $x^{n+1} \leq Cx^n$ whenever $x > k$. Suppose that there is a pair of witnesses and k for which $x^{n+1} \leq Cx^n$ whenever $x > k$. This, in turn, means that when $x > 0$, we have $x \leq C$ (by dividing the inequality by x^n). However, no matter what C and k are, the inequality $x \leq C$ cannot hold for all x with $x > k$. For instance, when x is greater than both k and C, it is not true that $x \leq C$ even though $x > k$. This contradiction shows that x^{n+1} is not $O(x^n)$.

When x is a very large positive integer, the order functions satisfy the following relationships:

$$O(1) < O(\log_2 x) < O(x) < O(x \log_2 x) < O(x^2) < O(2^x) < O(x!) < O(x^x).$$

Fig. 12.3 shows a display of the growth of functions commonly used in big-O estimates. Note that when we say the order of magnitude of an algorithm is a constant,

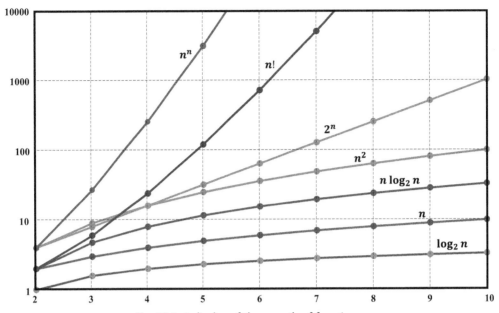

Fig. 12.3 A display of the growth of functions.

we mean that the execution time is bounded by a constant (i.e., it is independent of the input size x). If the order is linear, the execution time grows linearly (i.e., it is directly proportional to the input size x).

Example 12.6
Give a big-O estimate for $f(n) = an \log_a(n!) + (n^2 + a)\log_a(n^2 + a)$, where $a > 1$ is a real number and $n > 1$ is an integer.

Solution
We have

$$an \log_a(n!) = an \log_a(1 \times 2 \times \ldots \times n) \le an \log_a(n \times n \times \ldots \times n)$$
$$= an \log_a(n^n) = an^2 \log_a(n).$$

If we choose $n > a + 1$ (i.e., $k = a + 1$), we then have

$$(n^2 + a)\log_a(n^2 + a) \le (n^2 + a)\log_a(n^2 + an^2) = (n^2 + a)\log_a((1 + a)n^2)$$
$$= (n^2 + a)(\log_a(1 + a) + \log_a(n^2)) = (n^2 + a)(\log_a(1 + a) + 2\log_a(n))$$
$$\le (n^2 + a)(\log_a(n) + 2\log_a(n)) \le (n^2 + a)(3\log_a(n)) \le (2n^2)(3\log_a(n))$$
$$= 6n^2 \log_a(n).$$

We therefore have

$$f(n) = an \log_a(n!) + (n^2 + a)\log_a(n^2 + a) < an^2 \log_a(n) + 6n^2 \log_a(n)$$
$$= (an^2 + 6n^2)\log_a(n) = (a+6)n^2 \log_a(n) = O(n^2 \log_a(n)) \rightarrow C = a+6.$$

Let $f(x)$ and $g(x)$ be real-valued functions defined on the same set of nonnegative real numbers. Then $f(x)$ is $\Omega(g(x))$, read as $f(x)$ is **big–Omega** of $g(x)$ if there are real constants C and k such that $|f(x)| \geq C|g(x)|$ whenever $x > k$. This definition indicates $g(x)$ provides a lower bound for the size of $f(x)$ for large values of x. Note that $f(x)$ is $\Omega(g(x))$ if and only if $g(x)$ is $O(f(x))$. In short, if for sufficiently large value of x, the values of $f(x)$ are greater than those of a multiple of $g(x)$, then $f(x)$ is $\Omega(g(x))$.

Example 12.7

Give a $\Omega(g(x))$ estimate for $f(x) = 100x^3 - 10x^2 + 1$.

Solution

For $x \geq 1$, we have

$$x^3 \geq x^2 \rightarrow -x^3 \leq -x^2 \rightarrow -10x^3 \leq -10x^2.$$

Consequently, we have

$$100x^3 - 10x^2 + 1 \geq 100x^3 - 10x^3 = 90x^3.$$

The function $f(x) = 100x^3 - 10x^2 + 1$ is $\Omega(x^3)$ with $k = 1$ and $C = 90$.

Example 12.8

Give a big-Ω estimate for $f(x) = \frac{15\sqrt{x}(2x+9)}{x+1}$, if $x \geq 0$.

Solution

For $x > 1$, we have

$$\frac{15\sqrt{x}(2x+9)}{x+1} > \frac{15\sqrt{x}(2x+2)}{x+1} = \frac{30\sqrt{x}(x+1)}{x+1} > 30\sqrt{x}.$$

The function $f(x) = \frac{15\sqrt{x}(2x+9)}{x+1}$ is $\Omega(\sqrt{x})$ with $x > 1$ and $C = 30$.

Let $f(x)$ and $g(x)$ be real-valued functions defined on the same set of nonnegative real numbers. Then $f(x)$ is $\Theta(g(x))$, read as $f(x)$ is **big–Theta** of $g(x)$, if $f(x)$ is $O(g(x))$ and $f(x)$ is $\Omega(g(x))$. This definition indicates $g(x)$ provides a lower bound as well as an upper bound for the size of $f(x)$ for large values of x. Note that $f(x)$ is $\Theta(g(x))$ if and only if there are real numbers C_1 and C_2 and a positive real number such that $C_1|g(x)| \leq |f(x)| \leq C_2|g(x)|$, wherever $x > k$. In short, if for sufficiently large value of x, the values of $f(x)$ are bounded both above and below by those of a multiple of $g(x)$, then $f(x)$ is $\Theta(g(x))$.

Example 12.9
Give a big-Θ estimate for $f(x) = 5x^3 + 10x^2 \log_2 x$, if $x > 0$.

Solution
Assuming $x > 1$, we have the following:

$$x > \log_2 x > 0 \rightarrow 10x^3 > 10x^2 \log_2 x > 0 \rightarrow$$
$$5x^3 + 10x^3 > 5x^3 + 10x^2 \log_2 x > 5x^3 \rightarrow 15x^3 > f(x) > 5x^3.$$

Consequently, $5x^3 + 10x^2 \log_2 x$ is $\Theta(x^3)$, where $C_1 = 5$ and $C_2 = 15$, for $k = 1$.

Example 12.10
Give a big-Θ estimate for $f(n) = 1 + 2 + 3 + \dots + n$, where n is a positive integer.

Solution
We have

$$f(n) = 1 + 2 + 3 + \dots + n = \frac{n(n+1)}{2} = 0.5n^2 + 0.5n.$$

Assuming $n > 1$, we have the following:

$$0.5n^2 > 0.5n > 0 \rightarrow 0.5n^2 + 0.5n^2 > 0.5n^2 + 0.5n > 0.5n^2 \rightarrow n^2 > f(n) > 0.5n^2.$$

Consequently, $0.5n^2 + 0.5n$ is $\Theta(n^2)$, where $C_1 = 0.5$ and $C_2 = 1$, for $k = 1$.

If $f(x) = a_n x^n + a_{n-1} x^{n-1} + \dots + a_0$ is a polynomial of degree $n \geq 0$, where n is an integer, all coefficients are real numbers and $a_n \neq 0$, then for large values of x, we have the following:

$$\begin{cases} f(x) = O(x^s) & \text{for all integers } s \geq n \\ f(x) = \Omega(x^r) & \text{for all integers } r \leq n \\ f(x) = \Theta(x^n) \end{cases}$$

Note that big-Oh, big-Theta, and big-Omega notations can be extended to functions in more than one variable. For example, for functions in two variables and some witnesses, we have the following:

$$\begin{cases} f(x,\ y) \ = \ O(g(x,y)) \ \to \ |f(x,\ y)| \leq C_1 |g(x,\ y)| & \text{for } x > k_1 \text{ and } y > k_2 \\ f(x,\ y) \ = \ \Omega(g(x,y)) \ \to \ |f(x,\ y)| \geq C_2 |g(x,\ y)| & \text{for } x > k_1 \text{ and } y > k_2 \\ f(x,\ y) \ = \ \Theta(g(x,y)) \ \to \ C_2 |g(x,\ y)| \leq |f(x,\ y)| \leq C_1 |g(x,\ y)| & \text{for } x > k_1 \text{ and } y > k_2 \end{cases}$$

12.5 Sorting Algorithms

An entry in a database is a 2-tuple whose first component is a key and the second component is the corresponding data. In other words, a key in a database is a value from an ordered set, which is used to store and retrieve data. The process of arranging a collection of database entries into a sequence that conforms to the order of their keys is called *sorting*. A *sorting algorithm* is an algorithm that puts elements of a list in a certain order, such as an ascending order or a descending order.

A sorting algorithm puts the elements of a list in a certain place or rank according to kind, class, or nature. Such an arrangement in the context of discrete mathematics and computer science is generally in alphabetical or numerical order. Sorting examples may include words in a dictionary, names in a telephone directory, email addresses in a contact list, cities according to the sizes of their populations, and countries based on gross domestic product. A major advantage of using a sorted sequence of elements rather than an unsorted sequence of elements is that it is very much easier to find a particular element, especially when the sequence is quite long.

A significant portion of computing resources is often devoted to sorting problems. There are numerous sorting algorithms. Some algorithms are easier to implement, and some are more efficient either in general or for a set of particular inputs. Each algorithm has certain benefits and drawbacks. The focus of this section is on comparison-sorting algorithms, meaning sorting algorithms that can sort items of any type for which a less-than or equal-to relation is defined (i.e., which of two elements should occur first in the final sorted list). A comparison sort cannot perform better than $O(n \log_2(n))$ on average. Table 12.1 presents the complexities of various sorting methods.

The *bubble sort* is a simple but not very efficient sorting algorithm. In the bubble sort, smaller elements bubble to the top and larger elements sink to the bottom. It successively compares adjacent elements in the list and swaps them if they are not in the right order. The first element is compared with the second element, the elements are interchanged if the second element is smaller than the first one, otherwise no swapping is done. Then the current second element is compared with the third element, those elements are

TABLE 12.1 Complexities of various sorting methods.

Sorting methods	Worst-case time complexity	Average-case time complexity
Bubble sort	$O(n^2)$	$O(n^2)$
Insertion sort	$O(n^2)$	$O(n^2)$
Merge sort	$O(n \log n)$	$O(n \log n)$
Quicksort	$O(n^2)$	$O(n \log n)$
Selection sort	$O(n^2)$	$O(n^2)$

interchanged if the third element is smaller than the second one. This process of inter-changing a larger element with a smaller one following it starts with the first element and continues to the last element for a full pass. This procedure is repeated through several passes until the sort is complete. In a list of n elements, at the end of the ith pass, the i largest elements are correctly placed at the end of the list, where $1 \leq i \leq n$. The bubble sort takes $n - 1$ passes to sort a list of n elements. There are n elements, and for each element, $n - 1$ comparisons are made, this thus leads to a total time complexity of $O(n^2)$.

The *insertion sort* is a simple sorting algorithm that builds the final sorted list one element at a time. It is less efficient than most sorting algorithms unless the number of elements in the list is modest, say about 50 or less. The insertion sort begins with the second element and compares it with the first element and inserts it before the first element if it does not exceed the first element and after the first element otherwise. Then the third element is compared with the first element, and if it is larger than the first element, it is compared with the second element, and it is inserted in the correct position among the first three elements. The insertion sort iterates through the list and removes one element per iteration, finds the place that the element belongs to, and then places it there. The resulting list after k iterations has the property where the first $k + 1$ elements are sorted. The insertion sort requires at most $n - 1$ comparisons and $n - 1$ swaps. At each step, the number of comparisons and swaps decreases by one. Using the arithmetic progression, this gives rise to a total of $n(n-1)$ comparisons. This thus leads to a total time complexity of $O(n^2)$.

The *merge sort*, a recursive sorting algorithm, focuses on how to merge together two presorted lists such that the resulting list is also sorted. A merge sort proceeds by iteratively splitting lists into two sublists of equal or almost equal numbers of elements until each sublist contains one element; a list of one element is considered sorted. The merge sort then successively merges pairs of sublists, where both sublists are in increasing order, to produce a new larger sublist with elements in increasing order. This continues until the original list is put into increasing order. The merge sort is a recursive algorithm and time complexity can be expressed as the following recurrence relation $f(n) = 2f\left(\frac{n}{2}\right) + O(n)$, where $O(n)$ is what it takes to combine the sublists. The solution, that is, the time complexity of the merge sort, is then $O(n \log n)$.

The **quicksort** is an efficient sorting algorithm employing a divide-and-conquer strategy. It begins by selecting a pivot element, usually the first element from the list, and partitioning the other elements into two sublists. One sublist consists of all elements in the list less than the pivot and the other consists of all elements greater than the pivot. Then the pivot is put at the end of the first list as its final resting place. The two sublists are then recursively sorted until all sublists contain only one element. The sorted list can then be obtained by combining the sublists of one item in the order they occur. The quicksort requires $O(n)$ comparisons to scan through all n elements to divide them into two sublists. The quicksort is a recursive algorithm. In a balanced division, we have the average-case time complexity that can be expressed as the following recurrence relation $f(n) = 2f\left(\frac{n-1}{2}\right) + O(n)$, whose solution is then $O(n \log n)$. In the worst-case scenario (i.e., the pivot is the smallest or the largest in the list), the corresponding recurrence relation is $f(n) = 2f(n-1) + O(n)$, whose solution is then $O(n^2)$.

The **selection sort** is a very simple but inefficient sorting algorithm. The algorithm proceeds by finding the smallest element in the unsorted list and moves it to the top of the sorted list. Then the least element among the remaining elements of the unsorted list is found and moved to the second position of the ordered list. This procedure is repeated until the entire list has been sorted. The selection sort requires at $n - 1$ comparisons to select the smallest element and swap it to make it into the first position, and $n - 2$ comparisons to select the second smallest element and swap it, and so on. Using the arithmetic progression, this gives rise to a total of $\frac{n(n-1)}{2}$ comparisons. This thus leads to a total time complexity of $O(n^2)$.

Example 12.11
Sort the list $\{7, 6, 1, 9, 5\}$ using the following sorting algorithms:
(a) Bubble sort.
(b) Insertion sort.
(c) Merge sort.
(d) Quicksort.
(e) Selection sort.

Solution
(a) In each step of a pass, elements in boldface are being compared.

$$
\text{1st pass} \rightarrow
\begin{cases}
\mathbf{7,6},1,9,5 \\
6,\mathbf{7,1},9,5 \\
6,1,\mathbf{7,9},5, \\
6,1,7,\mathbf{9,5} \\
6,1,7,5,9
\end{cases}
\text{2nd pass} \rightarrow
\begin{cases}
\mathbf{6,1},7,5,9 \\
1,\mathbf{6,7},5,9 \\
1,6,\mathbf{7,5},9, \\
1,6,5,\mathbf{7,9} \\
1,6,5,7,9
\end{cases}
$$

$$
\text{3rd pass} \rightarrow
\begin{cases}
\mathbf{1,6},5,7,9 \\
1,\mathbf{6,5},7,9 \\
1,5,\mathbf{6,7},9, \\
1,5,6,\mathbf{7,9} \\
1,5,6,7,9
\end{cases}
\text{4th pass} \rightarrow
\begin{cases}
\mathbf{1,5},6,7,9 \\
1,\mathbf{5,6},7,9 \\
1,5,\mathbf{6,7},9 \\
1,5,6,\mathbf{7,9} \\
1,5,6,7,9
\end{cases}
$$

Note that the bubble sort needs one whole pass (i.e., the fourth pass) without any swap to know it is sorted.

(b) The insertion sort begins with 6 and compares it with 7. It inserts it before 7, as it does not exceed 7. Then, 1 is compared with 6, and as it is smaller than 6, it is inserted before 6. Next, 9 is compared with 1, as it is larger than 1, it is compared with 6, as it is larger than 6, it is compared with 7, as it is larger than 7, it is inserted after 7. Finally, 5 is compared with 1, as it is larger than 1, it is compared with 6, as it is smaller than 6, it is inserted after 1 and before 6. Note that the following shows the required steps, where the element under consideration in each step is in boldface:

$$7,\mathbf{6},1,9,5 \rightarrow 6,7,\mathbf{1},9,5 \rightarrow 1,6,7,\mathbf{9},5 \rightarrow 1,6,7,9,\mathbf{5} \rightarrow 1,5,6,7,9.$$

(c) Fig. 12.4 shows the merge sort, where the top part is the split and the bottom part is the merge.

(d) Employing the quicksort, the first element (i.e., 7) is selected, and we partition the other elements into two sublists. The first sublist consists of all elements in the list less than 7 (i.e., 6, 1, 5), and the other consists of all elements greater than 7 (i.e., 9). Then 7 is put at the end of the first sublist as its final resting place. The two sublists are then recursively sorted the same way, that is, 6 is selected in the first sublist and the other elements in that sublist are compared with 6 (i.e., 1, 5), and 9 is selected in the second sublist. Note that the following shows the required steps, where the element under consideration in each step is in boldface:

$$\mathbf{7},6,1,9,5 \rightarrow \mathbf{6},1,5,7,9 \rightarrow \mathbf{1},5,6,7,9 \rightarrow 1,5,6,7,9$$

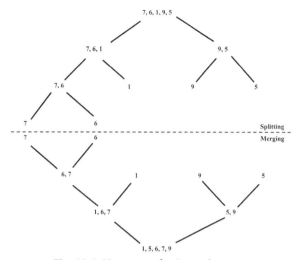

Fig. 12.4 Merge sort for Example 12.11.

(e) The selection sort finds 1, as it is the smallest element in the unsorted list, and moves it to the top of the sorted list. Then the least element among the remaining elements of the unsorted list (i.e., 5) is found and moved to the second position of the ordered list. This procedure is repeated until the entire list has been sorted. Note that the following shows the required steps, where the element under consideration in each step is in boldface:

$$7, 6, 1, 9, 5 \rightarrow \mathbf{1} \rightarrow 1, \mathbf{5} \rightarrow 1, 5, \mathbf{6} \rightarrow 1, 5, 6, \mathbf{7} \rightarrow 1, 5, 6, 7, \mathbf{9} \rightarrow 1, 5, 6, 7, 9.$$

12.6 Search Algorithms

Searching is the process of locating an element in a list. A *search algorithm* is an algorithm that involves a search problem. Searching a database employs a systematic procedure to find an entry with a key designated as the objective of the search. A search algorithm locates an element x in a list of distinct elements or determines that it is not in the list. The solution to the search is either the location of the element x in the list or 0 if x is not on the list. We now briefly introduce two well-known search algorithms whose worst-case and average time complexities are presented in Table 12.2.

TABLE 12.2 Complexities of various search methods.

Search methods	Worst-case time complexity	Average-case time complexity
Linear search	$O(n)$	$O(n)$
Binary search	$O(\log n)$	$O(\log n)$

The *linear search*, also known as the *sequential search*, is the simplest search algorithm. It is an algorithm, based on the brute-force algorithmic paradigm, that scans the elements of a list in sequence in search of x, the element that needs to be located. A comparison is made between x and the first element in the list, if they are the same, then the solution is 1. Otherwise, a comparison is made between x and the second element in the list, if they are the same, then the solution is 2. This process continues until a match is found and the solution is the location of the element sought. If no match is found, then the solution is 0. Linear search is applied on unsorted or unordered lists consisting of a small number of elements. Because n comparisons are required to find x, the linear search has a time complexity of $O(n)$, which means the time is linearly dependent on the number of elements in the list.

The list of data in a binary search must be in a sorted order for it to work, such as ascending order. This search algorithm, which is quite effective in large sorted array, is based on the divide-and-conquer algorithmic paradigm. A *binary search* works by comparing the element to be searched with the element in the middle of the array of elements. If we get a match, the position of the middle element is returned. If the target element is less than the middle element, the search continues in the upper half of the array (i.e., the target element is compared to the element in the middle of the upper subarray), and the process repeats itself. If the target element is greater than the middle element, the search continues in the lower half of the array (i.e., the target element is compared to the element in the middle of the lower subarray), and the process repeats itself. By doing this, the algorithm eliminates the half in which the target element cannot lie in each iteration. Assuming the number of elements is $n = 2^k$ (i.e., $k = \log_2 n$), at most $2k + 2 = 2\log_2 n + 2$ comparisons are required to perform a binary search. Binary search is thus more efficient than linear search, as it has a time complexity of $O(\log n)$. The worst case occurs when x is not in the list.

Example 12.12
Consider an array consisting of these integers: 2, 3, 5, 7, 23, 19, 17, 13, 11, 29, 31, 37.
(a) Use the linear search to find 17.
(b) Use the binary search to find 17.

Solution
(a) The first element of the array, that is, 2 is compared with 17, as they are not the same, 3 is compared with 17, as they are not the same, 5 is compared with 17; this process continues until 17, which is in the list, is found. Then the solution is 7, as 7 is the location of 17 in the list.
(b) In order to apply the binary search, we first need to sort the list. The sorted list, in ascending order, is then as follows: 2, 3, 5, 7, 11, 13, 17, 19, 23, 29, 31, 37. As the number of integers in the list is 12, there is no middle integer.

Therefore the sorted list is split into two sublists, each consisting of six integers, namely, 2, 3, 5, 7, 11, 13 and 17, 19, 23, 29, 31, 37. Then compare 17 and the largest integer in the first list. Because 17 > 13, the search for 17 must be restricted to the second sublist, namely, 17, 19, 23, 29, 31, 37. Next we split this sublist into two smaller sublists, each with three integers, namely, 17, 19, 23 and 29, 31, 37. Then, compare 17 and the largest integer in the first sublist. Because 23 > 17, the search for 17 can be restricted to the first sublist. Next we compare 17 to the middle integer in this sublist, namely, 19, as 19 > 17, we split this sublist into two integers, one smaller than 19 (i.e., 17) and the other larger than 19 (i.e., 23). The search has been narrowed down to one term, we thus compare 17 and 17, and 17 is thus located in the sorted list.

Exercises

(12.1)

Determine the increase in complexity for each of the following functions when the input size n is increased to $n + 1$, assuming n is a very large positive integer.

(a) $f(n) = \log n$.
(b) $f(n) = n$.
(c) $f(n) = n \log n$.
(d) $f(n) = n^m$, integer $m > 1$.
(e) $f(n) = 2^n$.
(f) $f(n) = n!$.

(12.2)

Apply the binary search algorithm to find 72 in the following sorted sequence:

$$1, \ 8, \ 9, \ 13, \ 22, \ 27, \ 36, \ 47, \ 49, \ 72, \ 81, \ 100, \ 121, \ 144, \ 150$$

(12.3)

Determine the big-Oh and big-Theta of the following function:

$$f(x) = \frac{x^7 + x^4 + x^3 + x^2 + x + 1}{x^3 + 1}.$$

(12.4)

Show that $8x^4$ is $O(x^5)$.

(12.5)

Determine the big-Oh of the following function, where n and m are both positive integers.

$$\left(n \log n + n^2\right)^m.$$

(12.6)

Describe how the number of comparisons used in the worst-case changes in each of the following search algorithms when the size of the list increases from n to mn, where $n > 0$ and $m > 0$ are positive integers.

(a) Linear search.

(b) Binary search.

(12.7)

Show that $f(x, y) = (x^2 + y^2 + xy + y \log x)^4$ is $O(x^8 y^8)$.

(12.8)

Describe how the number of comparisons used in the worst-case changes in each of the following sorting algorithms when the size of the list increases from n to mn, where $n > 0$ and $m > 0$ are positive integers.

(a) Bubble sort.

(b) Insertion sort.

(12.9)

Arrange the following functions in a list so that each function is big-O of the next function.

$$2^n, n^{100}, (\log n)^4, \sqrt[3]{n} \log n, 10^n, (n!)^2.$$

(12.10)

Noting that $f(n)$ is $O(g(n))$, determine $g(n)$, where $f(n)$ is as follows:

$$f(n) = 1^k + 2^k + \ldots + n^k,$$

where k and n are both positive integers.

CHAPTER 13

Induction

Contents

Induction is a very powerful method of proof. A multitude of mathematical claims can be validated using this technique. However, it is not a tool for discovering true claims or deriving correct formulas. Induction can be effectively used to prove a conjecture once it has been made (and is true) while not providing insights as to why it is true. Induction deals with families of statements that are generally indexed by the natural numbers (positive integers). In this chapter, mathematical induction, strong induction, and the well-ordering principle, which are all equivalent principles, will be discussed, where the validity of each can be proven from the validity of the other two techniques.

13.1 Deductive Reasoning and Inductive Reasoning

Deductive reasoning, which is top-down logic, contrasts with inductive reasoning, which is bottom-up logic. While the conclusion of a deductive argument is certain, based on the facts provided, the truth of the conclusion of an inductive argument may be probable based upon the evidence given.

Deductive reasoning refers to the process of concluding that something must be true because it is a specific case of a general principle that is already known to be true. Deductive reasoning is the process of reasoning from premises to reach a logically certain conclusion; it is logically valid and is the fundamental method in which mathematical facts are shown to be true. Deductive reasoning provides a guarantee of the truth of the conclusion if the premises (assumptions) are true. In other words, in a deductive argument, the premises are intended to provide such a strong support for the conclusion that, if the premises are true, then it would be impossible for the conclusion to be false. For example, a general principle in plane geometry states that the sum of the angles in any triangle is 180 degrees, then one can conclude that the sum of the angles in an isosceles right triangle is also 180 degrees. Another example is that the colonial powers systematically colonized countries and oppressed their people, then one can conclude that the

Discrete Mathematics
ISBN 978-0-12-820656-0, https://doi.org/10.1016/B978-0-12-820656-0.00013-7

British Empire, as it was a major colonial power, also colonized countries and oppressed people in a systematic manner. In summary, deductive reasoning requires one to start with a few general ideas, called premises, and apply them to a specific situation. Recognized rules, laws, theories, and other widely accepted truths are used to prove that a conclusion is right.

Inductive reasoning is the process of reasoning that a general principle is true because the special cases are true. Inductive reasoning makes broad generalizations from specific observations. Basically, there is data, and then conclusions are drawn from the data. Inductive reasoning is a process of reasoning in which the premises are viewed as supplying some evidence for the truth of the conclusion. It is also described as a method where one's experiences and observations, including what are learned from others, are synthesized to come up with a general truth. For example, if all the people one has ever met from a particular country have been racist, one might then conclude all the citizens of that country are racist. Inductive reasoning is not logically valid. Just because all the people one happens to have met from a country were racist is no guarantee at all that all the people from that country are racist. Therefore this form of reasoning has no part in a mathematical proof. Even if all of the premises are true in a statement, inductive reasoning allows for the conclusion to be false. For instance, my neighbor is a grandfather. My neighbor is bald. Therefore all grandfathers are bald. The conclusion does not follow logically from the statements. In summary, inductive reasoning uses a set of specific observations to reach an overarching conclusion. Therefore a few particular premises create a pattern that gives way to a broad idea that is possibly true.

Inductive reasoning is part of the discovery process whereby the observation of special cases leads one to suspect very strongly (though not know with absolute logical certainty) that some general principle is true. Deductive reasoning, on the other hand, is the method you would use to demonstrate with logical certainty that the special case is true. In other words, inductive reasoning is used to formulate hypotheses and theories, and deductive reasoning is employed when applying them to specific situations. The difference between the two kinds of reasoning lies in the relationship between the premises and the conclusion. If the truth of the premises definitely establishes the truth of the conclusion (due to definition, logical structure, or mathematical necessity), then it is deductive reasoning. If the truth of the premises does not definitely establish the truth of the conclusion but nonetheless provides a reason to believe the conclusion may be true, then the argument is inductive.

13.2 Mathematical Induction

The idea behind mathematical induction is in showing how each statement follows from the previous one, all that remains is to kick off this logical chain reaction from some starting point. Informal metaphors of mathematical induction may include climbing an infinite ladder step by step (i.e., every rung of the ladder could be reached after reaching the first rung) or the sequential falling of an infinite row of dominoes (i.e., every single domino could knock over after the first domino falls backward).

Let $P(n)$ be a propositional function, where n is a positive integer (i.e., $n \in N$). A proof by mathematical induction consists of two steps. The first is the **basis step**, where we show $P(1)$, called the **basis hypothesis**, is true. The second is the **inductive step**, where we show that if $P(k)$, called the **inductive hypothesis**, is true for an arbitrary integer $k \geq 1$, then $P(k+1)$ is also true. The principle of mathematical induction states that by proving these two key steps, we can conclude that the predicate $P(n)$ is true for every positive integer $n \in N$. The mathematical induction is also called **finite induction** or **weak induction**. It is also referred to as **incomplete induction**, which is a misnomer as not only is it complete, but it is a valid method of proof.

Expressing as a rule of inference, while assuming the domain is the set of positive integers N, the **principle of mathematical induction** can be stated as follows:

$$(P(1) \wedge \forall k(P(k) \rightarrow P(k+1))) \rightarrow \forall n P(n).$$

The **proof of the principle of mathematical induction** is based on a proof by contradiction. We thus assume that there is at least one positive integer for which $P(n)$ is false. In view of the well-ordering principle that states every nonempty subset of the set of positive integers N has a smallest element, there is the nonempty set T of positive integers for which $P(n)$ is false and its minimum element is denoted by the integer j. According to the basis hypothesis, $P(1)$ is true. This implies $j > 1$, and in turn $j - 1$ is a positive integer. Because $j - 1 < j$, we have $j - 1 \notin T$, as j is the smallest element in T. As a result, $P(j-1)$ is true. As $P(j-1)$ is true, then induction indicates that $P(j)$ is true. This means $j \notin T$, which contradicts j being the minimum element of T, thus resulting in an absurd conclusion. Therefore T being nonempty just cannot be true. If T being nonempty leads to a contradiction, then T must be empty. Hence $P(n)$ is true for every integer in N.

In mathematical induction, we first show that $P(1)$ is true, and then show that if $P(k)$ is true, then $P(k+1)$ is true, where $k \geq 1$. Therefore we know that $P(2)$ is true because $P(1)$ is true. Moreover, we know $P(3)$ is true because $P(2)$ is true. Continuing along these lines, we see that $P(n)$ is true for every integer $n > 1$. In a way, the principle of mathematical induction contains a technique based on a chain of deductive reasoning that infinitely many steps ($n \rightarrow \infty$) are true in just two steps.

It is important to highlight that mathematical induction can also be used to prove $P(n)$ is true, where the basis hypothesis is not $P(1)$. For instance, mathematical induction can prove the equality $2^0 + 2^1 + 2^2 + \dots + 2^n = 2^{n+1} - 1$, where the basis hypothesis is $P(0)$, or the inequality $2^n < n!$, where the basis hypothesis is $P(4)$. In principle, mathematical induction can be used to prove a propositional function $P(n)$ is true for every integer $n \geq a$, where a is a fixed integer.

Note that the basis and inductive steps are both essential in the principle of mathematical induction as the validity of each step is necessary, but not sufficient, to guarantee the proposition is true. As an example, consider the formula $2^n = 2n$, for every positive integer n. We can easily verify the basis hypothesis, as $2^1 = 2 \times 1$ is true. However, we cannot show the inductive hypothesis to be true because if we have $2^k = 2k$ for an integer $k > 1$, we then have $2^{k+1} \neq 2(k+1)$. As another example, consider the summation formula $1 + 3 + 5 + \ldots + (2n-1) = n^2 + 1$ for every positive integer n. The inductive hypothesis is true, simply because if we assume, we have $1 + 3 + 5 + \ldots + (2k-1) = k^2 + 1$ for an integer $k > 1$, we can then have $1 + 3 + 5 + \ldots + (2k-1) + (2k+1) = k^2 + 1 + (2k+1) = (k+1)^2 + 1$ for an integer $k+1$. However, the summation formula does not hold for any positive integer, as we cannot show the basis hypothesis to be true, simply because for $k = 1$, we have $1 \neq 1^2 + 1$.

Sometimes induction errors occur mainly due to the lack of precision. A case in point is that when we use mathematical induction carelessly to prove $a^n = 1$ for $a \neq 0$, $a \neq 1$, and $n \geq 0$. It is easy to verify the basis hypothesis, as $a^0 = 1$ is true. In the inductive step, we can show that for $k > 0$, the inductive hypothesis is true, that is, if we have $a^k = 1$, we then have $a^{k+1} = \frac{a^{k+1} \times a^k}{a^k} = \frac{a^k \times a^k}{a^{k-1}} = \frac{1 \times 1}{1} = 1$. The flaw in the proof lies in the inductive step because we did not include $k = 0$. Had we done that, it would have been obvious that in the denominator $a^{-1} \neq 1$, as $a \neq 1$.

Some mistakenly believe that a proof by mathematical induction is a fallacy known as a case of circular reasoning, in that what is assumed is what needs to be proven! The confusion stems from misinterpreting the inductive step for the conclusion. The inductive step involves showing that the implication $P(k) \rightarrow P(k+1)$ is a tautology. In other words, "if it is assumed" that $P(k)$ is true for a particular but arbitrarily chosen $k \geq 1$, then $P(k+1)$ is also true, whereas the conclusion is that $P(n)$ is true for every integer $n \geq 1$.

It is important to note that in order to prove a mathematical statement, verifying it for a number of different values of positive integer n does not form a formal proof, because there may exist a particular value of n for which the statement may not be true. In short, a proof for some values of n is not a proof for all values of n. To prove formally that a mathematical statement is true, we must show that it is true for all values of n greater than the initial value. However, it is impossible to verify it for every single value of n, as there are infinitely many values. For instance, in order to prove the function $n^2 - n + 41$, introduced by Euler, produces only prime numbers, it is not good enough to verify it for a few values of n. This is simply because for any positive integer $n < 41$, the function does produce a prime number, but for $n = 41$, it does not produce a prime number.

It is also possible to use mathematical induction with more than one variable, as the following ***two-dimensional mathematical induction principle*** reflects. Suppose $\{P(i, j)\}$ is a set of statements such that the basis step consisting of $P(0, j)$ for $j \geq 0$ and $P(i, 0)$ for $i \geq 0$ are true and the inductive step that if $P(i-1, j)$ and $P(i, j-1)$ are both true, then $P(i, j)$ is true for $i \geq 1$ and $j \geq 1$. By proving these two key steps, we can then conclude that $P(m, n)$ is true for all positive integers m and n. As an example, the principle of the two-dimensional mathematical induction can show that for all positive integers m and n, we have

$$P(m, n) = \sum_{r=1}^{n} r(r+1)(r+2)...(r+m-1) = \frac{n(n+1)(n+2)...(n+m)}{m+1}.$$

13.3 Applications of Mathematical Induction

Mathematical induction is a technique for proving statements of the form $\forall n P(n)$, where $P(n)$ is a propositional function and the domain is generally the set of positive integers. There exists a very wide set of applications of mathematical induction, such as summation and product formulas, inequalities, divisibility, regions in plane geometry, set identities, complexity of algorithms, theorems about graphs and trees, and correctness of computer programs and algorithms.

Example 13.1
Using mathematical induction, prove the following summation formulas for every integer $n \geq 1$:

(a) $1 + 2 + 3 + ... + n = \dfrac{n(n+1)}{2}$.

(b) $1 + r + r^2 + ... + r^{n-1} = \dfrac{r^n - 1}{r - 1}$, $r \neq 1$.

(c) $\displaystyle\sum_{i=0}^{n} \left(-\frac{1}{2}\right)^i = \frac{2}{3} + \frac{1}{3}\left(-\frac{1}{2}\right)^n$.

Solution
Let $P(n)$ be the propositional function of interest that must be proven to be true for every integer $n \geq 1$.

(a) Basis step: When $n = 1$, we have $1 = \frac{1(1+1)}{2}$, therefore $P(1)$ is true.

Inductive step: Assuming the inductive hypothesis

$$P(k) = 1 + 2 + ... + k = \frac{k(k+1)}{2}$$

is true, for an arbitrary integer $k \geq 1$, we must show

$$P(k+1) = 1 + 2 + \ldots + k + (k+1) = \frac{(k+1)(k+2)}{2}$$

is true. If we add $(k+1)$ to both sides of the equation representing $P(k)$, then we have

$$1 + 2 + \ldots + k + (k+1) = \frac{k(k+1)}{2} + (k+1) = \frac{(k+1)(k+2)}{2}.$$

This thus means that the statement $P(k+1)$ also holds true, as was to be shown.

(b) Basis step: When $n = 1$, we have $1 = \frac{r^1 - 1}{r - 1}$, therefore $P(1)$ is true.

Inductive step: Assuming the inductive hypothesis

$$P(k) = 1 + r + r^2 + \ldots + r^{k-1} = \frac{r^k - 1}{r - 1}$$

is true, for an arbitrary integer $k \geq 1$, we must show

$$P(k+1) = 1 + r + r^2 + \ldots + r^{k-1} + \left(r^k\right) = \frac{r^{k+1} - 1}{r - 1}$$

is true. If we add $\left(r^k\right)$ to both sides of the equation representing $P(k)$, then we have

$$1 + r + r^2 + \ldots + r^{k-1} + \left(r^k\right) = \frac{r^k - 1}{r - 1} + \left(r^k\right) = \frac{r^{k+1} - 1}{r - 1}.$$

This thus means that the statement $P(k+1)$ also holds true, as was to be shown.

(c) Basis step: $P(1)$ is true, as we have

$$\sum_{i=0}^{1} \left(-\frac{1}{2}\right)^i = \frac{2}{3} + \frac{1}{3}\left(-\frac{1}{2}\right)^1 \rightarrow \left(-\frac{1}{2}\right)^0 + \left(-\frac{1}{2}\right)^1 = \frac{2}{3} - \frac{1}{6} \rightarrow \frac{1}{2} = \frac{1}{2}.$$

Inductive step: Assuming the inductive hypothesis

$$P(k) = \sum_{i=0}^{k} \left(-\frac{1}{2}\right)^i = \frac{2}{3} + \frac{1}{3}\left(-\frac{1}{2}\right)^k$$

is true, for an arbitrary integer $k \geq 1$, we must show

$$P(k+1) = \sum_{i=0}^{k+1} \left(-\frac{1}{2}\right)^i = \frac{2}{3} + \frac{1}{3}\left(-\frac{1}{2}\right)^{k+1}$$

is true. Using the assumption that $P(k)$ is true, we have

$$\sum_{i=0}^{k+1}\left(-\frac{1}{2}\right)^{i} = \sum_{i=0}^{k}\left(-\frac{1}{2}\right)^{i} + \left(-\frac{1}{2}\right)^{k+1} = \frac{2}{3} + \frac{1}{3}\left(-\frac{1}{2}\right)^{k} + \left(-\frac{1}{2}\right)^{k+1}$$

$$= \frac{2}{3} + \left(-\frac{1}{2}\right)^{k}\left(\frac{1}{3}-\frac{1}{2}\right) = \frac{2}{3} + \left(-\frac{1}{2}\right)^{k}\left(-\frac{1}{6}\right)$$

$$= \frac{2}{3} + \left(-\frac{1}{2}\right)^{k} \times \left(-\frac{1}{2}\right) \times \frac{1}{3} = \frac{2}{3} + \frac{1}{3}\left(-\frac{1}{2}\right)^{k+1}.$$

This thus means that the statement $P(k+1)$ also holds true, as was to be shown.

Example 13.2

Using mathematical induction, prove the following inequalities for every integer $n \geq 1$:

(a) $(1 + x)^{n} \geq 1 + nx$, for every real number $x > -1$.

(b) $2^{n} > n$.

Solution

Let $P(n)$ be the propositional function of interest that must be proven to be true for every integer $n \geq 1$.

(a) Basis step: When $n = 1$, we have $(1 + x)^{1} = 1 + x$, therefore $P(1)$ is true.

   Inductive step: Assuming the inductive hypothesis

$$(1 + x)^{k} \geq 1 + kx$$

is true, that is $P(k)$ is true, for an arbitrary integer $k \geq 1$, we must show

$$(1 + x)^{k+1} \geq 1 + (k+1)x$$

is true, that is, $P(k+1)$ is true. Using the assumption that $P(k)$ is true, we multiply both sides of the inequality in $P(k)$ by $(1 + x)$, we thus obtain

$$(1 + x)^{k+1} \geq (1 + x)(1 + kx) = 1 + (k+1)x + kx^{2} \geq 1 + (k+1)x$$

as $kx^{2} \geq 0$. This thus means that the statement $P(k+1)$ also holds true, as was to be shown.

(b) Basis step: When $n = 1$, we have $2^{1} > 1$, therefore $P(1)$ is true.

   Inductive step: Assuming the inductive hypothesis

$$2^{k} > k$$

is true, that is $P(k)$ is true, for an arbitrary integer $k \geq 1$, we must show

$$2^{k+1} > k + 1$$

is true, that is, $P(k+1)$ is true. Using the assumption that $P(k)$ is true, we multiply both sides of the inequality in $P(k)$ by 2, we thus obtain

$$2 \times 2^k > 2 \times k \to 2^{k+1} > 2k \geq k+1$$

as $k \geq 1$. This thus means that the statement $P(k+1)$ also holds true, as was to be shown.

Example 13.3

Using mathematical induction, prove the following divisibility cases for every integer $n \geq 0$:
(a) $n^4 + 2n^3 - n^2 - 2n$ is divisible by 24.
(b) $7^n - 2^n$ is divisible by 5.
(c) $4 \times \left(2^{n-2} \times 3^n + 1\right)$ is divisible by 5.

Solution

Let $P(n)$ be the propositional function of interest that must be proven to be true for every integer $n \geq 0$.
(a) Basis step: When $n = 0$, the value of the function is 0, which is divisible by 24, therefore $P(0)$ is true.
  Inductive step: Assuming the inductive hypothesis

$$k^4 + 2k^3 - k^2 - 2k = 24m$$

is true for a nonnegative integer m, that is, $P(k)$ is true for an arbitrary integer $k \geq 0$, we must then show

$$(k+1)^4 + 2(k+1)^3 - (k+1)^2 - 2(k+1) = 24t$$

is true for a nonnegative integer t, that is, $P(k+1)$ is true. Using the assumption that $P(k)$ is true, we have

$$(k+1)^4 + 2(k+1)^3 - (k+1)^2 - 2(k+1) = k^4 + 6k^3 + 11k^2 + 6k$$
$$= \left(k^4 + 2k^3 - k^2 - 2k\right) + 4k^3 + 12k^2 + 8k = 24m + 4k(k+1)(k+2)$$
$$= 24t$$

as k, $k+1$, and $k+2$ are three consecutive integers, their product is thus a multiple of 6, and as a result, $4k(k+1)(k+2)$ is also a multiple of 24. This thus means that the statement $P(k+1)$ also holds true, as was to be shown.
(b) Basis step: When $n = 0$, the value of the function is 0, which is divisible by 5, therefore $P(0)$ is true.

Inductive step: Assuming the inductive hypothesis

$$7^k - 2^k = 5m$$

is true for a nonnegative integer m, that is, $P(k)$ is true for an arbitrary integer $k \geq 0$, we must show

$$7^{k+1} - 2^{k+1} = 5t$$

is true for a nonnegative integer t, that is, $P(k+1)$ is true. Using the assumption that $P(k)$ is true, we have

$$7^{k+1} - 2^{k+1} = 7 \times 7^k - 2^{k+1} = 7(5m + 2^k) - 2 \times 2^k = 35m + 5 \times 2^k$$

$$= 5(7m + 2^k) = 5t.$$

This thus means that the statement $P(k+1)$ also holds true, as was to be shown.

(c) Basis step: When $n = 0$, the value of the function is 5, which is divisible by 5, therefore $P(0)$ is true.

Inductive step: Assuming the inductive hypothesis

$$4 \times (2^{k-2} \times 3^k + 1) = 5m$$

is true for a nonnegative integer m, that is, $P(k)$ is true for an arbitrary integer $k \geq 0$, we must show

$$4 \times (2^{k-1} \times 3^{k+1} + 1) = 5t$$

is true for a nonnegative integer t, that is, $P(k+1)$ is true. Using the assumption that $P(k)$ is true, we have

$$4 \times (2^{k-1} \times 3^{k+1} + 1) = 4 \times (2^{k-2} \times 2 \times 3^k \times 3 + 1)$$

$$= 4 \times (6 \times 2^{k-2} \times 3^k + 1)$$

$$= 20 \times 2^{k-2} \times 3^k + 4 \times (2^{k-2} \times 3^k + 1)$$

$$= 20 \times 2^{k-2} \times 3^k + 5m = 5(4 \times 2^{k-2} \times 3^k + m)$$

$$= 5t.$$

This thus means that the statement $P(k+1)$ also holds true, as was to be shown.

Example 13.4

Using mathematical induction, show that n lines in the plane, where no two lines are parallel and no three lines meeting in a point, divide the plane into the following number of regions:

$$\frac{n^2 + n + 2}{2}.$$

Solution

Let $P(n)$ be the propositional function of interest that must be proven to be true for every integer $n \geq 0$.

Basis step: If we have no lines in the plane (i.e., $n = 0$), then there exists just one region, therefore $P(0)$ is true.

Inductive step: Suppose now that we have k lines dividing the plane into $\frac{k^2+k+2}{2}$ regions and we will add a $(k + 1)$th line. This extra line will meet each of the previous k lines because we have assumed it to be parallel with none of them, and also it meets each of these k lines in a distinct point, as we have assumed that no three lines are concurrent. These k points of intersection divide the new line into $k + 1$ segments. For each of these $k + 1$ segments there are now two regions, one on either side of the segment, where previously there had been only one region. Thus by adding this $(k + 1)$th line, we have created $k + 1$ new regions. In total, the number of regions we now have is $\frac{k^2+k+2}{2} + (k+1) = \frac{(k+1)^2+(k+1)+2}{2}$. Therefore the statement $P(k+1)$ also holds true, as was to be shown.

Example 13.5

Using mathematical induction, prove the following product formula for every integer $n \geq 2$:

$$\prod_{m=2}^{n}\left(1 - \frac{1}{m^2}\right) = \left(1 - \frac{1}{2^2}\right) \times \ldots \times \left(1 - \frac{1}{n^2}\right) = \frac{n+1}{2n}.$$

Solution

Let $P(n)$ be the propositional function of interest that must be proven to be true for every integer $n \geq 2$.

Basis step: When $n = 2$, we have $1 - \frac{1}{4} = \frac{3}{4}$, therefore $P(2)$ is true.

Inductive step: Assuming the inductive hypothesis

$$\prod_{m=2}^{k}\left(1 - \frac{1}{m^2}\right) = \frac{k+1}{2k}$$

is true, that is $P(k)$ is true, for an arbitrary integer $k \geq 2$, we must show

$$\prod_{m=2}^{k+1}\left(1 - \frac{1}{m^2}\right) = \frac{(k+1)+1}{2(k+1)}$$

is true, that is, $P(k+1)$ is true. Using the assumption that $P(k)$ is true, we multiply both sides of the equality in $P(k)$ by $\left(1 - \frac{1}{(k+1)^2}\right)$, we thus obtain

$$\left(1 - \frac{1}{(k+1)^2}\right) \prod_{m=2}^{k}\left(1 - \frac{1}{m^2}\right) = \left(1 - \frac{1}{(k+1)^2}\right)\left(\frac{k+1}{2k}\right)$$

$$\rightarrow \prod_{m=2}^{k+1}\left(1 - \frac{1}{m^2}\right) = \frac{(k+1)+1}{2(k+1)}.$$

This thus means that the statement $P(k+1)$ also holds true, as was to be shown.

Example 13.6

Using mathematical induction, prove the binomial theorem, for an integer $n \geq 0$, where x and y are variables and $\binom{n}{i}$ is n choose i, we have:

$$(x+y)^n = \sum_{i=0}^{n}\binom{n}{i}x^i y^{n-i}.$$

Solution

Let $P(n)$ be the propositional function of interest that must be proven to be true for every integer $n \geq 0$.

Basis step: When $n = 0$, both sides of the formula are equal to 1, therefore $P(0)$ is true.

Inductive step: If we assume the inductive hypothesis

$$(x+y)^k = \sum_{i=0}^{k}\binom{k}{i}x^i y^{k-i}$$

is true, we must then show

$$(x+y)^{k+1} = \sum_{i=0}^{k+1}\binom{k+1}{i}x^i y^{k+1-i}$$

is true. We have

$$(x+y)^{k+1} = (x+y)(x+y)^k = (x+y)\sum_{i=0}^{k}\binom{k}{i}x^i y^{k-i}$$

$$= \sum_{i=0}^{k}\binom{k}{i}x^{i+1} y^{k-i} + \sum_{i=0}^{k}\binom{k}{i}x^i y^{k+1-i}.$$

As we take out the last term from the first sum (i.e., x^{k+1}) and the first term from the second sum (i.e., y^{k+1}), we obtain

$$(x+y)^{k+1} = x^{k+1} + \sum_{i=0}^{k-1}\binom{k}{i}x^{i+1} y^{k-i} + \sum_{i=1}^{k}\binom{k}{i}x^i y^{k+1-i} + y^{k+1}.$$

We now make a change of variable in the first sum, as we set $i = i - 1$; we therefore have

$$(x+y)^{k+1} = x^{k+1} + \sum_{i=1}^{k}\binom{k}{i-1}x^i y^{k+1-i} + \sum_{i=1}^{k}\binom{k}{i}x^i y^{k+1-i} + y^{k+1}$$

$$= x^{k+1} + \sum_{i=1}^{k}\left(\binom{k}{i-1} + \binom{k}{i}\right)x^i y^{k+1-i} + y^{k+1}$$

$$= x^{k+1} + \sum_{i=1}^{k}\binom{k+1}{i}x^i y^{k+1-i} + y^{k+1} = \sum_{i=0}^{k+1}\binom{k+1}{i}x^i y^{k+1-i}.$$

This thus means that the statement $P(k+1)$ also holds true, as it was to be shown.

13.4 Strong Induction

When the truth of $P(k)$ might not be enough to establish the truth of $P(k+1)$, we need to use strong induction. Let $P(n)$ be a propositional function, where n is a positive integer. A proof by strong induction consists of two steps. First, the **basis step**, where we show $P(1)$, called the **basis** hypothesis, is true. Second, the **inductive step**, where we show that for all positive integers $k \geq 1$, if $P(1), P(2), \ldots, P(k-1), P(k)$, called the **inductive hypothesis**, are all true, then $P(k+1)$ is also true. The principle of strong induction states that by proving these two key steps we can conclude that $P(n)$ is true for every positive integer $n \geq 1$. The strong induction is also called **complete induction**.

Expressing as a rule of inference, while assuming the domain is the set of positive integers N, the **second principle of mathematical induction** can be stated as follows:

$$(P(1) \wedge \forall k((P(1) \wedge P(2) \wedge \ldots \wedge P(k-1) \wedge P(k)) \rightarrow P(k+1))) \quad \rightarrow \quad \forall n P(n).$$

Example 13.7

Using strong induction, show that for every integer $n \geq 2$, then either n is a prime or n can be written as the product of prime numbers, in other words, n is divisible by a prime number.

Solution

Let $P(n)$ be the proposition that n can be written as the product of primes or n is divisible by a prime number, for every integer $n \geq 2$.

Basis step: $P(2)$ is true, as 2 is a prime number.

Inductive step: Assuming $P(j)$ is true for all integers j with $2 \leq j \leq k$, we need to prove $P(k+1)$ is true. There are two mutually exclusive cases, namely when $k+1$ is prime and when $k+1$ is composite (not prime):

 (i) If $k+1$ is prime, then $P(k+1)$ is true.
 (ii) If $(k+1)$ is composite, then $k+1 = a \times b$, where a and b are integers with $2 \leq a \leq b < k+1$. Using the inductive hypothesis, both a and b can be written as product of primes. Thus $k+1$ can be written as the product of those primes in factorization of a and those in the factorization of b.

Example 13.8

Suppose $S(0) = 0$, $S(1) = 4$, and $S(n) = 6S(n-1) - 5S(n-2)$. Using strong induction, show that for every integer $n \geq 0$, $S(n) = 5^n - 1$.

Solution

Let $S(n)$ be the propositional function of interest that must be proven to be true for every integer $n \geq 0$.

Basis step: $S(0)$ and $S(1)$ are both true, as we have $5^0 - 1 = 0$ and $5^1 - 1 = 4$.

Inductive step: Assuming $S(j)$ is true for all integers j with $0 \leq j \leq k$, thus implying $S(k)$ and $S(k-1)$ are true, we then need to prove

$$S(k+1) = 5^{k+1} - 1$$

is true. We have

$$S(k+1) = 6S(k) - 5S(k-1) = 6\left(5^k - 1\right) - 5\left(5^{k-1} - 1\right) = 5^{k+1} - 1.$$

This thus means that the statement $S(k+1)$ also holds true, as was to be shown.

Example 13.9

Suppose $F(0) = 0$, $F(1) = 1$, and $F(n) = F(n-1) + F(n-2)$, for $n \geq 2$, where $F(0)$, $F(1)$, $F(2)$, ... are known as the **Fibonacci numbers**. Using strong induction, show $F(n) < 2^n$ for every integer $n \geq 0$.

Solution

Let $F(n)$ be the propositional function of interest that must be proven to be true for every integer $n \geq 0$.

Basis step: $F(0)$ and $F(1)$ are both true, as we have $F(0) = 0 < 2^0 = 1$ and $F(1) = 1 < 2^1 = 2$.

Inductive step: Assuming $F(j)$ is true for all integers j with $0 \leq j \leq k$, thus implying $F(k)$ and $F(k-1)$ are true, we then need to prove

$$F(k+1) < 2^{k+1}$$

is true. We have

$$F(k+1) = F(k) + F(k-1) < 2^k + 2^{k-1} < 2^k + 2^k < 2^{k+1}.$$

This thus means that the statement $F(k+1)$ also holds true, as was to be shown.

13.5 The Well-Ordering Principle

The **well-ordering principle** states that every nonempty set of positive integers has a least element. It is of great importance that the well-ordering principle can often be used directly in proofs.

Example 13.10

In each of the following cases, state the least element of the set, if it exists, and if not, explain why the well-ordering principle is not violated:

(a) The set of all positive real numbers less than 1.
(b) The set of all positive integers n such that $n^4 < n$.
(c) The set of all positive integers of the form $13 - 4k$, where k is an integer.

Solution

(a) There is no least positive real number. Because if x is a positive real number less than 1, then $\frac{x}{m}$, for every integer $m > 1$, is a positive real number less than x. There is no violation of well-ordering principle, because this set is not a set of integers.

(b) There is no such positive integer to satisfy the inequality. The set is thus empty. There is no violation of well-ordering principle, because this set has no element, let alone a least element.

(c) The smallest positive integer of the form $13 - 4k$ is 1, and that occurs when $k = 3$.

Example 13.11

Use the well-ordering principle to prove the division algorithm. The division algorithm states given any integer n and any positive integer d, there exist integers q and r such that $n = dq + r$ and $0 \leq r < d$. Note that we may call n a dividend, d a divisor, q a quotient, and r a remainder.

Solution

Let S be the set of all positive integers of the form $n - dk$, where k is an integer. The set S has at least one element. Note that if $n \geq 0$, then for $k = 0$, we have $n - d \times 0 = n \geq 0$, and if $n < 0$, then for $k = n$, we have $n - d \times n = n(1 - d) \geq 0$. Using the well-ordering principle for the integers, S thus contains a least element r. Then for some specific integer $k = q$, we have $n - dq = r$ or equivalently $n = dq + r$. Note that if $r \geq d$, then $n - d(q+1) = n - dq - d = r - d \geq 0$, and so $n - d(q+1)$ would be a positive integer in S that would be smaller than r. This is obviously a contradiction, as r is the smallest integer in S. Hence $0 \leq r < d$.

Exercises
(13.1)

(a) Using mathematical induction, prove the following arithmetic progression:

$$\sum_{i=1}^{n} a_i = a_1 + a_2 + \ldots + a_n = \frac{n(a_1 + a_n)}{2} = \frac{n(2a + (n-1)d)}{2},$$

where n is a positive integer, $a_i = a + (i-1)d$, while noting that a, known as the initial value, and d, called the common difference, are both finite constants.

(b) Using mathematical induction, prove the following geometric progression:

$$\sum_{i=1}^{n} a_k = a_1 + a_2 + \ldots + a_n = \frac{a(r^n - 1)}{r - 1},$$

where n is a positive integer, $a_i = ar^{i-1}$, while noting that a, known as the initial value, and $r \neq 1$, called the common ratio, are both finite constants.

(13.2)

(a) Using mathematical induction, prove the following:

$$P(n) = \sum_{p=1}^{n} p^2 = \frac{n(n+1)(2n+1)}{6}$$

for every integer $n \geq 1$.

(b) Using mathematical induction, prove the following:

$$P(n) = \sum_{p=1}^{n} p^3 = \frac{n^2(n+1)^2}{4}$$

for every integer $n \geq 1$.

(13.3)

Using mathematical induction, prove $2^n < n!$ for every integer $n \geq 4$.

(13.4)

Using mathematical induction, prove the following function is divisible by 120, for every integer $n \geq 0$.

$$P(n) = n^5 - 5n^3 + 4n.$$

(13.5)

Using mathematical induction, prove $P(n) = 2n^3 + 3n^2 + n$ is divisible by 6 for every integer $n \geq 1$.

(13.6)

Using mathematical induction, show that for all integers $P(n): n \geq 8$, $n\text{¢}$ can be obtained using 3¢ and 5¢ coins.

(13.7)

Using mathematical induction, for all integers $n \geq 0$, show that $P(n) = 2^{2n} - 1$ is divisible by 3.

(13.8)

Using mathematical induction, show that $P(n) = 5^n - 1$ is divisible by 4 for every integer $n \geq 1$.

(13.9)

Using mathematical induction, prove $P(n) = n^5 - n$ is a multiple of 10, for every integer $n \geq 1$.

(13.10)

(a) Suppose that at the start of a game, there are two players and two piles of matches, each pile contains the same number of matches. The player who removes the last match wins. Using strong induction, show that the second player can always win.

(b) Using the well-ordering principle, show that every decreasing sequence of nonnegative integers is finite.

CHAPTER 14

Recursion

Contents

Recursion is the process of defining a problem or the solution to a problem in terms of a simpler version of itself. It is used in a variety of disciplines ranging from linguistics to logic. Recursion is a powerful problem-solving technique extensively used in discrete mathematics and computer science, as such many programming languages support it. The application of recursion is where an entity being defined is applied within its own definition. Examples include a function (e.g., a factorial function), a set (e.g., the power set of a set), a tree (e.g., a full binary tree), a sequence (e.g., a geometric progression), and an algorithm (e.g., a sorting algorithm). This chapter briefly discusses various methods for solving recurrence relations, namely, by iteration, characteristic equations, and generating functions.

14.1 Sequences

A sequence with its discrete structure presents an ordered list of terms, where repetitions are allowed. Sequences are an important data structure in computer science and engineering. A function f whose domain is a subset of the set of integers, generally consecutive nonnegative integers, is called a ***sequence***. A sequence is a function whose domain is either all the integers between two given integers or all the integers greater than or equal to a given integer. The function $f(n)$, denoted by a_n and called a ***term*** of the sequence, is the image of the integer n. The integer n of the term a_n is called an ***index***. The nth term is the ***general term*** of a sequence, and it is often used to define a sequence. Note that the notation $\{a_n\}$ describes the sequence. For instance, the list of terms of the sequence $\{a_n\}$, where $a_n = n^2, n \geq 1$, namely $a_1, a_2, a_3, a_4, \ldots$, starts with 1, 4, 9, 16, ..., corresponding to $n = 1, 2, 3, 4, \ldots$. Hence the sequence is as follows: $\{1, 4, 9, 16, \ldots\}$.

Discrete Mathematics
ISBN 978-0-12-820656-0, https://doi.org/10.1016/B978-0-12-820656-0.00014-9

The terms of a sequence are usually ordered in increasing order of subscripts. The first term of the sequence is called the ***initial term***. A sequence is ***finite*** if its domain is finite; otherwise, it is ***infinite***. Note that a finite sequence of n terms is also called a ***string*** or an ***n-tuple***, its last term is known as the ***final term***, and the ***length of a string*** is the number of terms in the string. For instance, the set of two-digit integers that are positive represents a finite sequence of 90 terms, where 10 is its initial term and 99 is its final term; the set of prime numbers represents an infinite sequence, where the initial term is 2 and there is no final term, as there are infinitely many prime numbers.

The most widely known sequences are geometric and arithmetic progressions. A geometric progression is a sequence with the general term $a_n = ar^n$, where the ***initial term*** a and the ***common ratio*** r are real numbers, and $n \geq 0$ is an integer. The set of terms of a geometric progression is thus as follows: $\{a, \ ar, \ ar^2, \ ..., \ ar^n, \ ...\}$. An arithmetic progression is a sequence with the general term $a_n = a + nd$, where the ***initial term*** a and the ***common difference*** d are real numbers, and $n \geq 0$ is an integer. The set of terms of an arithmetic progression is thus as follows: $\{a, \ a+d, \ a+2d, \ ..., \ a+nd, \ ...\}$.

It is often required to add or multiply a number of terms in a sequence whose terms are as follows: $a_k, a_{k+1}, \ ... \ , a_m$, where m and k are both integers and $m \geq k$. The sum of the terms and the product of the terms can be written in a compact form as follows, respectively:

$$a_k + a_{k+1} + \ ... \ + a_m = \sum_{i=k}^{m} a_i$$

and

$$a_k \times a_{k+1} \times \ ... \ \times a_m = \prod_{j=k}^{m} a_j.$$

Note that the symbol Σ, the capital Greek letter sigma, is the ***summation notation***, the ***dummy variable*** i is the ***summation index***, the symbol Π, which is the capital Greek letter pi, is the ***product notation***, the ***dummy variable*** j is the ***product index***, and the integers k and m are the ***lower limit*** and the ***upper limit*** of each index, respectively. Note that a ***series*** is an extended sum of terms, such as the sum of the first thousand positive integers.

The terms of a sequence can be specified either by providing a formula for each term of the sequence as a function of its position or by expressing each term as a combination of the previous terms. The focus of this chapter is to find an ***explicit formula*** or a ***general formula***, called a ***closed-form formula***, for the terms of the sequence specified through a recurrence relation.

14.2 Recursively Defined Functions

Recursively defined functions are of great importance in the theory of computation in computer science. A *recursively defined function* $f(n)$ refers to itself, and its domain is a subset of the set of positive or nonnegative integers. In order for the definition not to be circular, the function definition must have the following two properties:

(i) **Basis clause**: A finite number of initial values of the function $f(n)$, known as the *initial conditions*, are specified, for which the function does not refer to itself. In other words, $f(b)$, $f(b+1)$, ..., $f(b+k-1)$, where integers $k \geq 1$ and $b \geq 0$ are given.

(ii) **Recursive clause**: The function $f(n)$ is defined in terms of the k preceding functional values $f(n-1)$, $f(n-2)$, ..., $f(n-k)$, with $k < n$, as the function refers to itself. An equation expressing a term of a sequence as a function of prior terms in the sequence is known as a *recurrence relation*.

Recursively defined functions are **well defined**, as the value of the function at every integer in the domain is determined in an unambiguous way. A **solution of a recurrence relation** is called a **sequence**, and a given recurrence relation may be satisfied by many different sequences. Once the initial conditions are included, there is then only one sequence satisfying the recurrence relation of interest. Note that any recurrence relation for a sequence can be written in terms of their differences, and the resulting equation involving its differences is called a **difference equation**. Although not all sequences can be represented by recurrence relations, many sequences that arise in the solution of most problems can be so represented. As an example, $a_n = 2a_{n-3} + 3a_{n-2}$ for $n = 4, 5, ..., 8$ is a recurrence relation, with $a_1 = 3$, $a_2 = 2$, and $a_3 = 1$ as the initial conditions. Thus we have the terms of the sequence as follows: $f(4) = a_4 = 12$, $f(5) = a_5 = 7$, $f(6) = a_6 = 38$, $f(7) = a_7 = 45$, and $f(8) = a_8 = 128$.

Example 14.1
Define recursively the following functions, and then for each function, find $f(4)$.
(a) The factorial function.
(b) The compound amount with annually compounded interest.
(c) The Fibonacci numbers: 0, 1, 1, 2, 3, 5, 8, 13, 21, 34, 55, ...
(d) The number of fist bumps in a group when each person fist bumps with everyone else only once.

Solution
(a) The factorial function f is defined by $f(n) = n!$, where $f(0) \triangleq 1$. Because $n! = n(n-1)!$, this function can be defined respectively as follows:
$f(0) = 1$ (initial condition)
$f(n) = n \times f(n-1)$, $n \geq 1$ (recurrence relation)

$$f(4) = 4 \times f(3) = 4 \times 3 \times f(2) = 4 \times 3 \times 2 \times f(1)$$
$$= 4 \times 3 \times 2 \times 1 \times f(0) = 24.$$

(b) The compound amount f is defined by $f(n) =$ (compound amount at the end of the $(n-1)$th year) + (interest earned during the nth year with an annual interest rate of i), where $f(0) = p$, known as the principal. Therefore f can be defined respectively as follows:

$f(0) = p$ (initial condition)
$f(n) = f(n-1) + i \times f(n-1) = (1+i) \times f(n-1)$, $n \geq 1$ (recurrence relation)

$$f(4) = (1+i) \times f(3) = (1+i) \times (1+i) \times f(2) = (1+i)^2 \times f(2)$$
$$= (1+i)^2 \times (1+i) \times f(1) = (1+i)^3 \times f(1)$$
$$= (1+i)^3 \times (1+i) \times f(0) = p(1+i)^4.$$

(c) Any Fibonacci number, except the first two, is the sum of the two immediately preceding it. Therefore f can be defined respectively as follows:

$f(1) = 0$ and $f(2) = 1$ (initial conditions)
$f(n) = f(n-1) + f(n-2)$, $n \geq 3$ (recurrence relation)

$$f(4) = f(3) + f(2) = f(2) + f(1) + f(2) = 2 \times f(2) + f(1) = 2 \times 1 + 0$$
$$= 2.$$

(d) With one person in the group, the number of fist bumps is obviously zero. With n persons in the group, one of them fist bumps with each of the remaining $n-1$ persons, resulting in $n-1$ fist bumps, and the number of fist bumps made by the other $n-1$ persons among themselves is $f(n-1)$. Therefore f can be defined respectively as follows:

$f(1) = 0$ (initial condition)
$f(n) = f(n-1) + (n-1)$, $n \geq 2$ (recurrence relation)

$$f(4) = f(3) + 3 = f(2) + 2 + 3 = f(1) + 1 + 2 + 3 = 6.$$

14.3 Recursive Algorithms

A recurrence relation that defines a sequence can be directly converted to an algorithm to compute the sequence, as recursive definitions consistently lead to recursive algorithms. A ***recursive algorithm*** is an algorithm that invokes itself during execution with a reduced version of itself, as it proceeds by reducing a problem to the same problem with smaller input. A recursive algorithm consists of the base case and the general case. The ***base case*** ensures the sequence of recursive calls will terminate after a finite number of steps, and the **general case** continues to call itself as long as the base case is not satisfied.

To develop a recursive algorithm, the following steps are of importance:

- Find ways to reduce the problem into smaller versions of itself. For instance, $(n-1)!$ is a smaller version of $n!$.
- Identify cases that can be directly solved to know when a base case is reached. For instance, calculate directly $0!$, $1!$, and $2!$ in $n!$.
- Determine how the solutions of smaller versions can lead to a larger version. For instance, $(n-1)!(n) = n!$.
- Make sure that each new invocation works on a smaller problem and that eventually a base case will be reached, thus allowing to terminate after a finite number of steps. Otherwise, the algorithm will never terminate.

Sometimes mathematical induction and often strong induction are needed to prove a recursive algorithm is correct, that is, for all possible input values, the correct output can be produced.

A class of recursive algorithms is called **divide–and–conquer algorithms** when they divide a problem into parts of the same problem of smaller size and they conquer the problem by using the solutions of the smaller problems, such as binary search and merge sort algorithms. Recurrence relation can be used to analyze the time complexity of divide-and-conquer algorithms. Consider a problem of size n that can be broken into a smaller subproblems, where each is the size of $\frac{n}{b}$, assuming n is a multiple of b, and suppose $f(n)$ is a nondecreasing function, where $f(1) = c$, noting that $a \geq 1$, $b \geq 2$, and $c \geq 1$ are all integers. Assuming $f(n)$ represents the number of operations required to solve the problem of size n, and $h(n)$ represents the maximum number of operations required to combine the solutions of the subproblems into a solution of the original problem, we then have the **divide–and–conquer recurrence relation** as follows:

$$f(n) = a \times f\left(\frac{n}{b}\right) + h(n).$$

A recursive algorithm is devised to evaluate the value of a function at a positive integer of interest in terms of the values of the function at smaller integers. On the other hand, an **iterative algorithm** is developed to evaluate the value of a function at the base cases and successively apply the recursive definition to find values of the function at larger integers. Oftentimes an iterative approach requires less computation than a recursive approach, but writing a software program for the nonrecursive version of a recursive algorithm is often a difficult and time-consuming task.

Example 14.2

Consider the Fibonacci sequence as recursively defined by $f(n) = f(n-1) + f(n-2)$, where $f(0) = 0$ and $f(1) = 1$. Find the number of additions required for $f(6)$ for each of the following two cases, and comment on the results.

(a) Using a recursive algorithm.

(b) Using an iterative algorithm.

Solution

(a) A recursive algorithm is based on a top-down approach. We evaluate the value of $f(6)$ in terms of the values of the function at smaller integers, namely, $f(5)$ and $f(4)$, and successively apply the recursive definition to find values of the function at smaller integers. Therefore we have the following:

$$
\begin{aligned}
f(6) = f(5) + f(4) &= f(4) + f(3) + f(3) + f(2) \\
&= f(3) + f(2) + f(2) + f(1) + f(2) + f(1) + f(1) + f(0) \\
&= f(2) + f(1) + f(1) + f(0) + f(1) + f(0) + f(1) + f(1) \\
&\quad + f(0) + f(1) + f(1) + f(0) \\
&= f(1) + f(0) + f(1) + f(1) + f(0) + f(1) + f(0) + f(1) \\
&\quad + f(1) + f(0) + f(1) + f(1) + f(0).
\end{aligned}
$$

Thus we need to carry out 12 additions for a recursive algorithm. In fact, it can be shown that the number of additions in a recursive algorithm to find the Fibonacci $f(n)$ is $f(n+1) - 1$.

(b) An iterative algorithm is based on a bottom-up approach. We use the given value of the function at the base cases, namely $f(0)$ and $f(1)$, and successively apply the recursive definition to find values of the function at larger integers until $f(6)$ is obtained. Therefore we have the following relations:

$$
\begin{aligned}
f(2) &= f(1) + f(0). \\
f(3) &= f(2) + f(1). \\
f(4) &= f(3) + f(2). \\
f(5) &= f(4) + f(3). \\
f(6) &= f(5) + f(4).
\end{aligned}
$$

Thus we need to carry out five additions for an iterative algorithm. In fact, it can be shown that the number of additions in an iterative algorithm to find the Fibonacci $f(n)$ is $n - 1$.

14.4 Solving Recurrence Relations by Iteration

A sequence is called a solution of a recursive relation if its terms satisfy the recurrence relation. We often want to know an explicit formula for a recurrence relation $f(n)$ or equivalently a sequence a_n, especially when we need to compute terms with very large n or assess general properties of a relation.

The most basic method for finding a closed-form formula for a recursively defined function is *iteration*. In the *iterative method*, we take the approach based on *back substitution* in order to see a pattern developing and then guess or discover an explicit formula. We can then use induction to verify the closed-form formula. Although it is a straightforward method, it can become algebraically unwieldy if the recurrence relation is too complex.

Example 14.3

Solve the following recurrence relations by iteration:

(a) $a_n = a_{n-1} + \frac{n(n+1)}{2}$, $n \geq 1, a_0 = 0.$
(b) $a_n = a_{n-1} + (n-1)$, $n \geq 2, a_1 = 0.$
(c) $a_n = a_{n-1} + m$, $n \geq 2, a_1 = p$, $m \neq 0.$
(d) $a_n = m a_{n-1}$, $n \geq 1, a_0 = p$, $m \neq 0.$

Solution

(a) Using iteration, we have

$$a_n = a_{n-1} + \frac{(n)(n+1)}{2} = a_{n-2} + \frac{(n-1)(n)}{2} + \frac{(n)(n+1)}{2}$$

$$= \ldots = a_0 + \frac{(1)(2)}{2} + \ldots + \frac{(n-1)(n)}{2} + \frac{(n)(n+1)}{2}$$

$$= a_0 + \sum_{i=1}^{n} \frac{i(i+1)}{2} = \sum_{i=1}^{n} \frac{i(i+1)}{2}$$

$$= \frac{1}{2} \left(\sum_{i=1}^{n} i^2 + \sum_{i=1}^{n} i \right)$$

$$= \frac{1}{2} \left(\frac{n(n+1)(2n+1)}{6} + \frac{n(n+1)}{2} \right)$$

$$= \frac{n(n+1)(n+2)}{6}, \quad n \geq 0.$$

(b) Using iteration, we have

$$
\begin{aligned}
a_n = a_{n-1} + (n-1) &= a_{n-2} + (n-2) + (n-1) \\
&= \ldots = a_1 + 1 + 2 + \ldots + (n-2) + (n-1) \\
&= 0 + 1 + 2 + \ldots + (n-2) + (n-1) \\
&= \frac{n(n-1)}{2}, \quad n \geq 1.
\end{aligned}
$$

(c) Using iteration, we have

$$
a_n = a_{n-1} + m = a_{n-2} + 2m = a_{n-3} + 3m = \ldots = a_1 + (n-1)m.
$$

We therefore have

$$
a_n = (n-1)m + p, \quad n \geq 1.
$$

(d) Using iteration, we have

$$
a_n = ma_{n-1} = m^2 \times a_{n-2} = \ldots = m^n \times a_0 = pm^n, \quad n \geq 0.
$$

It is important to highlight that in the iteration method when substitution is performed repeatedly, full simplification after a substitution should be in general avoided, so the terms left in expanded form can help recognize the pattern that develops.

Example 14.4

In a sports tournament, there are $n > 1$ teams. Suppose each team plays all other teams (i.e., it is a round-robin tournament in which each team plays in turn against every other only once). Using iteration, determine the total number of games a_n to be played in the tournament.

Solution

We have the following specific cases:

$$
k = 2 \ \rightarrow \ a_2 = 1.
$$

$$
k = 3 \ \rightarrow \ a_3 = 2 + 1.
$$

$$
k = 4 \ \rightarrow \ a_4 = 3 + 2 + 1.
$$

$$
k = 5 \ \rightarrow \ a_5 = 4 + 3 + 2 + 1.
$$

We can thus conclude

$$
k = n \ \rightarrow \ a_n = (n-1) + (n-2) + \ldots + 2 + 1.
$$

Using the arithmetic progression, we have

$$a_n = \frac{n(n-1)}{2}.$$

Note that using the method of iteration to find solutions to recurrence relations may sometimes fail; finding a closed-form formula for the Fibonacci sequence $(a_n = a_{n-1} + a_{n-2})$ using iteration is a case in point.

14.5 Solving Linear Homogeneous Recurrence Relations with Constant Coefficients

A recurrence relation of degree k is a function of the form

$$a_n = h(a_{n-1},\ a_{n-2}, ..., a_{n-k},\ n).$$

In other words, the nth term a_n of a sequence is a function of the preceding k terms $a_{n-1},\ a_{n-2}, ..., a_{n-k}$ and the integer n. However, one class of recurrence relations that can be explicitly solved in a systematic way and often occur in modeling of problems is the *linear homogeneous recurrence relation of degree k with constant coefficients*, which has the following form:

$$a_n = c_1 a_{n-1} + c_2 a_{n-2} + ... + c_k a_{n-k},$$

where $c_1,\ c_2,\ ...,\ c_k$ are real numbers and $c_k \neq 0$. The recurrence relation is *linear*, as there are no powers or products of the a_j's, where $j \in \{a_{n-1}, ..., a_{n-k}\}$. It is *homogeneous*, as no terms occur that are not multiples of the a_j's. The coefficients of the terms are all *constants*, as they do not depend on n. The *degree* is k as a_n is expressed in terms of the previous k terms of the sequence. Such a recurrence relation can be uniquely solved for a_n if the initial values of the first k terms of the sequence are specified. It is important to note that $a_n = 0$, called the *trivial solution*, is always a solution to any linear homogeneous recurrence relation with constant coefficients.

Example 14.5

Determine if each of the following recurrence relations is a linear homogeneous recurrence relation with constant coefficients (LHRRCC).

(a) $a_n = 4a_{n-1} - 3a_{n-2}$.

(b) $a_n = na_{n-1} - 3a_{n-2}$.

(c) $a_n = 4a_{n-2} - 3a_{n-4} + 9$.

(d) $a_n = n^2 a_{n-1} - 3a_{n-2} + 6$.

(e) $a_n = 4a_{n-1}a_{n-2} + a_{n-3}$.

(f) $a_n = 4a_{n-1} - (n-1)a_{n-2}a_{n-3}$.

(g) $a_n = 4a_{n-1} - 3a_{n-2}a_{n-3} + 2$.

(h) $a_n = n^3 a_{n-1}a_{n-3} - 3a_{n-2} + 5$.

Solution

In order to have a linear homogeneous recurrence relation with constant coefficients, the recurrence relation must be linear and homogeneous as well as have constant coefficients. If any one of these three requirements is not met, then the recurrence relation is not a linear homogeneous recurrence relation with constant coefficients, as presented in Table 14.1.

Suppose the sequence $\{a_n\}$ is generated by the linear homogeneous recurrence relation with constant coefficients $a_n = c_1 a_{n-1} + c_2 a_{n-2} + \ldots + c_k a_{n-k}$, where the coefficients are real numbers, and $c_k \neq 0$. We can then state that $a_n = r^n$, where r is a constant, is a solution of the recurrence relation if and only if we have

$$r^n = c_1 r^{n-1} + c_2 r^{n-2} + \ldots + c_k r^{n-k}.$$

After dividing both sides by r^{n-k} and taking all terms to one side, we have

$$r^k - c_1 r^{k-1} - c_2 r^{k-2} \ldots\ldots - c_{k-1}r - c_k = 0$$

which is known as the **characteristic equation** of the recurrence relation. The solutions of the characteristic equation are called **characteristic roots**, which can be used to give an explicit formula for all the solutions of the recurrence relation. Suppose the characteristic

Table 14.1 Identification of recurrence relations for Example 14.5.

Recurrence relation	Linear?	Homogeneous?	With constant coefficients?	LHRRCC?
(a)	Yes	Yes	Yes	Yes
(b)	Yes	Yes	No (n)	No
(c)	Yes	No (9)	Yes	No
(d)	Yes	No (6)	No (n^2)	No
(e)	No ($a_{n-1}a_{n-2}$)	Yes	Yes	No
(f)	No ($a_{n-2}a_{n-3}$)	Yes	No ($n-1$)	No
(g)	No ($a_{n-2}a_{n-3}$)	No (2)	Yes	No
(h)	No ($a_{n-1}a_{n-3}$)	No (5)	No (n^3)	No

equation of degree k has j distinct roots r_1, r_2, ..., r_j with multiplicity m_1, m_2, ..., m_j, respectively, where $m_1 + m_2 + \ldots + m_j = k$. Then, for any choice of constants $b_{i,p}$, where $1 \leq i \leq j$ and $0 \leq p \leq m_i - 1$, the following closed-form expression generates a solution to the recurrence relation:

$$a_n = \left(b_{1,0} + b_{1,1}n + \ldots + b_{1,m_1-1}n^{m_1-1} \right) r_1^n + \left(b_{2,0} + b_{2,1}n + \ldots \right.$$
$$\left. + b_{2,m_2-1}n^{m_2-1} \right) r_2^n +$$
$$\ldots + \left(b_{j,0} + b_{j,1}n + \ldots + b_{j,m_j-1}n^{m_j-1} \right) r_j^n.$$

If the initial values a_1, a_2, ..., a_{k-1} are specified, then unique values can be found for constants $b_{i,p}$ where $1 \leq i \leq j$ and $0 \leq p \leq m_i - 1$, so that the closed-form formula matches the sequence generated by the recursion relation.

Example 14.6

Solve the following linear homogeneous recurrence relations with constant coefficients:

$$a_n = 9a_{n-1} - 26a_{n-2} + 24a_{n-3}, \quad a_0 = 1, \; a_1 = 1, \; a_2 = -3.$$

Solution

We have $c_1 = 9$, $c_2 = -26$, $c_3 = 24$, and $k = 3$. Thus the characteristic equation and the characteristic roots are as follows:

$$r^3 - 9r^2 + 26r - 24 = 0 \; \rightarrow \; \begin{cases} r_1 = 2 \\ r_2 = 3 \\ r_3 = 4 \end{cases}$$

There are three distinct roots. We therefore have

$$j = 3 \; \& \; \begin{cases} m_1 = 1 \\ m_2 = 1 \\ m_3 = 1 \end{cases}$$

Note that $m_1 = 1$, $m_2 = 1$, and $m_3 = 1$ reflect the fact that each distinct root has a multiplicity of 1. Therefore the constants $b_{i,p}$, where $1 \leq i \leq 3$ and $p = 0$, are $b_{1,0}$, $b_{2,0}$, and $b_{3,0}$. Hence the closed-form solution is as follows:

$$a_n = \left(b_{1,0} \right) 2^n + \left(b_{2,0} \right) 3^n + \left(b_{3,0} \right) 4^n.$$

Using the initial values, we can find the constants of interest and solve the recurrence relation as follows:

$$\begin{cases} n = 0 \rightarrow a_0 = b_{1,0} + b_{2,0} + b_{3,0} = 1 \\ n = 1 \rightarrow a_1 = 2b_{1,0} + 3b_{2,0} + 4b_{3,0} = 1 \\ n = 2 \rightarrow a_2 = 4b_{1,0} + 9b_{2,0} + 16b_{3,0} = -3 \end{cases} \rightarrow \begin{cases} b_{1,0} = 1 \\ b_{2,0} = 1 \\ b_{3,0} = -1 \end{cases} \rightarrow$$

$a_n = 2^n + 3^n - 4^n.$

Example 14.7

Solve the following linear homogeneous recurrence relations with constant coefficients:

$$a_n = 7a_{n-1} - 16a_{n-2} + 12a_{n-3}, \quad a_0 = 0, \; a_1 = 1, \; a_2 = 3.$$

Solution

We have $c_1 = 7$, $c_2 = -16$, $c_3 = 12$, and $k = 3$. The characteristic equation and the characteristic roots are as follows:

$$r^3 - 7r^2 + 16r - 12 = 0 \rightarrow \begin{cases} r_1 = 2 \\ \\ r_2 = 3 \end{cases} \rightarrow$$

$$j = 2 \; \& \; \begin{cases} m_1 = 2 \\ \\ m_2 = 1 \end{cases} \rightarrow \begin{cases} i = 1, \; p = 0 \\ i = 1, \; p = 1 \\ i = 2, \; p = 0 \end{cases}$$

Noting that we have a root $r_2 = 3$ with multiplicity one and a root $r_1 = 2$ with multiplicity two, the closed-form solution is as follows:

$$a_n = \left(b_{1,0} + b_{1,1}n\right)2^n + \left(b_{2,0}\right)3^n.$$

Using the initial values, we can find the constants of interest and solve the recurrence relation as follows:

$$\begin{cases} n = 0 \rightarrow a_0 = b_{1,0} + b_{2,0} = 0 \\ n = 1 \rightarrow a_1 = 2b_{1,0} + 2b_{1,1} + 3b_{2,0} = 1 \\ n = 2 \rightarrow a_2 = 4b_{1,0} + 8b_{1,1} + 9b_{2,0} = 3 \end{cases} \rightarrow \begin{cases} b_{1,0} = 1 \\ b_{1,1} = 1 \\ b_{2,0} = -1 \end{cases} \rightarrow$$

$a_n = (1 + n)2^n - 3^n.$

Example 14.8

Solve the following linear homogeneous recurrence relations with constant coefficients:

$$a_n = 6a_{n-1} - 12a_{n-2} + 8a_{n-3}, \quad a_0 = 1, \ a_1 = 2, \ a_2 = -4.$$

Solution

We have $c_1 = 6$, $c_2 = -12$, $c_3 = 8$, and $k = 3$. The characteristic equation and the characteristic roots are as follows:

$$r^3 - 6r^2 + 12r - 8 = 0 \rightarrow r_1 = 2 \rightarrow j = 1 \ \& \ m_1 = 3 \rightarrow \begin{cases} i = 1, \ p = 0 \\ i = 1, \ p = 1 \\ i = 1, \ p = 2 \end{cases}$$

There is a root $r_1 = 2$ with multiplicity three. Hence the closed-form solution is as follows:

$$a_n = \left(b_{1,0} + b_{1,1}n + b_{1,2}n^2\right)2^n.$$

Using the initial values, we can find the constants of interest and solve the recurrence relation as follows:

$$\begin{cases} n = 0 \rightarrow a_0 = b_{1,0} = 1 \\ n = 1 \rightarrow a_1 = 2b_{1,0} + 2b_{1,1} + 2b_{1,2} = 2 \\ n = 2 \rightarrow a_2 = 4b_{1,0} + 8b_{1,1} + 16b_{1,2} = -4 \end{cases} \rightarrow \begin{cases} b_{1,0} = 1 \\ b_{1,1} = 1 \\ b_{1,2} = -1 \end{cases} \rightarrow$$

$$a_n = \left(1 + n - n^2\right)2^n.$$

14.6 Solving Linear Nonhomogeneous Recurrence Relations with Constant Coefficients

A linear nonhomogeneous recurrence relation with constant coefficients is in the following form:

$$a_n = c_1 a_{n-1} + c_2 a_{n-2} + \dots + c_k a_{k-n} + f(n),$$

where c_1, c_2, ..., c_k are real numbers, $c_k \neq 0$, and $f(n)$ is a function of the variable n only. The recurrence relation

$$a_n = c_1 a_{n-1} + c_2 a_{n-2} + \dots + c_k a_{k-n}$$

which is known in this context as the **associated homogeneous recurrence relation with constant coefficients**, provides a solution, denoted by $a_n^{(h)}$, to the linear nonhomogeneous recurrence relation with constant coefficients. The **general solution** of the linear nonhomogeneous recurrence relation with constant coefficients is then given by

$$a_n = a_n^{(h)} + a_n^{(p)},$$

where $a_n^{(p)}$ is called a **particular solution**. There is no general method to find a particular solution for every function $f(n)$. However, when $f(n)$ is a polynomial in n and the power of a constant, a particular solution can be rather easily found. Suppose $f(n)$ is as follows:

$$f(n) = \left(b_t n^t + b_{t-1} n^{t-1} + \ldots + b_1 n + b_0\right)\alpha^n,$$

where b_0, b_1, ..., b_t and α are real numbers. Then a particular solution has the following form:

$$a_n^{(p)} = g(n)\left(e_t n^t + e_{t-1} n^{t-1} + \ldots + e_1 n + e_0\right)\alpha^n,$$

where e_0, e_1, ..., e_t are some constants to be determined using the initial conditions. If α is not a root of the characteristic equation of the associated linear nonhomogeneous recurrence relation with constant coefficient, then $g(n) = 1$. If α is a root of this characteristic equation and its multiplicity is m, then $g(n) = n^m$.

Example 14.9
Solve the following linear nonhomogeneous recurrence relation with constant coefficients:

$$a_n = 5a_{n-1} - 6a_{n-2} + f(n), \quad n \geq 2, \ a_0 = 4, \ a_1 = 7.$$

(a) $f(n) = 8n^2$.
(b) $f(n) = 7^n$.

Solution
We first need to find the solution to the associated homogeneous recurrence relation with constant coefficients

$$a_n = 5a_{n-1} - 6a_{n-2}.$$

We have $c_1 = 5$, $c_2 = -6$, and $k = 2$. The characteristic equation and its distinct roots are as follows:

$$r^2 - 5r + 6 = 0 \rightarrow \begin{cases} r_1 = 2 \\ \\ r_2 = 3 \end{cases} \rightarrow j = 2 \ \& \begin{cases} m_1 = 1 \\ \\ m_2 = 1 \end{cases} \rightarrow \begin{cases} i = 1, \ p = 0 \\ \\ i = 2, \ p = 0 \end{cases}$$

Thus we have

$$a_n^{(h)} = b_{1,0}2^n + b_{2,0}3^n.$$

(a) As there is no constant to the power of n in $f(n)$, $\alpha = 1$, and because 1 is not a root of the characteristic equation, $g(n) = 1$, we have

$$a_n^{(p)} = e_2n^2 + e_1n + e_0$$

Thus we have the following three terms:

$$\begin{cases} a_n^{(p)} = e_2n^2 + e_1n + e_0 \\ a_{n-1}^{(p)} = e_2(n-1)^2 + e_1(n-1) + e_0 \\ a_{n-2}^{(p)} = e_2(n-2)^2 + e_1(n-2) + e_0 \end{cases}$$

After substituting these three terms into the relation

$$a_n = 5a_{n-1} - 6a_{n-2} + 8n^2$$

we have

$$e_2n^2 + e_1n + e_0 = 5\left(e_2(n-1)^2 + e_1(n-1) + e_0\right)$$
$$- 6\left(e_2(n-2)^2 + e_1(n-2) + e_0\right) + 8n^2.$$

We then simplify the terms on the right-hand side and group similar terms. Thus we obtain

$$e_2n^2 + e_1n + e_0 = (-e_2 + 8)n^2 + (14e_2 - e_1)n + (-19e_2 + 7e_1 - e_0).$$

By equating the coefficients of identical powers, we obtain

$$\begin{cases} e_2 = -e_2 + 8 \\ e_1 = 14e_2 - e_1 \\ e_0 = -19e_2 + 7e_1 - e_0 \end{cases} \rightarrow \begin{cases} e_2 = 4 \\ e_1 = 28 \\ e_1 = 60 \end{cases} \rightarrow a_n^{(p)} = 4n^2 + 28n + 60.$$

The general solution is then in the following form:

$$a_n = a_n^{(h)} + a_n^{(p)} = b_{1,0}2^n + b_{2,0}3^n + 4n^2 + 28n + 60.$$

Using the initial conditions, we have

$$\begin{cases} a_0 = b_{1,0} + b_{2,0} + 60 = 4 \\ \\ a_1 = 2b_{1,0} + 3b_{2,0} + 92 = 7 \end{cases} \rightarrow \begin{cases} b_{1,0} + b_{2,0} = -56 \\ \\ 2b_{1,0} + 3b_{2,0} = -85 \end{cases} \rightarrow \begin{cases} b_{1,0} = -83 \\ \\ b_{2,0} = 27 \end{cases}$$

Now that we know the values of all coefficients, the unique solution to the recurrence relation is as follows:

$$a_n = (-83)2^n + (27)3^n + 4n^2 + 28n + 60.$$

(b) As $\alpha = 7$, and 7 is not a root of the characteristic equation, we have $g(n) = 1$. Therefore

$$a_n^{(p)} = e_0 7^n.$$

Substituting $a_n^{(p)}$ into the linear nonhomogeneous recurrence relation, we can determine the coefficient of $a_n^{(p)}$:

$$a_n = 5a_{n-1} - 6a_{n-2} + 7^n \rightarrow e_0 7^n = 5e_0 7^{n-1} - 6e_0 7^{n-2} + 7^n \rightarrow e_0 = \frac{49}{20}.$$

Thus we have

$$a_n^{(p)} = \frac{49}{20} 7^n.$$

The general solution is then in the following form:

$$a_n = a_n^{(h)} + a_n^{(p)} = b_{1,0} 2^n + b_{2,0} 3^n + \frac{49}{20} 7^n.$$

Using the initial conditions, we have

$$\begin{cases} a_0 = b_{1,0} + b_{2,0} + \dfrac{49}{20} = 4 \\ \\ a_1 = 2b_{1,0} + 3b_{2,0} + \dfrac{343}{20} = 7 \end{cases} \rightarrow \begin{cases} b_{1,0} + b_{2,0} = \dfrac{31}{20} \\ \\ 2b_{1,0} + 3b_{2,0} = -\dfrac{203}{20} \end{cases} \rightarrow \begin{cases} b_{1,0} = \dfrac{296}{20} \\ \\ b_{2,0} = -\dfrac{265}{20} \end{cases}$$

Now that we know the values of all coefficients, the unique solution to the recurrence relation is as follows:

$$a_n = \left(\frac{296}{20}\right)2^n - \left(\frac{265}{20}\right)3^n + \left(\frac{49}{20}\right)7^n.$$

Example 14.10

Determine the form of a particular solution if the linear nonhomogeneous recurrence relation with constant coefficients is $a_n = 8a_{n-1} - 16a_{n-2} + f(n)$, when $f(n)$ is as follows:

(a) $f(n) = (n^2 + 2)3^n$.
(b) $f(n) = (n^3 + n + 5)4^n$.

Solution

We first need to find the solution to the associated homogeneous recurrence relation with constant coefficients

$$a_n = 8a_{n-1} - 16a_{n-2}.$$

We have $c_1 = 8$, $c_2 = 16$, and $k = 2$. The characteristic equation and its roots are as follows:

$$r^2 - 8r + 16 = 0 \rightarrow r_1 = 4 \rightarrow j = 1 \ \& \ m_1 = 2.$$

(a) Noting that $r = 4$, $\alpha = 3$, $g(n) = 1$, and $t = 2$, we have

$$a_n^{(p)} = (e_2 n^2 + e_1 n + e_0)3^n.$$

(b) Noting that $r = 4$, $m = 2$, $\alpha = 4$, $g(n) = n^2$, and $t = 2$, we have

$$a_n^{(p)} = n^2(e_2 n^2 + e_1 n + e_0)4^n.$$

14.7 Solving Recurrence Relations Using Generating Functions

Generating functions is a powerful and efficient tool to solve many types of problems, such as advanced counting, calculating the probability of discrete random variables, and solving recurrence relations. The solution to a recurrence relation with its initial conditions can be found when an explicit formula for the associated generating function is determined.

Generating functions can translate the terms of a sequence as coefficients of powers of a variable z in a formal power series. The **generating function** for the sequence $a_0, a_1, a_2, \ldots$ of real numbers is the following infinite series:

$$G(z) \triangleq a_0 + a_1 z + a_2 z^2 + \ldots = \sum_{n=0}^{\infty} a_n z^n.$$

Note that two generating functions can be added and multiplied as follows:

$$
\begin{cases}
G(z) = a_0 + a_1 z + a_2 z^2 + \ldots = \displaystyle\sum_{n=0}^{\infty} a_n z^n \\[4mm]
F(z) = b_0 + b_1 z + b_2 z^2 + \ldots = \displaystyle\sum_{n=0}^{\infty} b_n z^n
\end{cases}
\rightarrow
\begin{cases}
G(z) + F(z) = \displaystyle\sum_{n=0}^{\infty} (a_n + b_n) z^n \\[4mm]
F(z)\,G(z) = \displaystyle\sum_{n=0}^{\infty} \left(\sum_{j=0}^{n} b_j a_{n-j} \right) z^n
\end{cases}
$$

Table 14.2 presents some useful generating functions. In addition, the following highlights shifting a generating function and differentiating a generating function as two important mathematical operations that can help solve recurrence relations:

$$
z^m G(z) = \sum_{n=0}^{\infty} a_n z^{n+m} = \sum_{n=m}^{\infty} a_{n-m} z^n
$$

and

$$
G'(z) = \sum_{n=1}^{\infty} n a_n z^{n-1}.
$$

It is imperative to highlight that the goal is to use the generating function and its properties to a recurrence relation in order to find $G(z)$ in the form of a single summation, of course after some mathematical manipulation, and then to determine the sequence $\{a_n\}$ by using the definition of the generating function.

Table 14.2 Some useful generating functions.

$G(z)$	a_n
$\dfrac{1}{1-z^m} = \displaystyle\sum_{n=0}^{\infty} z^{mn}$	$\begin{cases} 1 & \text{if } m \text{ divides } n \\ 0 & \text{otherwise} \end{cases}$
$\dfrac{1}{(1-cz)^m} = \displaystyle\sum_{n=0}^{\infty} \left(\dfrac{(m+n-1)!}{n!\,(m-1)!} \right) c^n z^n$	$\left(\dfrac{(m+n-1)!}{n!\,(m-1)!} \right) c^n$
$\dfrac{z}{(1-z)^2} = \displaystyle\sum_{n=0}^{\infty} n z^n$	n
$\dfrac{z^2+z}{(1-z)^3} = \displaystyle\sum_{n=0}^{\infty} n^2 z^n$	n^2

Note: m is a positive integer and c is a real number.

Example 14.11

Solve the following recurrence relation using generating functions.

$$a_n = a_{n-1} + n, \quad n \geq 1, \ a_0 = -1.$$

Solution

We first multiply both sides of the recurrence relation by z^n to obtain

$$a_n z^n = a_{n-1} z^n + n z^n.$$

We then sum both sides of the equation starting with $n = 1$ to yield the following equation:

$$\sum_{n=1}^{\infty} a_n z^n = \sum_{n=1}^{\infty} a_{n-1} z^n + \sum_{n=1}^{\infty} n z^n.$$

We now change the indices and simplify the equation. Thus we have

$$\sum_{n=0}^{\infty} a_n z^n - a_0 = z \sum_{n=0}^{\infty} a_n z^n + \sum_{n=0}^{\infty} n z^n.$$

Using the definition of $G(z)$ and Table 14.2, we then have:

$$G(z) - a_0 = zG(z) + \frac{z}{(1-z)^2}.$$

Noting $a_0 = -1$, $G(z)$ can be obtained as follows:

$$G(z) = -\frac{1}{1-z} + \frac{z}{(1-z)^3}.$$

Using Table 14.2, we write each term in the preceding equation in terms of a summation:

$$G(z) = -\sum_{n=0}^{\infty} z^n + z \sum_{n=0}^{\infty} \left(\frac{(n+2)!}{n!2!} \right) z^n.$$

We now perform the following mathematical manipulation to obtain $G(z)$ in terms of a single summation:

$$G(z) = \left(-\sum_{n=0}^{\infty} z^n\right) + \left(z\sum_{n=0}^{\infty} \frac{(n+2)(n+1)}{2}z^n\right)$$

$$= \left(-\sum_{n=0}^{\infty} z^n\right) + \left(\sum_{n=0}^{\infty} \frac{(n+2)(n+1)}{2}z^{n+1}\right)$$

$$= \left(-1 - \sum_{n=1}^{\infty} z^n\right) + \left(\sum_{n=1}^{\infty} \frac{(n+1)(n)}{2}z^n\right)$$

$$= -1 + \sum_{n=1}^{\infty} \left((-1) + \frac{(n+1)(n)}{2}\right)z^n = \sum_{n=0}^{\infty} \left(-1 + \frac{(n+1)(n)}{2}\right)z^n$$

$$= \sum_{n=0}^{\infty} \left(\frac{(n+2)(n-1)}{2}\right)z^n.$$

Using the definition of the generating function, thus we have

$$G(z) = \sum_{n=0}^{\infty} a_n z^n = \sum_{n=0}^{\infty} \left(\frac{(n+2)(n-1)}{2}\right)z^n \rightarrow a_n = \frac{(n+2)(n-1)}{2}, \quad n \geq 0.$$

It is of utmost importance that after a recurrence relation of any type is solved by using any method, the solution is checked not only to make sure that it satisfies the recurrence relation but also that it fully meets the given initial conditions.

Exercises
(14.1)
(a) Determine the sum of the m terms in a geometric progression with the initial term a and the common ratio $r \neq 1$.
(b) Determine the sum of the m terms in an arithmetic progression with the initial term a and the common difference d.

(14.2)
Suppose that a person deposits $\$p$ in a saving account at a bank yielding i per year with compound interest, where i is represented in a percentage. Determine the amount of money in the account after k years for the following cases.
(a) The interest is paid m times a year.
(b) The interest is paid continuously.

(14.3)

Solve the following recurrence relations by iteration.

(a) $a_n = 2a_{n-1} + 1$, $n \geq 2$ and $a_1 = 1$.

(b) $a_n = 4a_{n-1} + 4$, $n \geq 1$ and $a_0 = 0$.

(14.4)

Find recurrence relations for sequences of n bits, along with the initial condition, for the following patterns:

(a) Containing three consecutive zeros.

(b) Not containing three consecutive zeros.

(14.5)

Solve the following linear homogeneous recurrence relations with constant coefficients.

(a) $a_n = 5a_{n-1} - 6a_{n-2}$, $a_0 = 7$, $a_1 = 16$.

(b) $a_n = 4a_{n-1} - 4a_{n-2}$, $a_0 = 1$, $a_1 = 1$.

(14.6)

Determine an explicit formula for the Fibonacci sequence.

$$f_n = f_{n-1} + f_{n-2}, \quad n \geq 2, \ f_0 = 1, \ f_1 = 1.$$

(14.7)

Determine an explicit formula for the following first-degree linear recurrence.

$$a_n = ma_{n-1} + g(n), \quad n > 0, \ m > 0, \ a_0 = k > 0.$$

(14.8)

Determine the general form of the solutions of a linear homogeneous recurrence relation if its characteristic equation has the following ten roots: 1, 1, 1, 1, -2, -2, -2, 3, 3, -4.

(14.9)

Solve the following recurrence relation using generation functions.

$$a_n = 6a_{n-1} - 9a_{n-2}, \quad n \geq 2, \ a_0 = 2 \text{ and } a_1 = 3.$$

(14.10)

Solve the following recurrence relation using generation functions.

$$a_n = a_{n-1} + n^2, \quad n \geq 1, \ a_0 = 0.$$

CHAPTER 15

Counting Methods

Contents

In a random experiment where the total number of possible outcomes is finite and the outcomes are equiprobable, we often need to know how many possible outcomes there are in total, as the probability of an event is the ratio of the number of outcomes in the event to the total number of possible outcomes. ***Counting methods*** are about how to determine the total number of equally likely outcomes in a random experiment without actually listing the outcomes. However, there are no absolute methods that can be used to solve all counting problems. This chapter discusses the basic rules of counting, permutations, and combinations along with some interesting applications.

15.1 Basic Rules of Counting

Suppose a task can be divided into a sequence of k independent subtasks, where $k > 0$ is an integer. A subtask is thus carried out regardless of how the other $k - 1$ subtasks are done. Assuming n_1, n_2, ..., and n_k are all positive integers, the first subtask can be carried out in n_1 ways, the second subtask in n_2 ways, ..., and the kth subtask in n_k ways. The ***fundamental principle of counting***, also known as the ***product rule for counting***, states that there is a total of $n_1 \times n_2 \times \ldots \times n_k$ distinct ways to carry out the task consisting of k independent subtasks.

Example 15.1
How many four-digit integers are there that are multiples of 20?

Solution
In a four-digit integer, the first (the most significant) digit cannot be zero. There are therefore nine possibilities for the first digit, from 1 to 9 inclusive. The second digit can be any one of the ten possible digits, from 0 to 9 inclusive. To be divisible by 20, the third digit cannot be an odd number, that is, it must be one of the five digits 0, 2, 4, 6, and 8, and the fourth (the least significant) digit must be 0. Using

Discrete Mathematics
ISBN 978-0-12-820656-0, https://doi.org/10.1016/B978-0-12-820656-0.00015-0

the product rule, there is then a total of $9 \times 10 \times 5 \times 1 = 450$ four-digit integers that are multiples of 20.

Suppose a task can be done in k mutually exclusive sets of ways, where $k > 0$ is an integer. The ways in any one set thus exclude the ways in the other $k - 1$ sets. Assuming n_1, n_2, ..., and n_k are all positive integers, the task can be carried out in one of n_1 ways in set 1, in one of n_2 ways in set 2, ..., and in one of n_k ways in set k, where the set of n_1 ways, the set of n_2 ways, ..., and the set of n_k ways are all pairwise disjoint (mutually exclusive) finite sets. The **sum rule for counting**, also known as the **addition rule for counting**, states that there is a total of $n_1 + n_2 + ... + n_k$ distinct ways to carry out the task.

Example 15.2
How many two-digit integers are there that are divisible by 11 or 13?

Solution
There is a set of 9 two-digit integers that are divisible by 11, and there is a set of 7 two-digit integers that are divisible by 13. These two sets are mutually exclusive, as there is no two-digit integer that is divisible by both 11 and 13. Using the sum rule, there are therefore $9 + 7 = 16$ two-digit integers that are divisible by 11 or 13.

Example 15.3
How many four-digit integers are there using the digits 0, 1, 2, 3, 4, and 5, where no digit is repeated and the integer is a multiple of 3?

Solution
For an integer to be a multiple of 3, the sum of its digits must be a multiple of 3. A digit can be used only once; therefore the smallest sum of all digits in a four-digit integer can be $0 + 1 + 2 + 3 = 6$, and the largest sum of all digits can be $2 + 3 + 4 + 5 = 14$. The sum that must be divisible by 3 can then be 6, 9, or 12. Note that the first digit cannot be 0, as it is a four-digit integer. We therefore have the following mutually exclusive cases:
- The sum of the integers is 6. The four integers are 0, 1, 2, and 3. Using the product rule, there are $3 \times 3 \times 2 \times 1 = 18$ integers.
- The sum of the integers is 9. The four integers are 0, 1, 3, and 5. Using the product rule, there are $3 \times 3 \times 2 \times 1 = 18$ integers.
- The sum of the integers is 9. The four integers are 0, 2, 3, and 4. Using the product rule, there are $3 \times 3 \times 2 \times 1 = 18$ integers.
- The sum of the integers is 12. The four integers are 0, 3, 4, and 5. Using the product rule, there are $3 \times 3 \times 2 \times 1 = 18$ integers.
- The sum of the integers is 12. The four integers are 1, 2, 4, and 5. Using the product rule, there are $4 \times 3 \times 2 \times 1 = 24$ integers.

By the sum rule, the total number of integers is thus $18 + 18 + 18 + 18 + 24 = 96$.

Suppose a task can be accomplished in k sets of ways, where $k > 0$ is an integer. Assuming n_1, n_2, …, and n_k are all positive integers, the task can be accomplished in one of n_1 ways in set 1, in one of n_2 ways in set 2, …, and in one of n_k ways in set k, where the set of n_1 ways, the set of n_2 ways, …, and the set of n_k ways are not pairwise disjoint finite sets. In other words, some of the ways in the sets to carry out the task are common. The **subtraction rule for counting**, also widely known as the **principle of inclusion–exclusion**, states that the number of distinct ways to accomplish the task is $n_1 + n_2 + … + n_k$ minus the number of common ways that have been overcounted, so no common way is counted more than once. Note that this rule was extensively discussed in the context of sets.

Example 15.4

How many two-digit integers are there that are multiples of at least one of these three integers 3, 4, and 5?

Solution

The set of two-digit integers that are multiples of 3, the set of integers that are multiples of 4, and the set of integers that are multiples of 5 are not mutually exclusive. There are 30 two-digit integers that are divisible by 3, there are 22 two-digit integers that are divisible by 4, and there are 18 two-digit integers that are divisible by 5. Note that there are 8 two-digit integers that are divisible by both 3 and 4, there are 6 two-digit integers that are divisible by both 3 and 5, and there are 4 two-digit integers that are divisible by both 4 and 5. Also, there is 1 two-digit integer that is divisible by all 3, 4, and 5. Using the subtraction rule, there are therefore $30 + 22 + 18 - 8 - 6 - 4 + 1 = 53$ two-digit integers that are multiples of 3, 4, or 5.

There are also some counting problems that cannot be directly solved using any basic rules of counting. For instance, certain problems that require tree diagrams with asymmetric structures to solve cannot easily use these rules because there are some conditions in these problems that must be met.

The outcomes of a finite sequential experiment can be represented by a **tree diagram**. A tree structure is a logical way to keep a systematic track of all possibilities in cases in which events occur in sequence but in a finite number of ways. In order to use trees in counting, a branch is used to represent each possible choice, and the leaves, which are the nodes not having other branches starting at them, are used to represent the possible outcomes. The number of branches that originate from a node represents the number of events that can occur, given that the event represented by that node occurs.

Example 15.5

How many two-digit integers are there that the least significant digit is greater than the most significant digit?

Solution

Fig. 15.1 shows the tree diagram for all two-digit integers with the least significant digit being greater than the most significant digit (MSD) while noting that the MSD cannot be 0. As shown on the tree, the total number of such two-digit integers is 36 $(= 8 + 7 + 6 + 5 + 4 + 3 + 2 + 1)$.

Example 15.6

Suppose that a random experiment begins by tossing two typical dice. If the dice match (i.e., the outcomes are one of the six possible doubles), the dice are rolled one more time, if not, they are not rolled any more. Determine the total number of possible outcomes.

Solution

Note that when a pair of typical dice is rolled, there are 15 possible nondoubles: $(1, 2)$, $(1, 3)$, $(1, 4)$, $(1, 5)$, $(1, 6)$, $(2, 3)$, $(2, 4)$, $(2, 5)$, $(2, 6)$, $(3, 4)$, $(3, 5)$, $(3, 6)$, $(4, 5)$, $(4, 6)$, and $(5, 6)$. There are also 6 possible doubles: $(1, 1)$, $(2, 2)$, $(3, 3)$, $(4, 4)$, $(5, 5)$, and $(6, 6)$. Fig. 15.2 shows the tree diagram for all possible outcomes, where there are three paths representing the three different kinds of outcome sequences. The total number of possible outcomes is thus the sum of the path products, namely 141 $(= 15 + 6 \times 6 + 6 \times 15)$.

15.2 The Pigeonhole Principle

It is common sense to say if there are more items than containers, then at least one container must contain more than one item. This obvious statement is a type of counting argument that can be used to demonstrate possibly interesting results. For example, if a mother has three children, at least two of them have the same sex, that is, at least two of them are girls or two of them are boys.

The *pigeonhole principle*, also known as the **Dirichlet drawer principle**, states that if $k > 0$ is an integer and k pigeonholes are occupied by $k + 1$ or more pigeons, then at least one pigeonhole is occupied by more than one pigeon.

Example 15.7

Determine the minimum number of digits in an integer to guarantee that at least two of the digits are the same digits.

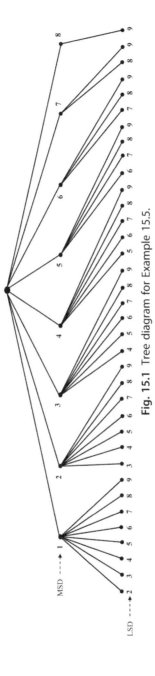

Fig. 15.1 Tree diagram for Example 15.5.

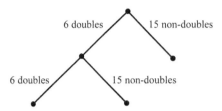

Fig. 15.2 Tree diagram for Example 15.6.

Solution

Using the pigeonhole principle, we have $k = 10$, as there are ten distinct digits in the Hindu–Arabic numeral system. In order to have at least two of the digits the same, the minimum number of the digits in the integer must then be $k + 1 = 11$.

The *generalized pigeonhole principle* states that if k and n are positive integers and k pigeonholes are occupied by $m = kn + 1$ or more pigeons, then at least one pigeonhole is occupied by $n + 1$ or more pigeons.

Example 15.8

A bag contains 18 red balls and 18 blue balls. A woman selects balls at random one at a time without looking at them. Determine the minimum number of balls that she must select from the bag to be sure of having at least nine balls of the same color.

Solution

Using the generalized pigeonhole principle, we have $k = 2$, as there are two different colors and we have $n = 8$, as $n + 1 = 9$. The minimum number of balls that she must select to be sure of having at least nine balls of the same color is thus $m = kn + 1 = 2 \times 8 + 1 = 17$.

It is important to note that the pigeonhole principles can be proven using a proof by contraposition.

15.3 Random Arrangements and Selections

Permutations and combinations arise when a subset is chosen from a set. In a counting problem, we may determine the number of ways to randomly choose a set of k objects from a set of n distinguishable objects.

An ordered arrangement of k distinguishable objects from a set of $n \geq k$ objects is called a *k-permutation*. In a k-permutation, different outcomes are distinguished by the order in which objects are chosen in a sequence. Therefore in a k-permutation, both the identity of the objects and their order of arrangement matter, as a permutation results in a list of objects.

As order counts with permutations, some real-life examples of permutations include the order of letters in a word in a natural language, the sequence of digits in a telephone number, books on a shelf in a library, the winners in an Olympic game, multiplication of matrices, and the order of alphanumeric characters in a password.

An unordered selection of k distinguishable objects from a set of $n \geq k$ objects is called a **k-combination**. In a k-combination, the identity of objects in a sequence matters but not their order of selection, as a combination results in a group of objects.

As order does not count with combinations, some real-life examples of combinations include multiplication of numbers, putting toppings on a pizza, counting subsets, buying groceries, handshakes among a group of people, taking attendance, voting in an election, answering questions on an exam, games in a round-robin tournament, dice rolled in a dice game, and cards dealt to form a hand in a card game. It is interesting to note that the term *combination lock* is a misnomer, as the sequence of numbers to unlock matters; in fact, a combination lock should be called a permutation lock.

In a **selection with replacement** (repetition, substitution), after an object out of n objects is selected, it is returned to the set, and it is thus possible that it will be selected again. In sampling with replacement, the total number of possible outcomes thus remains the same after each selection.

In a **selection without replacement**, an object, once selected, is not available for future selections. In sampling without replacement, the total number of possible outcomes of each selection depends on the outcomes of previous selections.

15.4 Permutations and Combinations

Before introducing the permutations and combinations formulas, we need to define the symbol $\binom{m}{r}$. The symbol $\binom{m}{r}$, read as **m choose r**, where m and r are both integers with $0 \leq r \leq m$, is called a **binomial coefficient**. Note that we have $\binom{m}{r} \triangleq \frac{m!}{r!(m-r)!}$, where $m!$, read as **m factorial**, is defined as $m! \triangleq m \times (m-1) \times (m-2) \times \ldots \times 2 \times 1$ and $0! \triangleq 1$. We now consider both permutations and combinations, each with and without replacement (i.e., all four cases), and highlight their applications through some examples.

If n and k are integers, such that $0 \leq k \leq n$, then the number of ways to make ordered arrangements of k objects from a set of n distinct objects but without repetition (i.e., **permutation without replacement**) is as follows:

$$n(n-1)(n-2)\ldots(n-k+1) = \frac{n!}{(n-k)!}.$$

Example 15.9

How many different ways are there to have a gold medalist, a silver medalist, and a bronze medalist from 32 national teams that have entered a sports competition?

Solution

Because it matters which team wins which medal, and no team can win more than one medal, it is a permutation without replacement. Noting that $n = 32$ and $k = 3$, the number of ways is thus $\frac{32!}{29!} = 32 \times 31 \times 30 = 29,760$.

If n and k are integers, such that $0 \leq k \leq n$, then the number of ways to make unordered selections of k objects from a set of n distinct objects but without repetition (i.e., **combination without replacement**) is as follows:

$$\binom{n}{k} = \frac{n!}{k!(n-k)!}.$$

Example 15.10

There are 22 players on a soccer team. The starting lineup consists of only 11 players. How many possible starting lineups are there, assuming what positions they play is of no concern?

Solution

Because the order of the selection of the players is immaterial and no player can be selected more than once, it is a combination without replacement. Noting that $n = 22$ and $k = 11$, the number of ways is thus $\frac{22!}{11!11!} = 705,432$.

If n and k are integers, such that $0 \leq k$ and $1 \leq n$, then the number of ways to make ordered arrangements of k objects from a set of n objects, when repetition of objects allowed (i.e., **permutation with replacement**) is as follows:

$$n \times n \times \ldots \times n = n^k.$$

Example 15.11

How many four-letter passwords from the capital letters A to Z inclusive can be made, noting that a letter can be repeated in a password?

Solution

This is a permutation with replacement, as the order of capital letters in a password matters and a capital letter can be used in a password more than once. Noting that $n = 26$ and $k = 4$, the number of passwords is thus $26^4 = 456,976$.

If n and k are integers, such that $0 \leq k$ and $1 \leq n$, then the number of ways to make unordered selections of k objects from a set of n objects, when repetition of objects allowed (i.e., **combination with replacement**) is as follows:

$$\binom{n + k - 1}{k} = \frac{(n + k - 1)!}{k!(n - 1)!}.$$

Example 15.12

There is a list of 25 different exotic foods on the menu of a special restaurant. For a flat price, a customer can select four foods. In how many ways can a selection of four foods be chosen?

Solution

This is a combination but with replacement, as the order of the selection does not matter and a food can be selected more than once. With $n = 25$ and $k = 4$, the number of ways is thus $\frac{28!}{4!24!} = 20,475$.

Table 15.1 summarizes the formulas for the numbers of ordered arrangements (i.e., permutations) and unordered selections (i.e., combinations) of k objects, with and without repetition (replacement) allowed, from a set of n distinct objects. Note that when $k = 0$, the number of permutations is 1 (i.e., the list has no objects) and the number of combinations is 1 (i.e., the group has no objects).

Suppose $n_1 \geq k_1$, $n_2 \geq k_2$, ..., and $n_m \geq k_m$ are all positive integers, and k_1 items from a group of n_1 items, k_2 items from a group of n_2 items, ..., and k_m items from a group of n_m items are selected in an unordered fashion without replacement while noting that $n = n_1 + n_2 + \ldots + n_m$ is the total number of items available and $k = k_1 + k_2 + \ldots + k_m$ is the total number of items selected. The number of ways to make such a particular selection is the product of m binomial terms and is as follows:

$$\binom{n_1}{k_1} \times \ldots \times \binom{n_m}{k_m} = \frac{n_1!}{k_1!(n_1 - k_1)!} \times \ldots \times \frac{n_m!}{k_m!(n_m - k_m)!}.$$

Table 15.1 Number of permutations and combinations.

	Permutations (ordered arrangements)	Combinations (unordered selections)
No replacement	$\frac{n!}{(n-k)!}$	$\frac{n!}{k!(n-k)!}$
With replacement	n^k	$\frac{(n+k-1)!}{k!(n-1)!}$

Note: k is the number of elements chosen from a set with n elements.

Example 15.13

Suppose we have 10 black balls, 20 white balls, and 18 yellow balls in a bag. Determine the number of ways to make a combination without replacement of five black balls, four white balls, and six yellow balls.

Solution

As we have $n_1 = 10$, $n_2 = 20$, $n_3 = 18$, $k_1 = 5$, $k_2 = 4$, and $k_3 = 6$, the number of ways is thus $\frac{10!20!18!}{5!5!4!16!6!12!} = 22,665,530,160$.

Example 15.14

Bridge is a popular card game in which 52 cards are dealt to four players, each having 13 cards. The order in which the cards are dealt is not important, only the final 13 cards each player ends up with are of importance. How many different ways are there to deal hands of 13 cards to each of four players?

Solution

Note that there are $\binom{52}{13}$ ways to choose the 13 cards of the first player,

$\binom{52-13}{13} = \binom{39}{13}$ ways to choose the 13 cards of the second player,

$\binom{39-13}{13} = \binom{26}{13}$ ways to choose the 13 cards of the third player, and

$\binom{26-13}{13} = \binom{13}{13}$ ways to choose the 13 cards of the fourth player. The total number of possible ways is thus as follows:

$$\binom{52}{13} \times \binom{39}{13} \times \binom{26}{13} \times \binom{13}{13} = \frac{52!}{13!13!13!13!} \cong 5.364 \times 10^{28}.$$

15.5 Applications

One paramount application of counting is the **Birthday Paradox**, which is about finding out the minimum number of people who need to be in a room so that it is more likely than not that at least two of them have the same birthday. It is imperative to note that the solution to the birthday problem leads to the solution of secure communications using message authentication.

This problem is now addressed in a more general way, that is, determine the probability that in a set of k randomly chosen people in a room, there is at least one pair of people who have the same birthday. To this end, we make the following assumptions:

- A year is not a leap year (i.e., there are 365 days in a year).
- Each day of the year is equally probable for a birthday.
- The birthdays of the people are independent (there are no twins, triplets, etc.).

The occurrence of one pair of people to have the same birthday seems unlikely unless k is quite large, and in fact, by the pigeonhole principle, the probability reaches 100% when the number of people reaches 366 (i.e., $k = 366$). We need to take the indirect approach, in which no two people share a common birthday. In other words, what needs to be done is to count the number of ways that k people can have distinct birthdays in a year.

The first selected birthday could be any day, with the probability of $\frac{365}{365}$. The probability that a randomly selected person whose birthday is different from the first birthday is $\frac{364}{365}$. The probability that a randomly selected person whose birthday is different from both birthdays is $\frac{363}{365}$. In general, the ith person, with $2 \leq i \leq 365$, has a birthday different from the birthdays of $i - 1$ people, already given that these $i - 1$ people have different birthdays, is $\frac{365-(i-1)}{365} = \frac{366-i}{365}$. We can thus conclude that the probability that k people have different birthdays is the multiplication of k independent probabilities. This probability is thus the number of ways of making a permutation of k days taken from 365 (i.e., $\frac{365!}{(365-k)!}$) without replacement divided by the number of ways making an ordered with replacement selection of k days from 365 (i.e., 365^k). We thus have the following:

$$p_k = \left(\frac{366 - 1}{365}\right)\left(\frac{366 - 2}{365}\right) \cdots \left(\frac{366 - k}{365}\right) = \frac{365!}{365^k(365 - k)!}.$$

Note that the probability that among k people at least two people having the same birthday is $1 - p_k$. It is interesting to note that the minimum number of people needed so that the probability that at least two people have the same birthday is greater than 50% is only 23. With only 50 people, the probability is greater than 97%, and with only 70 people, the probability is greater than 99.9%.

The probabilities are quite high simply because every pair of people are potential matches, and as the number of people increases, the number of pairs increases much faster. It is thus a key point to highlight the fact that in the birthday problem, neither of the two people is chosen in advance. The probabilities are for some collection of two or more people, and we cannot specify any of the people ahead of time.

Assuming we have n possibilities (instead of 365 days) and also $n \gg k$, it can be shown that the above probability of interest can be closely approximated as follows:

$$p_k = \frac{n!}{n^k(n - k)!} \cong \exp\left(-\frac{k(k - 1)}{2n}\right) \rightarrow k \cong \sqrt{-2 \ln p_k}\sqrt{n}.$$

We can thus determine the smallest value of n given a value of k such that the probability of no collision is greater than a particular threshold.

In the context of secure communications using message authentication code, we have $n = 2^m$, where m, the number of bits in the authenticator, is typically 128, 196, or 256.

Example 15.15

In a message authentication, assuming $m = 128$, determine k for any two messages that their authenticators match for each of these cases, $1 - p_k = 0.5$ and $1 - p_k = 10^{-12}$.

Solution

We can obtain k as follows:

$$m = 128 \rightarrow n = 2^{128} \cong 3.403 \times 10^{34} \rightarrow \begin{cases} 1 - p_k = 0.5 & \rightarrow \quad k \cong 2.171 \times 10^{19} \\ 1 - p_k = 10^{-12} & \rightarrow \quad k \cong 2.608 \times 10^{13} \end{cases}$$

Another application in counting is **quality control**. Suppose there are K items, out of which $k \leq K$ are defective, that is, $K - k$ items work properly. $M \leq K$ items are chosen at random and tested, that is, $K - M$ items remain untested. It is important to determine the probability that m of the M tested items are found defective, where we have $m \leq k$ and $m \leq M$.

It is an unordered sampling without replacement. There are $\binom{k}{m}$ ways to choose the m defective items from the total of k defective items and $\binom{K - k}{M - m}$ ways to choose the $M - m$ nondefective items from the total of $K - k$ nondetective items. Hence, there are $\binom{k}{m}\binom{K - k}{M - m}$ possible ways to make such a selection. However, the number of ways to select M items out of K items at random is $\binom{K}{M}$. The probability of such an occurrence, also known as the hypergeometric probability, is then as follows:

$$p = \frac{\binom{k}{m}\binom{K - k}{M - m}}{\binom{K}{M}}.$$

Example 15.16

Suppose in a factory where 10,000 items are built every day, on average, five of them are defective. Suppose 10 items are tested at random and one of them is defective. Determine the probability of such an occurrence.

Solution

We have $K = 10,000$, $M = 5$, $k = 10$, and $m = 1$. We thus have

$$p = \frac{\binom{10}{1}\binom{9990}{4}}{\binom{10000}{5}} \cong 0.5\%.$$

Exercises

(15.1)

From 21 consonants and five vowels in the English language, how many words can be formed consisting of two different consonants and two different vowels? Note that the words do not need to have meaning.

(15.2)

In a bag, there are three black balls, four white balls, and five red balls. Two balls are picked from the bag. How many ways are there if the two balls are not of the same color?

(15.3)

How many even four-digit integers are there if only the six digits 0, 1, 2, 3, 4, and 5 can be used, but no digit more than once?

(15.4)

Determine the total number of three-symbol passwords using the letters from A to Z inclusive and the digits 0 to 9 inclusive, while noting that a letter or a digit can be used more than once in a password.

(15.5)

Determine the total number of four-letter words using the letters from A to Z inclusive while noting that the words do not need to have meaning, and a letter cannot be used more than once in a word.

(15.6)

In the 5/52 lottery, the player picks five different numbers from a possible set of 52 different numbers, ranging from 1 to 52, inclusive. At a lottery drawing, five different balls are drawn at random (i.e., with equal probability) from a device containing the 52 balls representing the 52 different numbers. Note that no repetition is allowed, that is, once a ball is drawn, it is not put back in the device. Determine the total number of possible combinations in this lottery.

(15.7)

In a bag, there are four beige balls, three gray balls, and five brown balls. Two balls are picked from the bag. How many ways are there if the two balls are not of the same color?

(15.8)

Suppose the number of permutations without replacement is 210 and the number of combinations without replacement is 35. Determine n (the total number of objects) and k (the number of chosen objects).

(15.9)

A drawer has six black socks, four brown socks, and two gray socks. Determine the number of ways two socks can be picked from the drawer for each of the following cases:

(a) The two socks can be any color.
(b) The two socks must be the same color.

(15.10)

Teams A and B play against one another in a championship series. The first team that wins three games or wins two games in a row wins the tournament. Determine the number of ways the tournament can occur.

CHAPTER 16

Discrete Probability

Contents

There is nothing certain in life, as uncertainty exists in virtually every single aspect of life. Probability is a numerical measure of how likely an event is to occur or be the case, and thus it provides answers to questions that involve uncertainty. Probability has numerous applications in a host of diverse disciplines, including science and engineering. The focus of this chapter is on the fundamental concepts in discrete probability, along with some well-known applications.

16.1 Basic Terminology

An experiment is a measurement process where its end result is called an ***outcome***. An ***event*** is a collection of outcomes or consists of a single outcome. In a ***random experiment*** the outcome is always unpredictable and the conditions under which it is performed cannot be known in advance. A repetition of an experiment is called a ***trial***. In ***independent trials*** the outcome of a trial is independent of the outcomes of the past and future trials. In other words, a trial in a random experiment is independent if the likelihood of each possible outcome does not change from trial to trial, such as coin tossing and dice rolling.

The ***sample space*** S of a random experiment is defined as the set of all possible outcomes of an experiment. In a random experiment the outcomes, also known as ***sample points***, cannot occur simultaneously. An event is thus a subset of the sample space of an experiment. When no single outcome is any more likely than any other, we have equally likely outcomes, such as tossing a fair coin, where the probability of getting a head is equal to the probability of getting a tail.

Discrete Mathematics
ISBN 978-0-12-820656-0, https://doi.org/10.1016/B978-0-12-820656-0.00016-2

Two **mutually exclusive events**, also known as **disjoint events**, exist if the occurrence of one excludes the occurrence of the other. The **union** of two events A and B, denoted by $A \cup B$, is the set of all outcomes that are in either one of them or in both of them. The **intersection**, also known as the **joint event**, of two events A and B, denoted by $A \cap B$, is the set of all outcomes that are in both events.

The **complement** of event A consists of all outcomes that are not included in the event and is denoted by the event A^c. A **sure event**, generally denoted by S, consists of all outcomes and thus always occurs. A **null event**, denoted by $\varnothing$, contains not even one outcome and thus never occurs.

The sample space in discrete probability, known as the **discrete sample space**, is countable. In a discrete sample space the probability law for a random experiment can be specified by giving the probabilities of all possible outcomes. With a finite nonempty sample space of equally likely outcomes, the probability of an event that is a subset of the sample space is the ratio of the number of outcomes in the event to the number of outcomes in the sample space.

Example 16.1

Suppose three fair typical dice are rolled. Note that the outcome of a die, which may include 1, 2, 3, 4, 5, or 6, is the side of a die that is uppermost while resting on a flat surface, and a die is fair when the probability of each of its six outcomes is $\frac{1}{6}$.

(a) Determine the probability of the event when the sum of the three outcomes is 5.

(b) Provide examples of a sure event and a null event when the three dice are rolled.

Solution

(a) There are six possible outcomes for each die, where the outcome of a die is independent of the outcomes of the other two dice. We thus have $216 \, (= 6 \times 6 \times 6)$ possible outcomes in rolling three dice. As the dice are fair, we have 216 equally likely outcomes. There are six possible outcomes, each with a sum of five, namely, (1, 1, 3), (1, 2, 2), (1, 3, 1), (2, 1, 2), (2, 2, 1), and (3, 1, 1). Hence the probability that a sum of five comes up is $\frac{6}{216} = \frac{1}{36}$.

(b) An example of a sure event is that the sum of the three outcomes is less than 19 and an example of a null event is that the sum of the three outcomes is less than 3.

16.2 The Axioms of Probability

Axioms are self-evidently true statements that are unproven. In the *axiomatic definition of probability*, the probability of the event A, denoted by $P(A)$, in the sample space S is a real number assigned to A that satisfies the following *axioms of probability*:

Axiom I: $P(A) \geq 0$.
Axiom II: $P(S) = 1$.
Axiom III: If $A_1, A_2, \ldots$ is a countable sequence of events such that for all $i \neq j$, where $\varnothing$ is the null event, that is, they are pairwise disjoint (mutually exclusive) events, then $P(A_1 \cup A_2 \cup \ldots) = P(A_1) + P(A_2) + \ldots$.

These axioms meet the intuitive requirements of probability. *Axiom I of probability* points to the fact that the probability of an event is nonnegative, namely, the chance that something happens is always at least zero. *Axiom II of probability* states that the probability of all possible outcomes is one, namely, the chance that something happens is always 100%. *Axiom III of probability* highlights that the total probability of a number of nonoverlapping events is the sum of the individual probabilities. Based on the axioms of probability, Table 16.1 presents the corollaries of probability, which are quite useful in solving probability problems.

Table 16.1 Corollaries of probability.

$$P(\varnothing) = 0$$
The impossible event has probability zero; it provides a symmetry to Axiom II.

$$P(A) + P(A^c) = 1$$
The sum of the probabilities of two events that are partitioning the sample space is one.

$$P(A) \leq 1$$
The probability of an event is less than or equal to one; it is an upper bound on Axiom I.

$$P(A) = P(A \cap B) + P(A \cap B^c)$$
The probability of an event is the sum of the probabilities of two mutually-exclusive events.

$$A \subset B \rightarrow P(A) \leq P(B)$$
Probability is a nondecreasing function of the number of outcomes in an event.

$$P(A \cup B) = P(A) + P(B) - P(A \cap B)$$
It is the generalization of Axiom III, when the two events are not mutually exclusive.

$$P(A \cup B) \leq P(A) + P(B)$$
The sum of the probabilities of events is lower bounded by the probability of their union.

Example 16.2

Suppose for the events A and B, we have $P(A) = P(B) = 0.7$. Show that $P(A \cap B) \geq 0.4$.

Solution

Using Table 16.1, we have

$$P(A \cup B) \leq 1 \rightarrow P(A) + P(B) - P(A \cap B) \leq 1$$
$$\rightarrow 0.7 + 0.7 - P(A \cap B) \leq 1 \rightarrow P(A \cap B) \geq 0.4$$

Example 16.3

The police report that among drivers stopped on suspicion of impaired driving, 80% took test A, 20% test B, and 10% both tests A and B. Determine the probability for each of the following cases:

(a) A suspect is given test A or test B or both tests.

(b) A suspect is given either test A or test B, but not both tests.

(c) A suspect is given neither test A nor test B.

Solution

Let A be the event that a suspect is given test A and B be the event that a suspect is given test B. We thus have $P(A) = 0.8$, $P(B) = 0.2$, and $P(A \cap B) = 0.1$. Using the axioms and corollaries of probability, we have

(a) $P(A \cup B) = P(A) + P(B) - P(A \cap B) = 0.8 + 0.2 - 0.1 = 0.9$.

(b) $P(A \cup B) - P(A \cap B) = 0.9 - 0.1 = 0.8$.

(c) $P(A^c \cap B^c) = 1 - P(A \cup B) = 1 - 0.9 = 0.1$.

Example 16.4a

There are eight identical bags. In each bag there are eight balls, numbered from 1 to 8 inclusive. We randomly pick a ball from each bag. Determine the probability that the product of the eight numbers on the eight balls is a multiple of 3.

Solution

The product of eight numbers is not a multiple of 3 if none of them is a multiple of 3. The probability that a number from a bag is not a multiple of 3 is $\frac{6}{8} = \frac{3}{4}$, as out of the eight possible numbers, there are six numbers (i.e., 1, 2, 4, 5, 7, 8) that are not a multiple of 3. Therefore the probability that the product of eight numbers is not a multiple of 3 is $\left(\frac{3}{4}\right)^8$. We can then conclude that the probability that the product is a multiple of 3, using the complement law, is $1 - \left(\frac{3}{4}\right)^8 \cong 0.9$.

16.3 Joint Probability and Conditional Probability

The probability of the occurrence of a single event, such as $P(A)$ or $P(B)$, which takes a specific value irrespective of the probabilities of other events, is called a **marginal probability**. For instance, in rolling a typical die, the probability of getting a 2 represents a marginal probability. The probability that both events A and B simultaneously occur is known as the **joint probability** of events A and B, and is denoted by $P(A \cap B)$ or $P(A, B)$ and read as the probability of A and B. For instance, in rolling a pair of typical dice, the probability of getting a 1 on one die and a 6 on the other die represents a joint probability.

If we assume the probability of event B is influenced by the outcome of event A and we also know that event A has occurred, then the probability that event B will occur may not be the same as $P(B)$. The probability of event B when it is known that event A has occurred is defined as the **conditional probability**, denoted by $P(B|A)$ and read as the probability of B given A. The conditional probability $P(B|A)$ possesses the information that the occurrence of event A provides about event B. The conditional probability $P(B|A)$ and the conditional probability $P(A|B)$ are defined as follows:

$$
\begin{cases}
P(B|A) \triangleq \dfrac{P(A, B)}{P(A)} & P(A) > 0 \\[4mm]
P(A|B) \triangleq \dfrac{P(A, B)}{P(B)} & P(B) > 0
\end{cases}
\quad \rightarrow \quad P(A, B) = P(A)P(B|A) = P(B)P(A|B).
$$

Example 16.5

A box contains 25 cookies, of which x are bad. Two cookies are eaten one by one. Assuming the probability that both cookies are good is 0.4, determine the value of x.

Solution

Let A denote the event that the first cookie is good and B denote the event that the second cookie is good. Therefore the probability that the first cookie is good is $P(A) = \frac{25-x}{25}$, and the probability that the second cookie is also good is $P(B|A) = \frac{24-x}{24}$, as there are only 24 cookies left, out of which x are bad. We thus have

$$
P(A, B) = P(A)P(B|A) = \left(\frac{25 - x}{25} \right) \left(\frac{24 - x}{24} \right) = 0.4 \ \rightarrow \ x = 9.
$$

Example 16.6

A bag contains twelve red marbles, eight green marbles, and four blue marbles. A marble is drawn from the bag and it happens not to be a red marble. Determine the probability that it is a blue marble.

Solution

Let B denote the event that the selected ball is blue, and R^c is the event that the selected ball is not red. Therefore their probabilities are as follows:

$$P(B) = \frac{4}{12 + 8 + 4} = \frac{1}{6} \quad \& \quad P(R^c) = \frac{8 + 4}{12 + 8 + 4} = \frac{1}{2}.$$

We thus have

$$P(B|R^c) = \frac{P(B \cap R^c)}{P(R^c)} = \frac{P(B)}{P(R^c)} = \frac{\frac{1}{6}}{\frac{1}{2}} = \frac{1}{3}.$$

16.4 Statistically Independent Events and Mutually Exclusive Events

If the occurrence of event A has some bearing on the occurrence of event B, then the conditional probability of event B given event A, vis-à-vis the marginal probability of event B, may give rise to a larger probability (even one), yield a smaller probability (even zero), or even result in no change in the probability of event B.

Example 16.7

In rolling a fair die we define event B when the outcome is a multiple of 2. Determine the probability of B and the conditional probability $P(B|A)$ for the following cases:

(a) Event A represents an outcome that is a multiple of 2.
(b) Event A represents an outcome that is less than 3.
(c) Event A represents an outcome that is not a multiple of 2.

Solution

Because it is a fair die, we have the event B and its probability as follows:

$$B = \{2, 4, 6\} \quad \rightarrow \quad P(B) = \frac{3}{6} = \frac{1}{2}.$$

The conditional probabilities of interest can be thus obtained as follows:

(a) $\begin{cases} A = \{2,4,6\} \\ P(A) = \dfrac{1}{2} \end{cases}$ $\rightarrow$ $\begin{cases} A \cap B = \{2,4,6\} \\ P(A,B) = \dfrac{1}{2} \end{cases}$ $\rightarrow$

$P(B|A) = \dfrac{\frac{1}{2}}{\frac{1}{2}} = 1 \quad \rightarrow \quad P(B|A) > P(B)$

(b) $\begin{cases} A = \{1,2\} \\ P(A) = \dfrac{1}{3} \end{cases}$ $\rightarrow$ $\begin{cases} A \cap B = \{2\} \\ P(A,B) = \dfrac{1}{6} \end{cases}$ $\rightarrow$

$P(B|A) = \dfrac{\frac{1}{6}}{\frac{1}{3}} = \dfrac{1}{2} \quad \rightarrow \quad P(B|A) = P(B)$

(c) $\begin{cases} A = \{1,3,5\} \\ P(A) = \dfrac{1}{2} \end{cases}$ $\rightarrow$ $\begin{cases} A \cap B = \{\varnothing\} \\ P(A,B) = 0 \end{cases}$ $\rightarrow$

$P(B|A) = \dfrac{0}{\frac{1}{2}} = 0 \quad \rightarrow \quad P(B|A) < P(B)$

If the occurrence of event A has no statistical impact on the occurrence of event B, events A and B are then statistically independent. Statistical independence often arises from the physical independence of events and experiments. In random experiments it is common to assume that the events of separate trials are independent, like tossing a coin and rolling a die.

If the knowledge of event A does not change the probability of the occurrence of event B, then we have $P(B|A) = P(B)$. If the knowledge of event B does not change the probability of the occurrence of event A, then we have $P(A|B) = P(A)$. We can thus conclude that if we have $P(A \cap B) = P(A)P(B)$, the events A and B are then said to be **statistically independent**. It is important to emphasize that statistical independence between two events does not mean one event does not affect another event, but it merely means the probability of the joint event is equal to the product of the probabilities of individual events.

The concept of statistical independence can be extended to more than two events. In the case of n statistically independent events the probability of the intersection of events is equal to the product of the probabilities of n individual events, and also, a similar equality holds for every subset of the n events. Note that pairwise independence is not sufficient for n events to be statistically independent.

Example 16.8

The probability that the three participants A, B, and C in a TV game show each can answer the final question correctly is 80%. Assuming each of them answers the question once, determine the probability that all three of them fail to answer the question correctly.

Solution

Events A, B, and C are all statistically independent, that is, the probability of any two events is equal to the product of those two individual probabilities and also the probability of all three events is equal to the product of the three individual probabilities while noting that $P(A) = P(B) = P(C) = 0.8$. By applying the principle of inclusion–exclusion to the three events, the probability that at least one of them answers the question correctly is then as follows:

$$
\begin{aligned}
P(A \cup B \cup C) &= P(A) + P(B) + P(C) - P(A \cap B) - P(A \cap C) - P(B \cap C) \\
&\quad + P(A \cap B \cap C) \\
&= P(A) + P(B) + P(C) - P(A)P(B) - P(A)P(C) - P(B)P(C) \\
&\quad + P(A)P(B)P(C) \\
&= 0.8 + 0.8 + 0.8 - 0.8 \times 0.8 - 0.8 \times 0.8 - 0.8 \times 0.8 + 0.8 \times 0.8 \\
&\quad \times 0.8 = 0.992.
\end{aligned}
$$

The probability that all three of them fail to answer the question correctly is thus as follows:

$$
1 - P(A \cup B \cup C) = 1 - 0.992 = 0.008.
$$

If the joint probability of events A and B is zero, that is, $P(A \cap B) = 0$, these two events are then called **mutually exclusive** or **disjoint**. There is a clear distinction between the concept of statistically independent events and the concept of mutually exclusive events, even though both concepts seem to imply separation and distinctness. If the two events A and B are mutually exclusive, then events A and B cannot occur at the same time. Hence mutually exclusive events can be considered to be dependent events. If the two events A and B are both mutually exclusive and statistically independent, then it implies that at least one of the two events A and B has zero probability. Table 16.2 highlights the

Table 16.2 Statistically independent events and mutually exclusive events.

	$P(A \cap B)$	$P(A \cup B)$
Statistically independent events	$P(A)P(B)$	$P(A) + P(B) - P(A)P(B)$
Mutually exclusive events	0	$P(A) + P(B)$

requirements for two statistically independent events and those for two mutually exclusive events.

Example 16.9

Suppose for events A and B, we have $P(A \cup B) = 0.76$ and $P(A) - P(B) = 0.2$. Determine $P(A)$ and $P(B)$ for each of the following cases:

(a) Events A and B are statistically independent.
(b) Events A and B are mutually exclusive.

Solution

(a) Because A and B are statistically independent events, their joint probability is as follows: $P(A \cap B) = P(A)P(B)$. We thus need to solve the following system of two linear equations:

$$\begin{cases} P(A \cup B) = P(A) + P(B) - P(A)P(B) = 0.76 \\ P(A) - P(B) = 0.2 \end{cases} \rightarrow \begin{cases} P(A) = 0.6 \\ P(B) = 0.4 \end{cases}$$

(b) Because A and B are mutually exclusive events, their joint probability is as follows: $P(A \cap B) = 0$. We thus need to solve the following system of two linear equations:

$$\begin{cases} P(A \cup B) = P(A) + P(B) = 0.76 \\ P(A) - P(B) = 0.2 \end{cases} \rightarrow \begin{cases} P(A) = 0.48 \\ P(B) = 0.28 \end{cases}$$

16.5 Law of Total Probability and Bayes' Theorem

If events B_1, B_2, ..., B_n are all mutually exclusive events whose union forms the entire sample space S, that is, we have $S = B_1 \cup B_2 \cup ... \cup B_n$, we then refer to these events as a *partition* of S, as shown in Fig. 16.1.

In order to determine the probability of event A, it is sometimes best to separate all possible causes leading to event A. The *law of total probability*, also known as the *theorem on total probability*, is as follows:

$$P(A) = P(A \cap B_1) + ... + P(A \cap B_n) = P(A|B_1)\,P(B_1) + ... + P(A|B_n)\,P(B_n).$$

This *divide-and-conquer approach* is a practical tool used to determine the probability of A. This is due to the fact that the probability of event A can be expressed as a combination of the joint probabilities of event A and the mutually exclusive events B_1, B_2, ..., B_n.

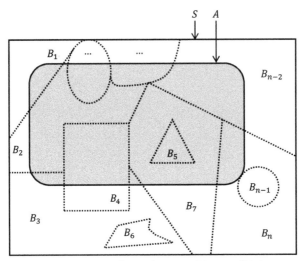

Fig. 16.1 A partition of sample space S into n disjoint sets.

Example 16.10

Suppose there are four companies supplying smartphones to the market. The smartphones built by companies A, G, M, and W have reliabilities of 99.9%, 99.5%, 99.0%, and 99.8%, respectively. It is known that companies A, G, M, and W supply 40%, 20%, 10%, and 30% of all smartphones in the market, respectively. Determine the reliability of a smartphone in the market.

Solution

Let $P(A)$ denote the reliability of a smartphone, B_A, B_G, B_M, and B_W denote the market shares of smartphones for companies A, G, M, and W, respectively. Noting that we have the following probabilities,

$$P(B_A) = 0.4, \quad P(B_G) = 0.2, \quad P(B_M) = 0.1, \quad P(B_W) = 0.3,$$

and

$$P(A|B_A) = 0.999, P(A|B_G) = 0.995, P(A|B_M) = 0.99, P(A|B_W) = 0.998,$$

the reliability of a smartphone in the market is thus as follows:

$$P(A) = 0.4 \times 0.999 + 0.2 \times 0.995 + 0.1 \times 0.99 + 0.3 \times 0.998 = 99.7\%.$$

When one conditional probability is given but the reversed conditional probability is required, the following relation, known as **Bayes' theorem** or **Bayes' rule**, which is based on the law of total probability, can be used:

$$P(B_1|A) = \frac{P(A, B_1)}{P(A)} = \frac{P(A|B_1)\,P(B_1)}{P(A|B_1)\,P(B_1) + P(A|B_2)\,P(B_2) + \ldots + P(A|B_n)\,P(B_n)},$$

where events B_1, B_2, ..., B_n are all mutually exclusive events whose union makes the entire sample space S. As an insight, the law of total probability is about effects from causes and Bayes' theorem is about causes from effects.

Note that *a priori* means derived by reasoning from self-evident propositions, and *a posteriori* means derived by reasoning from empirical evidence. To this effect, $P(B_i)$, the probability of an event B_i before the experiment is performed, is referred to as *a priori probability*, and $P(B_i|A)$, the probability of an event B_i after the experiment has been performed and the event A has occurred, is called *a posteriori probability*. Bayes' theorem connects the *a posteriori* probability with the *a priori* probability.

Example 16.11

In assessing the strength of evidence in a legal investigation a police detective always approaches his two informers to get information. The detective gets his information 80% of the time from informer A who tells a lie 75% of the time and 20% of the time from informer B, who tells a lie 40% of the time. Suppose the information the detective has received is truthful; determine the probability that the information was received from informer B.

Solution

Let T be the event that the received information is truthful. We thus have the following probabilities:

$$P(T) = P(T, A) + P(T, B) = P(T|A)P(A) + P(T|B)P(B)$$
$$= 0.25 \times 0.8 + 0.6 \times 0.2 = 0.32.$$

Using Bayes' rule, we get

$$P(B|T) = \frac{P(T, B)}{P(T)} = \frac{P(T|B)P(B)}{P(T)} = \frac{0.6 \times 0.2}{0.32} = \frac{3}{8} = 37.5\%.$$

Example 16.12

There are four bags. The first bag contains 10 black balls, 5 white balls, and 5 red balls; the second bag contains 3 black balls, 3 white balls, and 4 red balls; the third bag contains 18 white balls and 2 red balls; and the fourth bag contains 4 black balls and 16 white balls. The probability of randomly selecting the first bag is 40%, the second bag is 30%, the third bag is 20%, and the fourth bag is 10%. If a ball selected at random from one of the bags is white, determine the probability that it was drawn from the second bag.

Solution

There are four bags, A_1, A_2, A_3, and A_4, and the probabilities of randomly selecting them are as follows:

$$P(A_1) = 0.4, \quad P(A_2) = 0.3, \quad P(A_3) = 0.2, \quad P(A_4) = 0.1$$

Let W represent the event that the ball is white. We thus have

$$P(W|A_1) = \frac{5}{20}, \quad P(W|A_2) = \frac{3}{10}, \quad P(W|A_3) = \frac{18}{20}, \quad P(W|A_4) = \frac{16}{20}$$

Using Bayes' rule, the probability of interest is then as follows:

$$P(A_2|W) = \frac{P(A_2, W)}{P(W)} = \frac{P(W|A_2)P(A_2)}{P(W)}$$

$$= \frac{P(W|A_2)P(A_2)}{P(W|A_1)P(A_1) + P(W|A_2)P(A_2) + P(W|A_3)P(A_3) + P(W|A_4)P(A_4)}$$

$$= \frac{0.3 \times 0.3}{0.25 \times 0.4 + 0.3 \times 0.3 + 0.9 \times 0.2 + 0.8 \times 0.1} = \frac{0.09}{0.45} = 20\%.$$

The probability that a ball selected from the second bag reduced from 0.3, when no extra information was available, to 0.2, once we knew that the ball selected was white.

Example 16.13

In a country with a population of hundreds of millions, where capital punishment is legal and carried out, research points that 10,000 individuals are charged and tried for murder every year, and those convicted of murder are put to death. Past records consistently indicate that out of those tried, 95% are truly guilty and 5% are truly innocent. Out of those who are truly guilty, 95% are convicted and 5% are wrongly set free, and out of those who are truly innocent, 95% are set free and 5% are wrongly convicted. Determine the probability that a person who is tried is truly innocent but wrongly convicted of murder.

Solution

Suppose A is defined as the event that a person is truly innocent and B is defined as the event that the court finds that person guilty. Using Bayes' rule, we have the following:

$$P(A|B) = \frac{P(A) \times P(B|A)}{P(A) \times P(B|A) + P(A^c) \times P(B|A^c)} = \frac{5\% \times 5\%}{5\% \times 5\% + 95\% \times 95\%}$$
$$= \frac{1}{362} \cong 0.276\%.$$

Obviously, this is a very low probability. Nevertheless, it amounts to an incredible loss of the precious lives of 27 innocent individuals in that country. Hence capital punishment is an awful miscarriage of justice!

16.6 Applications in Probability

There are numerous applications in probability, but our focus here is limited to a few applications, including systems reliability, medical diagnostic testing, quality control using the Monte Carlo algorithm, and Bayesian spam filtering.

Reliability is an important aspect of the analysis, design, and operation of systems. With redundant components in a system, the probability of system failure can be minimized. To assess the reliability of a system with a number of components, we assume that the components fail independently, and the probability of failure in the ith component is p_i, where $0 \leq p_i \leq 1$, that is, its probability of functioning is $1 - p_i$.

Consider a system that consists of k components in series. Such a system functions if all k components are functioning or it fails if any one of the k components fails. Note that the probability that the system functions is lower than the functioning probability of the weakest component. Table 16.3 presents the probabilities that a system with k components in series functions or fails. In a system with components in series it is significantly easier to first determine the probability of functioning and then the probability of failure.

Consider a system that consists of k components in parallel. Such a system functions if at least a component is functioning or it fails if all k components fail. Note that the probability that the system functions is higher than the functioning probability of the strongest component. Table 16.3 presents the probabilities that a system with k components in

Table 16.3 Reliability probabilities.

System configuration	Probability of functioning	Probability of failure
Series	$\prod_{i=1}^{k} (1 - p_i)$	$1 - \prod_{i=1}^{k} (1 - p_i)$
Parallel	$1 - \prod_{i=1}^{k} p_i$	$\prod_{i=1}^{k} p_i$

parallel fails or functions. In a system with components in parallel it is significantly easier to first determine the probability of failure and then the probability of functioning.

It is important to note that sometimes there are some components in a system due to which the system does not consist of only series and parallel components. In such a complex system we can take a conditional probability approach through which mutually exclusive conditions are considered and then apply the law of total probability.

Example 16.14

Assume that the probability of failure of a component in each of the following two systems is p and the components fail independently. Noting that n, n_1, n_2, ..., n_n are all positive integers, determine the probability that each of the following two systems functions.

(a) A system consists of n subsystems in series. Each of the n subsystems consists of parallel components, where the first subsystem consists of n_1 parallel components, the second subsystem consists of n_2 parallel components, and finally the nth subsystem consists of n_n parallel components.

(b) A system consists of n subsystems in parallel. Each of the n subsystems consists of components in series, where the first subsystem consists of n_1 components in series, the second subsystem consists of n_2 components in series, and finally the nth subsystem consists of n_n components in series.

Solution

(a) The probability that the first subsystem consisting of n_1 parallel components fails is p^{n_1}, so it functions with the probability of $1 - p^{n_1}$. By the same line of reasoning, the probability of functioning for each of the other subsystems can be obtained. The probability that the system with n subsystems in series functions is thus as follows:

$$(1 - p^{n_1})(1 - p^{n_2})...(1 - p^{n_n}).$$

(b) The probability that the first subsystem consisting of n_1 series components functions is $(1 - p)^{n_1}$, so it fails with the probability of $1 - (1 - p)^{n_1}$. By the same line of reasoning, the probability of failure for each of the other subsystems can be obtained. The probability that the system with n subsystems in parallel functions is thus as follows:

$$1 - ((1 - (1 - p)^{n_1})(1 - (1 - p)^{n_2})...(1 - (1 - p)^{n_n})).$$

A major application of Bayes' theorem lies in **medical diagnosis testing**. Suppose that there is a particular rare disease that is independently and identically distributed throughout the general population. We further assume that genetics and environmental factors do not play any role, and a randomly selected individual can thus have it.

Assuming A is defined as the event that a person selected at random has the disease, and B is defined as the event that the test result is positive, Table 16.4 presents the relevant probabilities of interest. Note that for a perfect medical test, the *a posteriori* probabilities $P(A|B)$ and $P(A^c|B^c)$ must be 1, or equivalently, the *a posteriori* probabilities $P(A^c|B)$ and $P(A|B^c)$ must be zero. However, even for the most accurate medical diagnostic test, these ideal *a posteriori* probabilities cannot be achieved, as tests are always flawed. Using Bayes' rule, we can get the following results:

$$
\begin{cases}
P(A|B) = \dfrac{(1-\alpha)\rho}{(1-\alpha)\rho + \beta(1-\rho)} \\[3mm]
P(A^c|B) = \dfrac{\beta(1-\rho)}{(1-\alpha)\rho + \beta(1-\rho)}
\end{cases}
\quad \text{and} \quad
\begin{cases}
P(A^c|B^c) = \dfrac{(1-\beta)(1-\rho)}{(1-\beta)(1-\rho) + \alpha\rho} \\[3mm]
P(A|B^c) = \dfrac{\alpha\rho}{(1-\beta)(1-\rho) + \alpha\rho}
\end{cases}
$$

Table 16.4 Medical diagnostic testing probabilities.

$P(B^c|A) = \alpha$ (false–negative probability)
The probability of a negative test result given the person has the disease.

$P(B|A) = 1 - \alpha$ (true–positive probability)
The probability of a positive test result given the person has the disease.

$P(B|A^c) = \beta$ (false–positive probability)
The probability of a positive test result given the person does not have the disease.

$P(B^c|A^c) = 1 - \beta$ (true–negative probability)
The probability of a negative test result given the person does not have the disease.

$P(A) = \rho$
The probability that a person selected at random has the disease.

$P(A^c) = 1 - \rho$
The probability a person in the general population does not have the disease.

$P(A|B)$
The probability a person who tests positive for the disease has the disease.

$P(A^c|B^c)$
The probability a person who tests negative for the disease does not have the disease.

$P(A^c|B) = 1 - P(A|B)$
The probability a person who tests positive for the disease does not have the disease.

$P(A|B^c) = 1 - P(A^c|B^c)$
The probability a person who tests negative for the disease has the disease.

Example 16.15

Suppose that 2% of women in the general population have breast cancer. In total, 90% of mammograms detect breast cancer when it is there, and 5% of mammograms detect breast cancer when it is not there. Determine the probability that a woman who tests positive for the cancer has cancer.

Solution

Assuming A is defined as the event that a woman has cancer and B is defined as the event that the mammogram test result is positive, we have $P(A) = \rho = 2\%$, $P(B|A) = 1 - \alpha = 90\%$, and $P(B|A^c) = \beta = 5\%$. It is thus important to note that 98% of women do not have cancer, as we have $P(A^c) = 1 - \rho = 98\%$, 10% of mammograms miss cancer, as we have $P(B^c|A) = \alpha = 10\%$, and 95% correctly return a negative test result, as we have $P(B^c|A^c) = 1 - \beta = 95\%$. Table 16.5 captures the available information. The top row of Table 16.5 shows the true positive and false positive probabilities, while its bottom row shows the true negative and false negative probabilities. In addition, the two columns of Table 16.5 highlight the test results for women who have cancer and those who do not have cancer.

Table 16.6 provides all four possible cases with their corresponding probabilities, where the sum of all four probabilities is 1. We can now get the probability of interest as follows:

$$P(A|B) = \frac{0.9 \times 0.02}{0.9 \times 0.02 + 0.05 \times 0.98} \cong 26.9\%.$$

Algorithms that make random choices at one or more steps are called probabilistic algorithms. A particular class of probabilistic algorithms is the Monte Carlo algorithms. Monte Carlo algorithms always produce answers to decision problems, but a small probability remains that these answers may be incorrect. A ***Monte Carlo algorithm*** uses a sequence of tests and the probability that the algorithm answers the decision problem correctly increases as more tests are carried out. Suppose there are $n \gg 0$ items in a batch and the probability that an item is defective is p when random testing is done. To decide all items are good, n tests are required to guarantee that none of the items are defective. However,

Table 16.5 Available information for Example 16.15.

	With cancer (2%)	No cancer (98%)
Test positive	90%	5%
Test negative	10%	95%

Table 16.6 Summarized information for Example 16.15.

	With cancer	No cancer
Test positive	90% × 2% = 1.8% (true positive)	5% × 98% = 4.9% (false positive)
Test negative	10% × 2% = 0.2% (false negative)	95% × 98% = 93.1% (true negative)

a Monte Carlo algorithm can determine whether all items are good as long as some probability of error is acceptable.

A Monte Carlo algorithm proceeds by successively selecting items at random and testing them one by one, where the maximum number of items being tested is a predetermined $k \ll n$. When a defective item is encountered, the algorithm stops to indicate that out of the n items in a batch, there is at least one defective. If a tested item is good, the algorithm goes on to the next item. If after testing k items, no defective item is found, the algorithm concludes that all n items are good but with a modest probability of error. Note that the probability of finding not even a defective one is $(1 - p)^k$, which interestingly does not depend on n.

Example 16.16

Suppose there are $1,000,000$ cell phones in a factory warehouse, where the probability that a cell phone is in perfect condition is 0.999. Based on the Monte Carlo algorithm, determine the minimum number of cell phones that needs to be tested so the probability of finding not even a defective cell phone among those tested is less than one in a million.

Solution

With k as the number of cell phones tested, the probability of finding not even a defective one is $(1 - p)^k$. We thus have

$$(1 - p)^k = (0.999)^k \leq 10^{-6} \quad \rightarrow \quad \log(0.999)^k \leq \log 10^{-6}$$
$$\rightarrow \quad k \log 0.999 \leq -6 \quad \rightarrow \quad k \geq 13,809.$$

This probability is independent of 1,000,000, the total number of cell phones in the warehouse. The Monte Carlo algorithm saves a lot of testing $(1,000,000 - 13,809 = 986,191)$. We can thus conclude that when $13,809$ tested cell phones (i.e., just less than 1.4% of all cell phones) are all good, where the probability of such an occurrence is less than one in a million, the entire batch of 1,000,000 cell phones, even those not tested, is good, of course, with some probability of error.

Another major application of Bayes' theorem is Bayesian spam filtering. **Bayesian spam filtering** is a popular spam-filtering technique that relies on word probabilities and can tailor itself to the email needs of individual users. Bayesian spam filters look for occurrences of particular words. For instance, we may say if an email contains certain words or group of words, such as *urgent, bank, no catch, act now, risk free, account, money, special deal, million, virus, update*, that are often used in spam and some words, such as *food, car, see, dinner, kids, book, hope, good, glasses, love, hotel*, that are hardly used in spam, it is then likely to be spam. Such words have particular probabilities of occurring in a spam email vis-à-vis in a nonspam email.

Let S be the event that the email is spam and S^c be the event that the email is not spam. Suppose there is an email with a particular word, say X_1. Let W_1 be the event that the email contains the word X_1. Note that the empirical probability for a spam email containing the word X_1 is $P(W_1|S) = a(X_1)$, and the empirical probability for a nonspam email containing the word X_1 is $P(W_1|S^c) = b(X_1)$. A Bayesian spam detection software generally assumes that there is no *a priori* reason for an incoming email to be spam rather than nonspam and considers both cases to have equal probabilities (i.e., $P(S) = P(S^c) = 0.5$). Using Bayes' theorem, p_1, the probability that the email is spam, given that it contains X_1 is as follows:

$$P(S|W_1) = \frac{P(W_1|S)P(S)}{P(W_1)} = \frac{P(W_1|S)P(S)}{P(W_1|S)P(S) + P(W_1|S^c)P(S^c)} \rightarrow$$

$$p_1 = \frac{a(X_1)}{a(X_1) + b(X_1)}.$$

If p_1 is greater than a certain threshold, then the software classifies it as spam. Using a single word for the detection of spam may not be very effective, as it can lead to excessive false positives and false negatives.

If we use $k \geq 1$ words, say $X_1, X_2, \ldots, X_k$, we can then increase significantly the probability of detecting spam correctly. Assuming that $W_1, W_2, \ldots, W_k$ are the events that the email contains the words $X_1, X_2, \ldots, X_k$, the probability that an email is spam is the same as the probability that an email is nonspam (i.e., $P(S) = P(S^c) = 0.5$), and the events $W_1|S, W_2|S, \ldots, W_k|S$ are all independent, by using Bayes' theorem, p_k, the probability that an email containing all the words $X_1, X_2, \ldots, X_k$ is spam, is thus as follows:

$$p_k = P(S|W_1, \ldots, W_k) = \frac{P(W_1|S) \times \ldots \times P(W_k|S)}{P(W_1|S) \times \ldots \times P(W_k|S) + P(W_1|S^c) \times \ldots \times P(W_k|S^c)}$$

$$= \frac{a(X_1) \times \ldots \times a(X_k)}{(a(X_1) \times \ldots \times a(X_k)) + (b(X_1) \times \ldots \times b(X_k))}.$$

Example 16.17

Suppose a Bayesian spam filter is trained on a set of 100,000 spam emails and 4,000 emails that are not spam. The word *money* appears in 2500 spam emails and 200 nonspam emails, the word *urgent* appears in 8000 spam emails and 800 nonspam emails, the word *attention* appears in 20,000 spam emails and 80 nonspam emails, and the word *account* appears in 10,000 spam emails and 100 nonspam emails. Estimate the probability that a received email containing all four words of *money, urgent, attention,* and *account* is spam. Will the email be rejected as spam if the threshold for rejecting spam is set at 80%?

Solution

Assuming X_1, X_2, X_3, and X_4 refer to the words *money, urgent, attention,* and *account*, respectively, we have the following probabilities:

$$a(X_1) = \frac{2500}{100000} = 0.025 \quad \& \quad b(X_1) = \frac{200}{4000} = 0.05$$

$$a(X_2) = \frac{8000}{100000} = 0.08 \quad \& \quad b(X_2) = \frac{800}{4000} = 0.2$$

$$a(X_3) = \frac{20000}{100000} = 0.2 \quad \& \quad b(X_3) = \frac{80}{4000} = 0.02$$

$$a(X_4) = \frac{10000}{100000} = 0.1 \quad \& \quad b(X_4) = \frac{100}{4000} = 0.025.$$

We can thus obtain

$$p_4 = \frac{0.025 \times 0.08 \times 0.2 \times 0.1}{0.025 \times 0.08 \times 0.2 \times 0.1 + 0.05 \times 0.2 \times 0.02 \times 0.025} = \frac{8}{9} \cong 88.9\%.$$

As $p_4 > 80\%$, an incoming email containing all these words will be rejected.

Exercises

(16.1)

There are two events, A and B, such that $P(A \cap B) = 0.1$ and $P(A|B) = 0.25$. Determine $P(B|A)$.

(16.2)

There are two bags of balls; the first bag contains five red balls and four blue balls and the second bag contains three red balls and six blue balls. One ball is taken from the first bag and put in the second bag without seeing what the color of the ball is. Determine the probability that a ball now drawn from the second bag is blue.

(16.3)

Suppose we have two coins. One is fair, with $P(H) = P(T) = 0.5$, and one is unfair, with $P(H) = 0.51$ and $P(T) = 0.49$, where H stands for heads and T for tails. In case 1 we pick one of the two coins at random (i.e., with the probability of 0.5), toss it once, put it back, pick one of the two coins at random again, and toss it. In case 2 we pick one of the two coins at random and toss it twice. In each of these two cases determine the probability that both tosses are heads. In the second case determine which one of the two coins is more likely to have been picked.

(16.4)

In a certain city 20% teenage drivers text while driving. The research record indicates that 40% of those who text have a car accident and 1% of those who do not text have a car accident. If a teenage driver has an accident, what is the probability that he was texting?

(16.5)

Suppose that we have found out that the word *money* occurs in 500 of 4000 messages known as spam and in x of 2000 messages known not to be spam. Suppose our threshold for rejecting a message as spam is 90%, and it is equally likely that an incoming message is spam or not spam. Determine the range of values of x for which the messages will be rejected.

(16.6)

The probability that a student passes an exam is 0.95, given that he studied. The probability that he passes the exam without studying is 0.15. As he is a lazy student, we know that the probability that the student studies for an exam is just 50%. Given the student passed the exam, determine the probability that he studied.

(16.7)

A man and a woman get married today. The probabilities that the husband and the wife will be alive in 50 years from today are 0.7 and 0.8, respectively. Assume the events that the husband and the wife will be alive in 50 years from now are independent.
(a) Determine the probability that in 50 years both will be alive.
(b) Determine the probability that in 50 years at least one will be alive.

(16.8)

There are three boxes. The first box contains 4 brown balls and 2 white balls, the second box contains 11 brown balls and 11 white balls, and the third box contains 5 brown balls and 15 white balls. The probability of selecting any one box is 1 in 3. If a ball selected at random from one of the boxes is white, determine the probability that it was drawn from the first box.

(16.9)

Suppose there are a million cars made by a car company across the globe, where the probability that a car has a defective brake part is 0.0001. Based on the Monte Carlo algorithm, determine the minimum number of cars that need to be tested so that the probability of finding not even a defective car among those tested is less than one in a million.

(16.10)

Consider the system shown in Fig. 16.2, where the probability of failure of a component is $0 \leq p \leq 1$ and the components fail independently. Determine the probability that the system connecting points A and B fails, in terms of the probability p. Then, assuming the system failure is 3×10^{-12}, determine the value of p (i.e., the probability of failure of a component).

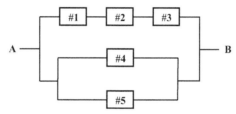

Fig. 16.2 System of components for Exercise 16.10.

CHAPTER 17

Discrete Random Variables

Contents

A numerical representation of the outcome of a random experiment is referred to as a random variable. Interestingly enough, the term *random variable* is a misnomer as it is not random, nor is it a variable; in fact, it is a deterministic function. The focus of this chapter is on probability models that assign real numbers to the random outcomes in the discrete sample spaces. A detailed discussion of all major concepts associated with a discrete random variable is provided. In addition, some well-known random variables along with their applications are introduced.

17.1 The Cumulative Distribution Function

The name of a random variable is represented by an uppercase letter, such as X, and the corresponding lowercase letter, such as x, represents a possible value of the random variable. A **random variable** X is a deterministic function that assigns a real number to each outcome in the sample space S. The sample space S is the **domain of the random variable**, and the set of all values taken on by the random variable, denoted by S_X, is the **range of the random variable**. The range S_X is a subset of all real numbers $(-\infty, \infty)$.

The range of a discrete random variable assumes values from a countable set. The defining characteristic of a discrete random variable is that the set of possible values in the range can all be listed, where it may be a finite list or a countably infinite list. Examples of discrete random variables include the outcomes resulting from rolling a pair of dice, which form a finite list, and the outcomes resulting from randomly selecting a positive integer, which form a countably infinite list.

Every random variable has a cumulative distribution function (cdf). The cdf of a random variable contains all the information required to calculate the probability for any event involving the random variable. The notation for cdf is $F_X(x)$, where we use the uppercase letter F with a subscript corresponding to the name of the random variable X as a

Discrete Mathematics
ISBN 978-0-12-820656-0, https://doi.org/10.1016/B978-0-12-820656-0.00017-4

function of a possible value of the random variable, represented by the lowercase letter x. The *cumulative distribution function* of a random variable X expresses the complete probability model of a random experiment as the following mathematical function:

$$F_X(x) = P(X \le x) \quad -\infty < x < \infty.$$

The event $\{X \le x\}$ and its probability may vary, as x is varied. For any real number x, the cdf is the probability that the random variable X is no larger than x, that is, the cdf is the probability that the random variable X takes on a value in the interval $(-\infty, x]$. Note that the cdf is a function that is continuous from the right. The properties of the cdf of a discrete random variable are presented in Table 17.1.

For a discrete random variable X, $F_X(x)$ has zero slope everywhere except at values of x with nonzero probabilities, and at these points, the cdf has a discontinuity in the form of a jump of magnitude $P(X = x) \ne 0$. The cdf for a discrete random variable is thus in the form of a staircase of finite or countably infinite number of steps. Using the unit step function $u(x)$, defined as $u(x) = 0$ for $x < 0$ and $u(x) = 1$ for $x \ge 0$, the cdf for a discrete random variable can be written in terms of unit step functions, where the number of the unit step functions corresponds to the number of nonzero probabilities.

17.2 The Probability Mass Function

A discrete random variable assumes values from a countably infinite set $S_X = \{x_1, x_2, x_3, \ldots\}$ or a finite set $S_X = \{x_1, x_2, \ldots, x_n\}$, where n is a positive integer. The notation for probability mass function (pmf) is $p_X(x)$, where we use the lowercase letter p, with a subscript corresponding to the name of the random variable X, as a function of a possible value of the random variable, represented by the lowercase letter x. The *pmf* of a discrete random variable X expresses the complete probability model of a random experiment as the following mathematical function:

$$p_X(x) = P(X = x) \quad x \in S_X.$$

Table 17.1 Properties of the cdf of a discrete random variable.

$0 \le F_X(x) \le 1$
$a < b \quad \rightarrow \quad F_X(a) \le F_X(b)$
$\lim_{x \to -\infty} F_X(x) = 0$
$\lim_{x \to \infty} F_X(x) = 1$
$F_X(b) = F_X(b^+)$
$P(a < X \le b) = F_X(b) - F_X(a)$
$P(X = b) = F_X(b^+) - F_X(b^-)$
$P(X > x) = 1 - F_X(x)$

Note that $p_X(x)$ is a function ranging over real numbers x and that $p_X(x)$ can be nonzero only at the values $x \in S_X$. For any value of x, the function is the probability of the event $\{X = x\}$ The pmf $p_X(x)$ contains all the information required to calculate the probability for any event involving the discrete random variable X. The pmf of a discrete random variable must satisfy the following properties:

(a) $p_X(x) \geq 0 \quad \forall x$

(b) $\displaystyle\sum_{x \in S_X} p_X(x) = 1$

(c) $P(X \in B) = \displaystyle\sum_{x \in B \subseteq S_X} p_X(x),$

where B is an event. All these three properties are consequences of the three axioms of probability. The graph of pmf $p_X(x)$ of a discrete random variable has vertical lines of height $p_X(x)$ at the values x in S_X. The cdf and pmf of a discrete random variable X are related as follows:

$$F_X(x) = \sum_{u \leq x} p_X(u) \quad x \in S_X.$$

In other words, the value of $F_X(x)$ is evaluated by simply adding together the probabilities $p_X(u)$ for all values of u that are no larger than x.

Example 17.1

Suppose we have a fair coin. Let X be the number of heads in tossing a coin four times. Find and sketch the pmf $p_X(x)$ and cdf $F_X(x)$ of the discrete random variable X.

Solution

In a coin toss there are two possibilities, namely, a head and a tail. In tossing a coin four times there are thus a total of $16 \left(= 2^4\right)$ different possible outcomes, as presented below:

$$\{TTTT, TTTH, TTHT, TTHH, THTT, THTH, THHT, THHH, HTTT,$$
$$HTTH, HTHT, HTHH, HHTT, HHTH, HHHT, HHHH\}.$$

As the coin is fair, the likelihood of getting a tail is the same as the likelihood of getting a head. It is a reasonable assumption that the coin tosses are independent. After tossing a fair coin four times, the probability of each of the 16 outcomes is $\frac{1}{16} \left(= \frac{1}{2} \times \frac{1}{2} \times \frac{1}{2} \times \frac{1}{2}\right)$. In short, we have 16 equally likely outcomes, and the pmf of the random variable X is therefore as follows:

$$P(X = 0) = P(TTTT) = \frac{1}{16}.$$

$$P(X = 1) = P(TTTH) + P(TTHT) + P(THTT) + P(HTTT) = \frac{4}{16}.$$

$$P(X = 2) = P(TTHH) + P(HTTH) + P(THTH) + P(HTHT) + P(THHT)$$
$$+ P(HHTT) = \frac{6}{16}.$$

$$P(X = 3) = P(HHHT) + P(HHTH) + P(HTHH) + P(THHH) = \frac{4}{16}.$$

$$P(X = 4) = P(HHHH) = \frac{1}{16}.$$

The cdf $F_X(x)$ is thus as follows:

$$F_X(x) = P(X \leq x) = \begin{cases} 0 & x < 0 \\ \dfrac{1}{16} & 0 \leq x < 1 \\ \dfrac{5}{16} & 1 \leq x < 2 \\ \dfrac{11}{16} & 2 \leq x < 3 \\ \dfrac{15}{16} & 3 \leq x < 4 \\ 1 & 4 \leq x \end{cases}$$

Using the unit step function $u(x)$, the cdf can also be written in the following compact form:

$$F_X(x) = \frac{1}{16}u(x) + \frac{4}{16}u(x-1) + \frac{6}{16}u(x-2) + \frac{4}{16}u(x-3) + \frac{1}{16}u(x-4).$$

Fig. 17.1 shows the pmf and cdf of this discrete random variable.

17.3 Expected Values

Expectation provides meaningful insight into the behavior of a random variable. The expected value or the mean of a random variable X represents a real number $(-\infty, \infty)$. The **expected value** of a discrete random variable X, denoted by $E[X]$ or μ_X, is defined as follows:

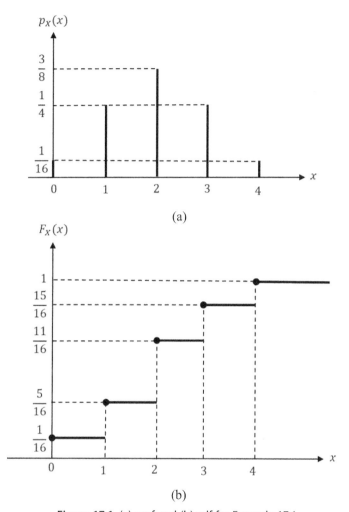

Figure 17.1 (a) pmf and (b) cdf for Example 17.1.

$$E[X] = \mu_X = \sum_{k=-\infty}^{\infty} x_k \, P(X = x_k) = \sum_{x \in S_X} x \, p_X(x).$$

The expected value is thus obtained by multiplying each possible value by its respective probability and then summing these products over all the values that have nonzero probabilities. For a discrete random variable whose expected value is defined, it may not be possible to observe an outcome that is equal to its expected value. For instance, the expected value of the random outcomes of a fair typical die is 3.5, but none of its outcomes can ever be 3.5.

Let X be a random variable and let $Y = g(X)$ denote a real-valued deterministic function of X. Therefore Y is also a random variable. The **expected value of a function** of a discrete random variable is defined as follows:

$$E[g(X)] = \sum_{k=-\infty}^{\infty} g(x_k)\, P(X = x_k).$$

The expected value of a derived random variable, such as $g(X)$, can be calculated without having its distribution. In general, the expected value of a function of a random variable is not equal to the function of the expected value of the random variable, that is, $E[g(X)] \neq g(E[X])$.

Note that the statistical expectation is a linear operation; that is, the mean value of a weighted sum of functions of a random variable equals the weighted sum of the mean values of individual functions of the random variable. Therefore we have

$$E\left[\sum_{i=1}^{n} a_i g_i(X)\right] = \sum_{i=1}^{n} a_i E[g_i(X)],$$

where $\{a_1, a_2, \ldots, a_n\}$ are some nonrandom constants.

Example 17.2

Let X be the random outcome in rolling a fair typical die. Suppose Y is a discrete random variable, where $Y = \sin\left(\frac{X\pi}{6}\right)$. Determine the expected values of the discrete random variables X and Y.

Solution
We have

$$E[X] = \mu_X = \sum_{k=1}^{6} \frac{1}{6} \times k = \frac{1}{6}(1+2+3+4+5+6) = 3.5$$

and

$$E[Y] = \sum_{k=1}^{6} \left(\frac{1}{6}\right) \times \sin\left(\frac{k\pi}{6}\right) = \frac{1}{6}\left(\frac{1}{2} + \frac{\sqrt{3}}{2} + 1 + \frac{\sqrt{3}}{2} + \frac{1}{2} + 0\right)$$

$$= \frac{1}{6}(2 + \sqrt{3}) \cong 0.622.$$

As the expected value of a random variable fails to show the spread of random values in its distribution, a measure to highlight its dispersion is essential and insightful. If we have $g(X) \triangleq (X - E[X])^2$, $E[g(X)]$ is then called the **variance** of the random variable X.

The variance of X is thus the mean square of the difference between a random variable X and its mean $E[X]$. As the statistical expectation is a linear operation, the variance of the random variable X, denoted by σ_X^2, can be equivalently expressed as follows:

$$\sigma_X^2 \triangleq E\left[(X - E[X])^2\right] = E\left[X^2\right] - (E[X])^2 = E\left[X^2\right] - \mu_X^2.$$

The variance of a random variable, if it is defined, is always nonnegative. The variance of a random variable is a measure of the variable's randomness, as it indicates the variability of the outcomes. For instance, a large variance indicates the random variable is quite spread out, and it is thus more unpredictable, whereas a small variance shows the random variable is concentrated around its mean and it is thus less random. In fact, when the variance is zero, the variable is no longer random, meaning there is no uncertainty at all. In addition, the variance, in contrast to the mean, is a nonlinear operator. In other words, the variance of a weighted sum of random variables, in general, is not equal to the weighted sum of the variances of the random variables.

The square root of the variance of X, denoted by σ_X, is called the **standard deviation** of the random variable X and is a positive quantity with the same unit as X. For instance, the random variable X and its standard deviation σ_X may be both in meters, volts, degrees Celsius, or kilograms. The great importance of the standard deviation of a random variable lies in the fact that it brings context to the mean value. For instance, a student who has a test mark of 10 points above the test mean is likely to be in the middle of the class, if the standard deviation of test marks is 20 points, however, the student is likely to be near the top of the class if the standard deviation is 5 points.

Example 17.3

Suppose we have a fair coin. Let X be the number of heads in tossing a coin five times. Determine the standard deviation of the discrete random variable X.

Solution

The pmf of the discrete random variable X is as follows:

$$p_X(0) = \frac{1}{32}, \; p_X(1) = \frac{5}{32}, \; p_X(2) = \frac{10}{32}, \; p_X(3) = \frac{10}{32}, \; p_X(4) = \frac{5}{32}, \; \text{and}$$

$$p_X(5) = \frac{1}{32}.$$

The mean is thus as follows:

$$E[X] = \mu_X = (0)\left(\frac{1}{32}\right) + (1)\left(\frac{5}{32}\right) + (2)\left(\frac{10}{32}\right) + (3)\left(\frac{10}{32}\right) + (4)\left(\frac{5}{32}\right)$$

$$+ (5)\left(\frac{1}{32}\right) = 2.5,$$

which intuitively makes sense, as out of five tosses of a fair coin, the number of heads on average is 2.5. The mean square is obtained as follows:

$$E[X^2] = (0)^2\left(\frac{1}{32}\right) + (1)^2\left(\frac{5}{32}\right) + (2)^2\left(\frac{10}{32}\right) + (3)^2\left(\frac{10}{32}\right) + (4)^2\left(\frac{5}{32}\right)$$
$$+ (5)^2\left(\frac{1}{32}\right) = 7.5.$$

Therefore the variance and standard deviation are as follows, respectively:

$$\sigma_X^2 = E[X^2] - (E[X])^2 = 7.5 - (2.5)^2 = 1.25 \rightarrow \sigma_X = \sqrt{1.25} \cong 1.118.$$

Example 17.4
Suppose $Y = aX + b$, where X is a random variable with mean μ_X and variance σ_X^2, and a and b are both real nonrandom constants. Determine the variance of the random variable Y.

Solution
The mean of the random variable Y is as follows:

$$\mu_Y = E[Y] = E[aX + b] = aE[X] + b.$$

This in turn means the mean value of Y is linearly related to the mean value of X in the same way that the random variables Y and X are linearly related. The mean square value of Y is as follows:

$$E[Y^2] = E[(aX + b)^2] = E[a^2X^2 + b^2 + 2abX] = a^2E[X^2] + b^2 + 2abE[X].$$

We can now determine the variance of Y:

$$\sigma_Y^2 = E[Y^2] - (E[Y])^2 = (a^2E[X^2] + b^2 + 2abE[X]) - (aE[X] + b)^2$$
$$= a^2(E[X^2] - (E[X])^2) = a^2\sigma_X^2.$$

There are two other simple measures that can provide further insights into the possible values of a discrete random variable, namely, the mode and median. The mode and median both make the most sense when a very large discrete sample space must be described. The *mode* of a random variable is that value that occurs most often (i.e., it has the greatest probability of occurring). Sometimes, a random variable has more than one mode and is thus named a *multimodal* random variable. The *median* of a random variable is that particular value for which the sum of the probabilities of all values greater than the median and the sum of the probabilities of all values less than the median are equal (i.e., each sum is equal to 0.5). The median of a discrete random variable may not exist. Note that median is generally employed when there are **outliers** in the sample data (i.e., when there are data values that appear remote from all or most of the other data values).

Example 17.5

Determine the median, mode, and mean of the discrete random variable X, whose pmf is presented in Table 17.2.

Solution

The median is 6, as the sum of the probabilities of all values greater than 6 and the sum of the probabilities of all values less than 6 each is equal to 50%. The mode is 8, as it is the most likely to occur. The mean is obtained as follows:

$$E[X] = 0 \times 2\% + 1 \times 3\% + 2 \times 5\% + 3 \times 10\% + 4 \times 10\% + 5 \times 20\%$$
$$+ 6 \times 0\% + 7 \times 20\% + 8 \times 25\% + 9 \times 3\% + 10 \times 2\% = 5.7.$$

17.4 Conditional Distributions

A conditional distribution can incorporate partial knowledge about the outcome of an experiment in the evaluation of probabilities of events. If there is some information about a random variable, then its conditional distribution needs to incorporate that. Suppose event B, defined as $\{X \le b\}$, is given, and we have $P(B) = P(X \le b) > 0$. Using Bayes' theorem, the conditional cdf of X given event B is then defined as follows:

$$F_X(x|B) = P(X \le x|B) = \frac{P(X \le x, X \le b)}{P(B)}.$$

Table 17.2 Pmf values for Example 17.5.

x	$p_X(x)$
0	2%
1	3%
2	5%
3	10%
4	10%
5	20%
6	0%
7	20%
8	25%
9	3%
10	2%

There are now two mutually exclusive cases, depending on whether x or b is larger. As a result, the **conditional cdf** and **conditional pmf** of X given event B can be simplified as follows:

$$F_X(x|B) = \begin{cases} \dfrac{P(B)}{P(B)} = 1 & x > b \\[2ex] \dfrac{P(X \leq x)}{P(B)} & x \leq b \end{cases} \quad \rightarrow \quad p_X(x|B) = \begin{cases} 0 & x \notin B \\[2ex] \dfrac{p_X(x)}{P(B)} & x \in B \end{cases}$$

Note that dividing the pmf $p_X(x)$ by $P(B) < 1$ ensures that for $x \leq b$, the summation of the conditional pmf $p_X(x|B)$, over the entire range of interest, is 1.

Example 17.6

Consider rolling an unfair (loaded) die whose probabilities of its possible outcomes are as follows:

$$P(1) = \frac{2}{7} \text{ and } P(2) = P(3) = P(4) = P(5) = P(6) = \frac{1}{7}.$$

Determine the conditional $p_X(x|B)$ pmf, where B is the set of outcomes that are prime numbers, that is, $B = \{2, 3, 5\}$

Solution

The probability of event B is as follows:

$$P(B) = P(X = 2) + P(X = 3) + P(X = 5) = \frac{1}{7} + \frac{1}{7} + \frac{1}{7} = \frac{3}{7}.$$

The conditional pmf is thus as follows:

$$p_X(x|B) = \begin{cases} 0 & x = 1, 4, 6 \\[2ex] \dfrac{\left(\dfrac{1}{7}\right)}{\left(\dfrac{3}{7}\right)} = \dfrac{1}{3} & x = 2, 3, 5 \end{cases}$$

17.5 Upper Bounds on Probability

A bound, by definition, encompasses all cases, including the worst case; therefore, when it is applied to a particular case, it may not be very tight. Our focus is on the Markov and Chebyshev inequalities, which both provide upper bounds on the probability that a value of a random variable is greater than some number. In principle, using more information about the random variable brings about tighter bounds. The Markov bound uses only the expected value and the Chebyshev bound uses both the expected value and the variance of the random variable.

The Markov inequality provides an upper bound on the probability that a value of a nonnegative random variable is greater than or equal to some positive constant. More specifically, for a random variable X such that $P(X < 0) = 0$ and thus $E[X] > 0$, the **Markov inequality** is as follows:

$$P(X \geq c) \leq \frac{E[X]}{c} \quad \rightarrow \quad P(X \geq kE[X]) \leq \frac{1}{k},$$

where $c > 0$, and thus $k = \frac{c}{E[X]} > 0$. A simple example of an application of the Markov inequality is that no more than 50% of the population can have more than two times the average income, this result is due to $k = 2$ and $c = 2E[X]$. Another example is when the average mark on an exam is 60%, an upper bound on the proportion of students who score at least 80% is then $P(X \geq 80\%) \leq \frac{E(X)}{80\%} = \frac{60\%}{80\%} = \frac{3}{4}$; thus at most, 75% of students can possibly score this high.

The Chebyshev inequality states that the probability of a large deviation from the expected value is inversely proportional to the square of the deviation. More specifically, for an arbitrary random variable X, the **Chebyshev inequality** is as follows:

$$P(|X - E[X]| \geq c) \leq \frac{\sigma_X^2}{c^2} \quad \rightarrow \quad P(|X - E[X]| \geq k\sigma_X) \leq \frac{1}{k^2},$$

where $E[X]$ and σ_X^2 are the mean and variance of the random variable X, respectively, and we have $c > 0$ and $k = \frac{c}{\sigma_X} > 0$. The Chebyshev inequality thus indicates that the probability that a random variable deviates from its mean by more than c in either direction is less than or equal to its variance divided by c^2. This confirms the fact that the probability of an outcome departing from the mean becomes smaller as the variance decreases. The Chebyshev inequality is valid for any negative or nonnegative random variable, and it is obviously useful when $c > \sigma_X$, that is, $k > 1$. A simple example of an application of the Chebyshev inequality is when a fair coin is flipped 100 times, where the expected value and variance are 50 and 25, respectively. By the Chebyshev inequality, the probability that the number of times the coin lands on heads is greater than 60 or less than 40 is then bounded as follows:

$$P((X < 40) \cup (X > 60)) = P(|X - 50| \geq 10) \leq \frac{25}{10^2} = 25\%.$$

Example 17.7

Suppose the pmf of the discrete random variable X is as follows: $p_X(x) = \binom{n}{x} p^x (1-p)^{n-x}$ for $x = 0, 1, 2, \ldots, n$, where n is a positive integer and $p = \frac{1}{10}$. Noting the expected value and variance of the random variable X are $\frac{n}{10}$ and $\frac{9n}{100}$, respectively, determine an upper bound on $P\left(X \geq \frac{n}{5}\right)$ using the Markov inequality and the Chebyshev inequality.

Solution

The Markov inequality is as follows:

$$P\left(X \geq \frac{n}{5}\right) \leq \frac{\left(\frac{n}{10}\right)}{\left(\frac{n}{5}\right)} = \frac{1}{2}.$$

In order to determine the Chebyshev inequality, we need to find $P\left(X \geq \frac{n}{5}\right)$ as follows:

$$P\left(X \geq \frac{n}{5}\right) = P\left(X \geq \frac{n}{10} + \frac{n}{10}\right) = P\left(X - \frac{n}{10} \geq \frac{n}{10}\right) = P\left(\left|X - \frac{n}{10}\right| \geq \frac{n}{10}\right).$$

The right-hand side of the preceding equation was obtained, as X is nonnegative. The Chebyshev inequality is thus as follows:

$$P\left(X \geq \frac{n}{5}\right) = P\left(\left|X - \frac{n}{10}\right| \geq \frac{n}{10}\right) \leq \frac{\dfrac{9n}{100}}{\dfrac{n^2}{100}} = \frac{9}{n}.$$

The Markov bound is a constant, whereas the Chebyshev bound is a function of n and converges to zero as n approaches infinity.

17.6 Special Random Variables and Their Applications

There are infinitely many discrete random variables, defined by their cdfs or equivalently pmfs, out of which a few have real-life applications. Some of the probability distributions that are important enough to have been given names, where there is a random experiment behind each of these distributions, are now discussed. The mean μ_X and variance σ_X^2 of each of these special discrete random variables are presented in Table 17.3.

The Bernoulli random variable X takes the value of 1 with probability p (also known as the probability of success) and the value of 0 with probability $1 - p$ (also known as the

Table 17.3 Means and variances of some discrete random variables.

Distribution	Mean (Expected value)	Variance
Bernoulli	p	$p(1-p)$
Binomial	np	$np(1-p)$
Geometric	$\frac{1}{p}$	$\frac{1-p}{p^2}$
Pascal	$\frac{k}{p}$	$\frac{k(1-p)}{p^2}$
Hypergeometric	$\frac{nK}{N}$	$\left(\frac{nK}{N}\right)\left(\frac{N-K}{N}\right)\left(\frac{N-n}{N-1}\right)$
Poisson	λ	λ
Uniform	$m+\frac{L+1}{2}$	$\frac{L^2-1}{12}$

probability of failure), where $0 < p < 1$. It is therefore a discrete random variable with the range $\{0, 1\}$. The pmf of a **Bernoulli random variable** is defined as follows:

$$p_X(x) = \begin{cases} 1-p & x = 0 \\ p & x = 1 \\ 0 & x \neq 0, x \neq 1 \end{cases} \quad 0 < p < 1.$$

A **Bernoulli trial**, which corresponds to sampling from the Bernoulli distribution, is a random experiment with exactly two possible outcomes, in which the probability of each of the two outcomes remains the same every time the experiment is conducted. The Bernoulli trial is equivalent to the tossing of a biased coin or examining if a component is defective in a system. The Bernoulli trial is a basic building block for some well-known discrete distributions, such as the binomial, geometric, and Pascal distributions.

Example 17.8

Determine the maximum value of the variance of the Bernoulli random variable.

Solution

The maximum value of the variance is obtained as follows:

$$\sigma_X^2 = p - p^2 \quad \rightarrow \quad \begin{cases} \dfrac{d\sigma_X^2}{dp} = 1 - 2p = 0 \\ \dfrac{d\sigma_X^2}{dp^2} = -2 < 0 \end{cases} \quad \rightarrow \quad p = \frac{1}{2}.$$

Therefore the maximum value of the variance occurs when $p = \frac{1}{2}$, which in turn corresponds to the highest level of uncertainty (randomness), as there are two equally likely outcomes.

The binomial random variable X is the number of times 1 (i.e., successes) occurs in n independent Bernoulli trials, where n is a positive integer, and each occurrence of 1 is assumed to have probability p, where $0 < p < 1$. Applications of binomial distribution may include the estimation of the probabilities of the number of times hitting the target or the number of erroneous bits when a packet of data is transmitted over a noisy communication channel. The pmf of a **binomial random variable** is defined as follows:

$$p_X(x) = \binom{n}{x} p^x (1-p)^{n-x} \qquad x = 0, 1, \ldots, n \text{ and } 0 < p < 1,$$

where $\binom{n}{x} \triangleq \frac{n!}{x!(n-x)!}$ is known as the **binomial coefficient**. The binomial distribution is frequently used to model the number of successes when the sampling is performed with replacement, as the draws are independent. Note that if the sampling is carried out without replacement from a finite population, the draws are not independent and the hypergeometric distribution must be employed.

Example 17.9

Suppose a packet of data consisting of 10,000 bits is independently transmitted over a channel in which the bit error rate (probability of an erroneous bit) is 0.0001. Using the binomial distribution, calculate the probability when the total number of errors is less than or equal to 2.

Solution

Noting that the transmission of a bit can be viewed as a Bernoulli trial, the probability of interest is as follows:

$$P(X \leq 2) = \sum_{x=0}^{2} \binom{10000}{x} (0.0001)^x (0.9999)^{10000-x} \cong 0.91971.$$

In a sequence of independent Bernoulli trials with a success probability p, with $0 < p < 1$, the random variable X that denotes the number of trials performed until the first success occurs is said to have the geometric distribution with probability p. For instance, the geometric distribution could be used to describe the number of candidates to be interviewed until a candidate is accepted or the number of cars to be test-driven until a car is bought. The pmf of a **geometric random variable** is defined as follows:

$$p_X(x) = (1-p)^{x-1} p \qquad x = 1, 2, \ldots \text{ and } 0 < p < 1.$$

Example 17.10

Tickets for the final match of the 2026 World Cup are sold exclusively online on a single website. Suppose that the chance of successfully accessing the website by a fan to buy a ticket is 0.01. Determine the probability that a fan has to attempt 100 or more times to get through.

Solution

The probability that a fan has to attempt 100 or more times to get through is as follows:

$$P(X \geq 100) = 1 - P(X \leq 99) = 1 - \left(1 - (0.99)^{99}\right) = (0.99)^{99} \cong 0.36973.$$

This is simply the probability that the first 99 attempts are unsuccessful.

The Pascal distribution is commonly used in quality control for a product and represents the number of Bernoulli trials that take place until one of the two outcomes is observed a certain number of times. For instance, in choosing 12 citizens to serve on a jury, the Pascal distribution could be applied to estimate the number of rejections before the jury selection is completed. The number of trials up to and including the success in a sequence of independent Bernoulli trials with a constant success probability p, where $0 < p < 1$, has a Pascal distribution. The pmf of a **Pascal random variable** is defined as follows:

$$p_X(x) = \binom{x-1}{k-1} p^k (1-p)^{x-k} \quad 0 < p < 1, \quad x = k, \ k+1, \ ..., \ \text{for } k \in \{1, 2, \ ...\}.$$

Example 17.11

Suppose a law firm is recruiting two lawyers, and each applicant interviewed has a probability of 0.8 to be hired. Determine the probability that up to and including three applicants need to be interviewed.

Solution

With $p = 0.8$ and $k = 2$, the probability that up to and including three applicants need to be interviewed (i.e., $x = 2$ and $x = 3$) is as follows:

$$P(x = 2) + P(x = 3) = \sum_{x=2}^{3} \binom{x-1}{1} (0.8)^2 (0.2)^{x-2} = 0.896.$$

The hypergeometric distribution has many applications in quality control, such as the chance of picking a defective part from a production line. Suppose there is a finite population of N items of which K possess a certain attribute, the hypergeometric distribution then describes the probability that a sample of n items, without replacement, is selected of which x possess the attribute. The pmf of a **hypergeometric random variable** is defined as follows:

$$p_X(x) = \frac{\binom{K}{x}\binom{N-K}{n-x}}{\binom{N}{n}} \qquad \max(0,\ n+K-N) \le x \le \min(n,\ K),$$

$$0 \le n \le N, \quad \text{and} \quad 0 \le x \le K.$$

Example 17.12
Suppose there are 40 students in a party, 16 of them are female, and the remaining 24 are male. A photo of two students is taken. Determine the probability that both of them in the photo are female.

Solution
With $N = 40$, $K = 16$, $n = 2$, and $x = 2$, the probability of interest is thus as follows:

$$P(X = 2) = \frac{\binom{16}{2} \times \binom{24}{0}}{\binom{40}{2}} = \frac{2}{13} \cong 0.1538.$$

Example 17.13
A 12-person jury is to be selected from a group of 24 potential jurors, of which 16 are men and 8 are women. Determine the probability of selecting 6 men and 6 women to form a 12-person jury.

Solution
It is an unordered sampling without replacement. There are $\binom{16}{6} = 8008$ ways to choose the six men and $\binom{8}{6} = 28$ ways to choose the six women. There are thus $\binom{16}{6}\binom{8}{6} = 224,224$ possible ways to make such a selection. However, the number of ways to select a 12-person jury at random, regardless of the gender

of the jurors, is $\binom{24}{12} = 2,704,156$. The probability of interest is the ratio of these two numbers of ways, so we have

$$\frac{\binom{16}{6}\binom{8}{6}}{\binom{24}{12}} = \frac{8008 \times 28}{2,704,156} \cong 0.083.$$

The Poisson distribution represents the number of occurrences of events occurring within certain specified boundaries, such as the number of text messages received by a mobile user during an hour, the number of potholes in a road, and the number of defective units in a sample taken from a production line. The Poisson random variable arises in situations where the events occur completely at random in time or space. These events occur with a constant average rate and independently of the time or space associated with the last event. The pmf of a **Poisson random variable** is defined as follows:

$$p_X(x) = \frac{e^{-\lambda}\lambda^x}{x!} \quad x = 0, 1, 2, \dots \quad \lambda > 0,$$

where λ is the parameter of the distribution reflecting the average rate of occurrence.

Example 17.14

Suppose that the number of errors in a book has a Poisson distribution with parameter $\lambda = 7$. Determine the probability that a book has no errors and the probability that the number of errors are three or more.

Solution

The probability that there is no error is as follows:

$$P(X = 0) = \frac{e^{-7}7^0}{0!} \cong 0.0009.$$

The probability that the number of errors is three or more is as follows:

$$P(X \geq 3) = 1 - \sum_{i=0}^{2} P(X = i) = 1 - \sum_{i=0}^{2} \frac{e^{-7}7^i}{i!} \cong 0.0296.$$

The discrete uniform random variable occurs when outcomes are equally likely, such as rolling a fair die. It takes on values in a set of L positive integers with equal probability. This distribution is generally employed when there is no information available regarding the possible outcomes; as such, the outcomes are all assumed to have the same probability. The pmf of a *discrete uniform random variable* is defined as follows:

$$P(X = k) = p_X(k) = \frac{1}{L} \quad k = m+1, \ldots, m+L, \quad -\infty < m < \infty, \quad L \in \{1, 2, \ldots\}.$$

Example 17.15

Suppose the mean and variance of a discrete uniform random variable X are both 4. Determine the set of consecutive integers that this discrete uniform random variable can take on.

Solution

We have

$$\sigma_X^2 = \frac{L^2 - 1}{12} = 4 \quad \rightarrow \quad L = 7.$$

Therefore we have

$$\mu_X = m + \frac{L+1}{2} = 4 \quad \rightarrow \quad m + \frac{7+1}{2} = 4 \quad \rightarrow \quad m = 0.$$

Hence X takes on values in the set of integers $\{1, 2, 3, 4, 5, 6, 7\}$, each with probability $\frac{1}{7}$.

Exercises

(17.1)

In a bag there are 12 identical balls numbered 1 to 12 inclusive. Let X be the discrete random variable denoting the ball drawn from the bag. Determine the conditional pmf of X given B, where B is the event representing balls with prime numbers.

(17.2)

The number of text messages sent by a teenager during an hour is a random variable. The mean and variance of this random variable are 15 and 9, respectively. Using the Chebyshev inequality, estimate the probability that the number of text messages is more than 5 from the mean.

(17.3)

Assume 1000 bits are independently transmitted over a digital communications channel in which the bit error rate is 0.001. Determine the probability when the total number of errors is greater than or equal to 998.

(17.4)

Suppose phone call arrivals at a call center are Poisson and occur at an average rate of 50 per hour. The call center has only one operator. If all calls are assumed to last 1 minute, determine the probability that a waiting line will occur.

(17.5)

The variance of the discrete uniform random variable Z, which takes on values in a set of n consecutive integers, is 4. Determine the mean of this random variable.

(17.6)

The number of major earthquakes in the world is represented by a Poisson distribution with a rate of $\lambda = 7.4$ earthquakes in a year. Determine the probability that there are exactly four earthquakes in a year. What is the probability that there are no earthquakes given that there are at most two earthquakes in a year?

(17.7)

Consider a six-sided cube-shaped die that is not fair. Let X be the discrete random variable that represents the outcome of a roll of the die. Determine the variance of the random variable X whose pmf is as follows:

$$P(X = 1) = P(X = 2) = P(X = 3) = P(X = 5) = P(X = 6) = 0.1$$
$$\&\ P(X = 4) = 0.5.$$

(17.8)

Suppose the average age of people in a town is 49 years. If all people over 70 years old should be vaccinated against a certain disease, use the Markov inequality to determine the maximum fraction of people of the town who should be vaccinated.

(17.9)

In a typical lottery game a player chooses n distinct numbers from 1 to N inclusive, where n and N are both positive integers. Determine the probability that $0 \leq k \leq n$ of the n balls picked match the player's choices. Assuming $N = 49$ and $n = 6$, determine the specific probabilities of winning for various values of k.

(17.10)

Suppose X is a discrete random variable represented by the outcome of rolling a fair typical die (i.e., its six outcomes 1, 2, 3, 4, 5, and 6 are equally likely), and the random variable Y is related to X by the deterministic function $Y = \cos\left(\frac{X\pi}{3}\right)$. Determine the sample space, pmf, and mean value of the random variable Y.

CHAPTER 18

Graphs

Contents

In graph theory the term *graph* does not refer to data charts, such as line graphs or bar graphs. In fact, graph theory, which was invented by Leonard Euler, is a branch of mathematics concerned with networks of points connected by lines. Graphs are used in a wide variety of models, as they are a powerful problem-solving tool to represent a complex situation with a single image that can be effectively analyzed both visually and with the aid of a computer. Graph theory is a broad topic, but in this chapter, we introduce only some basic concepts of graph theory.

18.1 Basic Definitions and Terminology

A *graph* $G = (V, E)$ consists of V, a nonempty finite set of *vertices* (*points* or *nodes*), and E, a finite set of *edges* (*lines*, *links*, or *arcs*), where an edge either joins one vertex to another or joins a vertex to itself. Note that the number of distinct vertices in the set V and the number of distinct edges in the set E are represented by $|V|$ and $|E|$, respectively. Geometrically, vertices are shown by small solid circles, and edges are represented by curves or straight-line segments. The number of vertices of a graph is called its *order*, and the number of edges of a graph is called its *size*.

Each edge of a graph has one or two vertices associated with it, called its *endpoints*. An *undirected edge* is said to connect its endpoints u and v and is denoted by $\{u, v\}$. An *undirected graph* consists of a set of vertices and a set of undirected edges, each of which is associated with a set of one or two vertices. Two vertices u and v in an undirected graph are called *adjacent* (or *neighbors*) in the graph G if u and v are endpoints of an edge e of the graph G, and an edge e is then called *incident with* the vertices u and v. The set of all neighbors of a vertex is called the *neighborhood* of the vertex. An *isolated vertex* is not adjacent to any vertex (i.e., it is not an endpoint of an edge). A *trivial graph* has just one vertex and no edges.

Discrete Mathematics
ISBN 978-0-12-820656-0, https://doi.org/10.1016/B978-0-12-820656-0.00018-6

An edge emanating from and terminating at the same vertex u is called a **loop** and is denoted by $\{u, u\}$. A loop thus has only one endpoint. It is possible to have more than one loop at a vertex. **Adjacent edges** have a common vertex. **Parallel edges** (or **multiple edges**) have the same two vertices.

A **simple graph** contains no loops or parallel edges. In a simple graph, each edge is associated to an unordered pair of vertices. Graphs that may have multiple edges connecting the same vertices are called **multigraphs**. When there are m different edges associated to the same unordered pair of vertices $\{u, v\}$, then it is an edge of **multiplicity** m. Graphs with loops and multiple edges may be called **pseudographs** (or **general graphs**). A simple graph in which each edge is assigned a positive number, called the **weight**, is a **weighted graph**.

A **directed edge** (or **arc**) is associated with an ordered pair of vertices (u, v), where it starts at u and ends at v. A directed edge is depicted by an arc with an arrow from u, the **initial vertex**, to v, the **terminal vertex**, to indicate the direction of an edge. The initial vertex and the terminal vertex of a loop are the same. A **directed graph** (or **digraph**) consists of a nonempty, finite set of vertices and a set of finite directed edges. When a directed graph has no loops and no multiple directed edges, it is called a **simple directed graph**. Directed graphs with **multiple directed edges** are called **directed multigraphs**. A graph with both directed and undirected edges is called a **mixed graph**. In general, we use the term graph to refer only to the undirected graph.

The **degree of a vertex** u in an undirected graph G, denoted by $\deg(u)$, is the number of edges incident with (meeting at or ending at) u. The **degree sequence of a graph** is the sequence of the degrees of its vertices, usually given in increasing or decreasing order. The **total degree** of the graph G is the sum of the degrees of all the vertices of G. Note that for an isolated vertex u, $\deg(u) = 0$, and with a loop at a vertex u, $\deg(u) = 2$. A vertex u with $\deg(u) = 1$ is called a **pendant vertex** (or **leaf vertex**).

The **handshaking theorem** states that the sum of the degrees of all vertices in an undirected graph is twice the total number of edges (i.e., $2|E|$), which also includes multiple edges and loops. Because the total degree of an undirected graph is even, it is possible to determine if a given number of edges and vertices with known degrees can generate an undirected graph. For instance, the sequence degree $\{1, 2, 3, 4, 7\}$ is not graphical because its sum is odd. In addition, an undirected graph has an even number of vertices of odd degree.

In a graph with directed edges, the **in-degree of a vertex** v, denoted by $\deg^-(v)$, is the number of edges with v as their terminal vertex (the number of arcs having v as a head). The **out-degree of a vertex** u, denoted by $\deg^+(u)$, is the number of edges with u as their initial vertex (the number of arcs having v as a tail). Note that the sum of the in-degrees of all vertices and the sum of the out-degrees of all vertices in a graph with directed edges are equal, and in turn is the same as the number of edges $|E|$ in the graph.

Example 18.1

(a) Determine the number of vertices, the number of edges, and the degree of each vertex in the undirected graph shown in Fig. 18.1a.

(b) Determine the number of vertices, the number of edges, and the in-degree and out-degree of each vertex in the directed graph shown in Fig. 18.1b.

Solution

(a) There are 5 vertices and 11 edges. The degrees are as follows: $\deg(a) = 4$, $\deg(b) = 6$, $\deg(c) = 1$, $\deg(d) = 5$, and $\deg(e) = 6$. Note that according to the handshaking theorem, the sum of the degrees of all vertices is twice the number of edges, that is we have $4 + 6 + 1 + 5 + 6 = 2 \times 11 = 22$.

(b) There are 4 vertices and 8 edges. The degrees are as follows: $\deg^-(a) = 2$, $\deg^+(a) = 2$, $\deg^-(b) = 3$, $\deg^+(b) = 4$, $\deg^-(c) = 2$, $\deg^+(c) = 1$, $\deg^-(d) = 1$, and $\deg^+(d) = 1$. Note that the sum of the in-degrees of all vertices and the sum of the out-degrees of all vertices are equal, and in turn is the same as the number of edges in the graph, that is we have $2 + 3 + 2 + 1 = 2 + 4 + 1 + 1 = 8$.

Graph models are mathematical representations that involve graphs. There are many real-world problems in most disciplines that can be solved using graph models. In fact, it is not very easy to find an area that graph theory has not been applied. Table 18.1 highlights some real-life examples of graph models. In order to build a graph model, some fundamental questions need to be answered, such as the number of vertices, the degree of each vertex, the number of edges (including loops and multiple edges), and the directivity and weight of each edge.

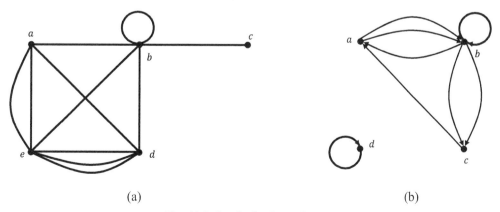

(a)   (b)

Fig. 18.1 Graphs for Example 18.1.

Table 18.1 Real-life examples of graph models.

Graph model	Vertices	Edges
Drug interactions	Medications	Interactions
Electric circuits	Circuit components	Wires
Family trees in genealogy	Family members	Parenthood
Flowcharts in computer programming	Steps to do	Flow
Food webs in ecology	Species	Who eats whom
Information management	Data records	Decisions
Module dependency in software design	Modules	Dependency
Personnel assignment	People/tasks	Job capabilities
Physical chemistry	Atoms	Molecular bonds
Protein interaction in biology	Proteins	Interactions
Scheduling in operations research	Activities	Activity conflicts
Semantic in linguistics	Words	Connections among words
Set theory	Elements	Relatedness
Shortest path in network optimization	Locations	Distances
Social networks	Individuals/organizations	Relationships
Sport tournaments	Teams	Who beats whom
Supply chain management	Supply and demand	Supply lines
Telecommunications networks	Transceivers	Wired/wireless links
The World Wide Web	Computers/cell phones	Wired/wireless links
Transportation networks	Intersections	Roads

18.2 Types of Graphs

A *subgraph* of a simple graph $G = (V, E)$ is a graph $G_1 = (V_1, E_1)$ where $V_1 \subseteq V$ and $E_1 \subseteq E$. In other words, a graph G_1 is a subgraph of G if and only if every vertex in G_1 is also a vertex in G, every edge in G_1 is also in G, and every edge in G_1 has the same endpoints as it has in G. A *spanning subgraph* of a graph G is a subgraph that contains all the vertices of G.

The *complement* or *inverse* of a simple graph $G = (V, E)$ is the *complementary graph* $G_1 = (V_1, E_1)$ that has the same vertices as G and has edges joining every pair of vertices that are not joined in G, and vice versa, that is, two distinct vertices in G_1 are adjacent if and only if they are not adjacent in G.

The *converse* of a directed graph $G_1 = (V_1, E_1)$ is the directed graph $G_2 = (V_2, E_2)$, where the set E_2 of edges of G_2 is obtained by reversing the direction of each edge in the set E_1 of edges of G_1.

The *graph union* of two simple graphs $G_1 = (V_1, E_1)$ and $G_2 = (V_2, E_2)$, denoted by $G_1 \cup G_2$, is the simple graph with the vertex set $V_1 \cup V_2$ and the edge set $E_1 \cup E_2$. The *graph intersection* of two simple graphs $G_1 = (V_1, E_1)$ and $G_2 = (V_2, E_2)$, denoted by $G_1 \cap G_2$, is the simple graph with the vertex set

$V_1 \cap V_2$ and the edge set $E_1 \cap E_2$. Note that the graph intersection of two simple graphs is a subgraph of each of them, and each graph is a subgraph of the graph union.

Example 18.2

Consider the simple graphs G and H, as shown in Fig. 18.2.

(a) List all the subgraphs of the graph G that each has two edges and three vertices.

(b) Determine the complement of the graph G.

(c) Find the graph union of the graphs G and H.

(d) Find the graph intersection of the graphs G and H.

Solution

The subgraphs, complement, union, and intersection graphs are all shown in Fig. 18.3.

(a) There are three subgraphs of G meeting the requirements, as shown in Fig. 18.3a.

(b) The complement of the graph G is shown in Fig. 18.3b.

(c) The graph union of the graphs G and H is shown in Fig. 18.3c.

(d) The graph intersection of the graphs G and H is shown in Fig. 18.3d.

Fig. 18.4 shows several classes of simple graphs, including complete, cycle, star, wheel, and linear graphs, as they have many applications in telecommunications and computer networks.

A **complete graph** (or **mesh topology**) is a simple graph with an edge between every two distinct vertices. A complete graph with $n \geq 3$ vertices, denoted by K_n, has $\frac{n(n-1)}{2}$ edges, and each vertex has degree $n - 1$. A simple graph for which there is at least one pair of distinct vertices not connected by an edge is called a **noncomplete graph**.

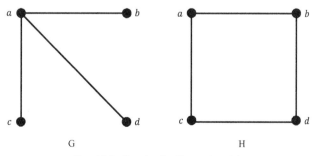

Fig. 18.2 Graphs for Example 18.2.

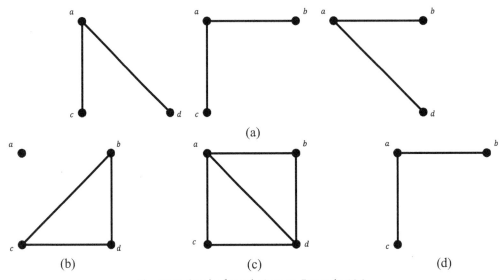

Fig. 18.3 Graphs for solutions to Example 18.2.

A **cycle graph** (or **ring topology**), denoted by C_n, with $n \geq 3$, consists of n vertices $v_1, v_2, ..., v_n$ as well as n edges $\{v_1, v_2\}, \{v_2, v_3\}, ..., \{v_{n-1}, v_n\}, \{v_n, v_1\}$, while noting that each vertex has degree 2.

A **star graph**, denoted by S_{n+1}, with $n \geq 3$, has $n + 1$ vertices $v_1, v_2, ..., v_{n+1}$ as well as n edges $\{v_1, v_{n+1}\}, \{v_2, v_{n+1}\}, ..., \{v_n, v_{n+1}\}$, that is, the vertices $v_1, v_2, ..., v_n$ are all

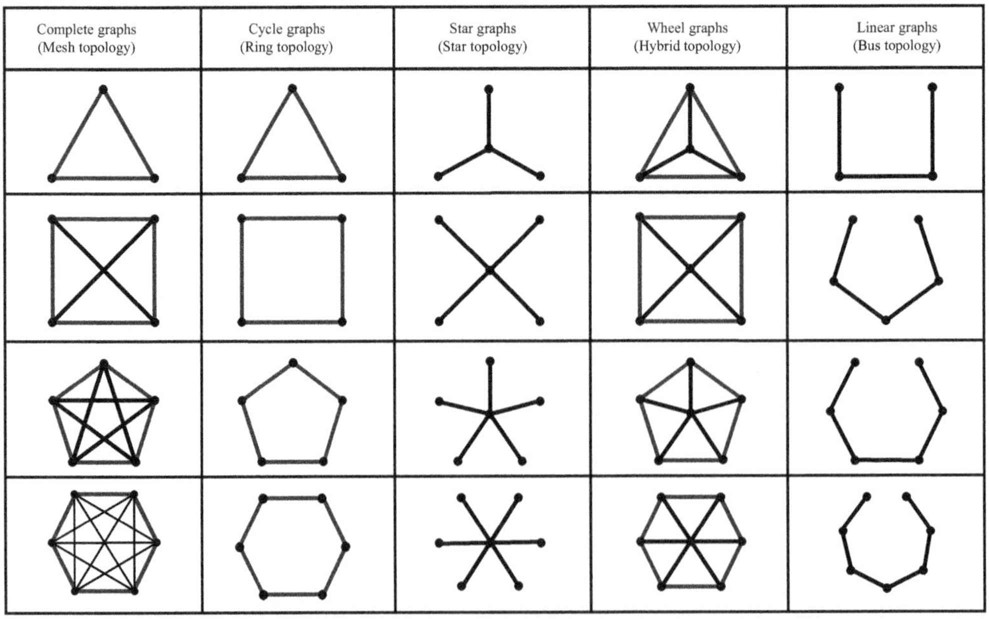

Complete graphs (Mesh topology)	Cycle graphs (Ring topology)	Star graphs (Star topology)	Wheel graphs (Hybrid topology)	Linear graphs (Bus topology)

Fig. 18.4 Special simple graphs.

connected to a central vertex v_{n+1} with degree n, but each of the other n vertices has degree 1.

A *wheel graph* (or *hybrid topology*), denoted by W_{n+1}, with $n \geq 3$, is the union of C_n and S_{n+1} graphs; it thus has $n + 1$ vertices as well as $2n$ edges, where each vertex around the perimeter has degree 3, but the central vertex has degree n.

A *linear graph* (or *bus topology*), denoted by L_n, has n vertices v_1, v_2, ..., v_n and with $n - 1$ edges $\{v_1, v_2\}$, $\{v_2, v_3\}$, ..., $\{v_{n-1}, v_n\}$, while noting that the last two vertices at the two ends of the bus each has degree 1, but each of the other vertices has degree 2.

A *grid graph* $G_{m,n}$ is a simple graph with vertices arranged in an m by n grid. Edges join vertices that are vertically or horizontally adjacent. An *n-dimensional hypercube* (or *n-cube*) Q_n is a simple graph that has vertices representing the 2^n bit strings of length n. Two vertices are adjacent if and only if the bit strings that they represent differ in exactly one-bit position. For instance, when $n = 1$, it is a single edge connecting two vertices each with degree 1, when $n = 2$, edges form a square to connect the four vertices each with degree 2, and when $n = 3$, edges form a cube to connect the eight vertices, each with degree 3.

A *regular graph* is a simple graph in which every vertex of the graph has the same degree. For instance, the complete graph K_n is a regular graph where the degree of each vertex is $n - 1$, the cycle graph C_n is a regular graph where the degree of each vertex is 2, and the n-cube is a regular graph where the degree of each vertex is n. However, the star, wheel, and linear graphs are not regular graphs.

If the vertex set V of a simple graph $G(V, E)$ can be partitioned into two disjoint (mutually exclusive), nonempty subsets V_1 and V_2, known as *bipartite sets*, such that every edge in the graph connects a vertex in V_1 and a vertex in V_2, that is, no edge in the graph G connects either two vertices in V_1 or two vertices in V_2, then we call the graph G the *bipartite graph* and the pair (V_1, V_2) a *bipartition* of the vertex set V of the graph G. Let G be a bipartite graph with m vertices in V_1 and n vertices in V_2. If an edge exists between every vertex in V_1 and every vertex in V_2, the graph G is then called a *complete bipartite graph*, and has $m \times n$ edges.

Example 18.3
Determine which of the simple graphs shown in Fig. 18.5 are bipartite.

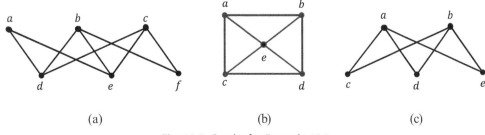

Fig. 18.5 Graphs for Example 18.3.

Solution

Fig. 18.5a is a bipartite graph, as the vertex set $V = \{a, b, c, d, e, f\}$ consists of the two disjoint sets $V_1 = \{a, b, c\}$ and $V_2 = \{d, e, f\}$, and every edge has one vertex in V_1 and the other in V_2. Note that because there is no edge connecting vertices a and f, Fig. 18.5a is not a complete bipartite. Fig. 18.5b is not a bipartite graph, and Fig. 18.5c is a complete bipartite graph.

18.3 Graph Representation and Isomorphism

Representations of graphs by the dot and line diagrams are quite appealing and informative but may often lead to visual misrepresentations. There are therefore other useful ways to represent graphs, such as lists and matrices. However, complex graphs can be better represented by matrices and thus more easily manipulated by computers.

Lists can be used to represent simple graphs, including edge lists and adjacency lists. An **edge list** is a table with rows indexed by the vertices, where each row provides a list of all edges incident with the row's indexing vertex. An **adjacency list** is a table with rows indexed by the vertices, where each row provides a list of all vertices adjacent with the row's indexing vertex. When there are many vertices and edges, then the representation of graphs by edge list or adjacency list can prove to be impractical. To simplify computation, graphs can be represented using matrices, such as adjacency matrices and incidence matrices.

Suppose $G = (V, E)$ is a simple undirected graph with n vertices $v_1, v_2, ..., v_n$. The **adjacency matrix** A of the graph G is an $n \times n$ zero-one matrix, with rows and columns indexed by the n vertices (in the same order), where the entry a_{ij} is 1 if the vertices v_i and v_j are adjacent (i.e., $\{v_i, v_j\}$ is an edge of G), and it is 0 otherwise.

Adjacency matrices can also be used to represent undirected graphs with loops and multiple edges. The entry a_{ij} of the matrix equals the number of multiple edges associated to $\{v_i, v_j\}$. Hence the resulting adjacency matrix is not a zero-one matrix. The entry a_{ii} of the matrix equals the number of loops associated to the vertex v_i, and the nonzero entries along the main diagonal of an adjacency matrix thus indicate the presence of loops. Note that all undirected graphs have symmetric matrices. In the case of a directed graph, the entry a_{ij} represents the number of arrows from v_i to v_j, for all $i \& j = 1, 2, ..., n$. Note that by reordering the vertices of a directed graph, the rows and columns are then moved around.

Suppose $G = (V, E)$ is a simple undirected graph with n vertices $v_1, v_2, ..., v_n$ and m edges $e_1, e_2, ..., e_m$. The **incidence matrix** M of the graph G is an $m \times n$ zero-one matrix, with rows indexed by the vertices and columns indexed by the edges, where the entry m_{ij} is 1, if the edge e_j is incident with the vertex v_i, and it is 0 otherwise. A loop is represented using a column with exactly one entry equal to 1 corresponding to the vertex that is incident with this loop, and multiple edges are represented using columns with identical entries, as these edges are incident with the same pair of vertices.

Example 18.4

Consider the undirected pseudograph shown in Fig. 18.6.

(a) Use an adjacency matrix to represent it.

(b) Use an incidence matrix to represent it.

Solution

(a) The adjacency matrix is as follows:

$$
A = \begin{array}{c} \\ v_1 \\ v_2 \\ v_3 \\ v_4 \\ v_5 \end{array} \begin{array}{c} \begin{array}{ccccc} v_1 & v_2 & v_3 & v_4 & v_5 \end{array} \\ \left(\begin{array}{ccccc} 1 & 2 & 0 & 0 & 0 \\ 2 & 0 & 1 & 0 & 1 \\ 0 & 1 & 0 & 0 & 1 \\ 0 & 0 & 0 & 1 & 1 \\ 0 & 1 & 1 & 1 & 0 \end{array} \right) \end{array}.
$$

(b) The incidence matrix is as follows:

$$
M = \begin{array}{c} \\ v_1 \\ v_2 \\ v_3 \\ v_4 \\ v_5 \end{array} \begin{array}{c} \begin{array}{cccccccc} e_1 & e_2 & e_3 & e_4 & e_5 & e_6 & e_7 & e_8 \end{array} \\ \left(\begin{array}{cccccccc} 1 & 1 & 1 & 0 & 0 & 0 & 0 & 0 \\ 0 & 1 & 1 & 1 & 0 & 1 & 0 & 0 \\ 0 & 0 & 0 & 1 & 1 & 0 & 0 & 0 \\ 0 & 0 & 0 & 0 & 0 & 0 & 1 & 1 \\ 0 & 0 & 0 & 0 & 1 & 1 & 1 & 0 \end{array} \right) \end{array}.
$$

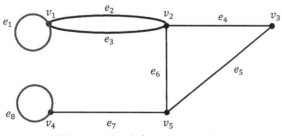

Fig. 18.6 Graph for Example 18.4.

Some graphs may seem to differ but can be essentially the same, as graphs can have multiple representations. Two simple graphs with the same structure and hence the same properties (i.e., they are the same except for labeling of their vertices and edges) are called *isomorphic*. There is a one-to-one correspondence between vertices of two isomorphic graphs preserving the adjacency relationship. Isomorphism of simple graphs is an equivalent relation. More formally, the simple graphs $G_1 = (V_1, E_1)$ and $G_2 = (V_2, E_2)$ are isomorphic if there exists a one-to-one correspondence (i.e., a one-to-one and onto function) f from V_1 to V_2 such that a and b are adjacent vertices in G_1 if and only if $f(a)$ and $f(b)$ are adjacent vertices in G_2 for all a and b in V_1.

Isomorphism has a wide range of applications, such as molecular graphs to model chemical compounds, design of electronic circuits, bioinformatics, and computer vision. It is thus important to show whether two simple graphs are isomorphic or not. However, it is often a very difficult task to do so, as there are no known simple tests for determining graph isomorphism.

A property shared by isomorphic graphs is called an *isomorphism invariant*. For instance, the number of vertices, the number of edges, and the number of vertices of each degree are all isomorphism invariants. If these invariants differ in two simple graphs, then they are not isomorphic. However, the converse is not true, that is, when these invariants are the same, it does not necessarily mean that the two graphs are isomorphic. There is no particular set of invariants that can guarantee isomorphism.

There is no efficient method to determine whether two simple graphs, each with n vertices and m edges, are isomorphic or not while noting that the number of one-to-one correspondences from vertices to vertices is $n!$ and the number of one-to-one correspondences from edges to edges is $m!$. As a result, the total number of pairs of functions to test is $(n!) \times (m!)$, which is simply impossible to calculate when n and m are not small integers. For instance, with $m = n = 6$, which reflects a set of graphs of modest order and size, the number of pairs of functions to be tested is over half a million.

Example 18.5

Consider the pentagon $G_1 = (V_1, E_1)$ and the pentagram $G_2 = (V_2, E_2)$, as shown in Fig. 18.7. Show that the two simple graphs G_1 and G_2 are isomorphic.

Solution

Note that both graphs have five vertices and five edges, with degree two for each vertex. By defining a function $f : V_1 \rightarrow V_2$, as we have $f(a) = u$, $f(b) = w$, $f(c) = x$, $f(d) = y$, and $f(e) = z$, it becomes obvious that it is a one-to-one correspondence. The function f preserves the adjacency relationship because

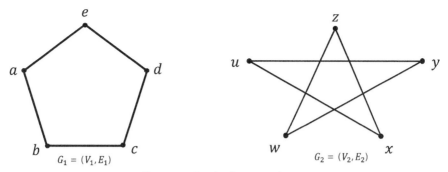

Fig. 18.7 Graphs for Example 18.5.

the adjacency matrix of G_1 is identical to the adjacency matrix of G_2, as reflected below:

$$A_{G_1} = \begin{pmatrix} 0 & 1 & 0 & 0 & 1 \\ 1 & 0 & 1 & 0 & 0 \\ 0 & 1 & 0 & 1 & 0 \\ 0 & 0 & 1 & 0 & 1 \\ 1 & 0 & 0 & 1 & 0 \end{pmatrix} \quad \& \quad A_{G_2} = \begin{pmatrix} 0 & 1 & 0 & 0 & 1 \\ 1 & 0 & 1 & 0 & 0 \\ 0 & 1 & 0 & 1 & 0 \\ 0 & 0 & 1 & 0 & 1 \\ 1 & 0 & 0 & 1 & 0 \end{pmatrix}.$$

Therefore $G_1 = (V_1, E_1)$ and $G_2 = (V_2, E_2)$ are isomorphic.

Example 18.6
Identify all nonisomorphic graphs that have two vertices and two edges.

Solution
There are four nonisomorphic graphs with two vertices and two edges, as shown in Fig. 18.8. The graph (i) has no loops, the graph (ii) has one loop, the graph (iii) has two loops on the same vertex, and the graph (iv) has two loops on separate vertices. Table 18.2 shows the degrees of all vertices. Note that although the number of vertices of each degree in the graph (i) is the same as that in the graph (iv), these two graphs are not isomorphic.

18.4 Connectivity

Going from one place to another in a graph is accomplished by going from one vertex to another along a sequence of adjacent edges. Certain types of sequences of adjacent

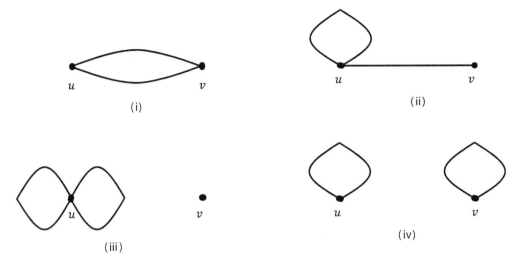

Fig. 18.8 Graphs for Example 18.6.

vertices and edges are of special importance, such as those that do not have a repeated edge or a repeated vertex and those that start and end at the same vertex. It is of great importance to note that there is considerable variation of terminology concerning the concepts regarding connectivity, as the terminology used in graph theory has not been standardized.

A *walk* is a finite nonempty sequence of alternating vertices and edges of the graph, that is, it is any route through a graph from vertex to vertex along edges. Note that a walk can travel over any edge and any vertex any number of times, and the length of a walk is simply the number of edges passed in that walk. A *closed walk* is when the starting vertex is the same as the ending vertex; otherwise, it is an *open walk.* If a graph does not have any parallel edges, then any walk in the graph is uniquely determined by its sequence of vertices.

A *trail* is a walk that does not pass over the same edge twice, that is, it does not contain a repeated edge. A trail might visit the same vertex twice, but only if it comes and goes from a different edge each time. A *path* is a trail that does not include any vertex twice.

Table 18.2 Degrees of all vertices for Example 18.6.

Graphs	Degree of vertex u	Degree of vertex v
(i)	2	2
(ii)	3	1
(iii)	4	0
(iv)	2	2

Table 18.3 Requirements for various connectivities.

	Repeated edge?	Repeated vertex?	Starts and ends at the same point?
Walk	Allowed	Allowed	Allowed
Trail	No	Allowed	Allowed
Path	No	No	No
Cycle	No	No	Yes
Circuit	No	Allowed	Yes

A *cycle* is a path (closed trail) that begins and ends on the same vertex and no other vertices are repeated. A *circuit* is a closed trail that begins and ends on the same vertex and does not contain a repeated edge but may have repeated vertices. Table 18.3 summarizes the relevant requirements for various connectivities.

Note that for simple graphs, it is unambiguous to specify a walk by naming only the vertices that it crosses or giving a sequence of edges, but for pseudographs and multi-graphs, the edges must be specified because there might be multiple edges connecting vertices. For digraphs, walks can travel edges only in the direction of the arrows.

Example 18.7
Consider the graph shown in Fig. 18.9 and identify the following connectivities:
(a) *abcefcbd*.
(b) *abcefcd*.
(c) *abcefcdba*.
(d) *bcefcdb*.

Solution
(a) It is a walk of length 7, but it is not a trail, nor is it a path.
(b) It is a trail, but it is not a path.

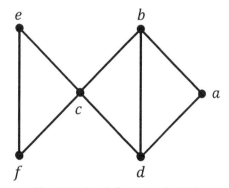

Fig. 18.9 Graph for Example 18.7.

(c) It is a closed walk, but it is not a circuit.

(d) It is a circuit, but it is not a cycle.

A graph is **connected** if there is a walk between every two distinct vertices of the graph; otherwise, it is disconnected. The length of a walk between any two distinct vertices of a connected graph with n vertices is at most $n - 1$.

If G is a connected graph with n vertices v_1, v_2, ..., v_n, and has the adjacency matrix A, then a_{ij} of the matrix A^k, with $1 \leq k \leq n - 1$, is the number of walks of length k from vertex v_i to vertex v_j. The graph G is connected if and only if the following has only nonzero entries.

$$\sum_{k=1}^{n-1} A^k.$$

The length of the shortest walk between the vertices v_i and v_j is the smallest l such that there is at least one walk of length l from v_i to v_j. We can find the length l by computing successively A, A^2, A^3, ..., until we find the smallest positive integer l such that the a_{ij} entry of A^l is not 0.

Example 18.8

Show the graph shown in Fig. 18.10 is connected.

Solution

Noting $n = 4$, the adjacency matrix A of the graph is as follows:

$$A = \begin{pmatrix} 0 & 1 & 1 & 0 \\ 1 & 0 & 2 & 0 \\ 1 & 2 & 0 & 1 \\ 0 & 0 & 1 & 0 \end{pmatrix}.$$

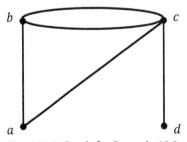

Fig. 18.10 Graph for Example 18.8.

Obviously, there are zero entries in A, as they reflect there are no loops at the vertices a, b, c, or d, nor are there any lines between a and d, or b and d. We therefore determine A^2 as follows:

$$A^2 = \begin{pmatrix} 2 & 2 & 2 & 1 \\ 2 & 5 & 1 & 2 \\ 2 & 1 & 6 & 0 \\ 1 & 2 & 0 & 1 \end{pmatrix}.$$

There are 2-stage links (walks of length 2) between any two vertices, except between the vertices c and d. For instance, there are exactly five different 2-stage links between the vertex b and itself, which are as follows: (i) from b to c and c to b using the upper link exclusively, (ii) from b to c and c to b using the lower link exclusively, (iii) from b to c using the upper link and c to b using the lower link, (iv) from b to c using the lower link and c to b using the upper link, and (v) from b to a and from a to b using the same link. By adding A and A^2, we get the following matrix:

$$A + A^2 = \begin{pmatrix} 0 & 1 & 1 & 0 \\ 1 & 0 & 2 & 0 \\ 1 & 2 & 0 & 1 \\ 0 & 0 & 1 & 0 \end{pmatrix} + \begin{pmatrix} 2 & 2 & 2 & 1 \\ 2 & 5 & 1 & 2 \\ 2 & 1 & 6 & 0 \\ 1 & 2 & 0 & 1 \end{pmatrix} = \begin{pmatrix} 2 & 3 & 3 & 1 \\ 3 & 5 & 3 & 2 \\ 3 & 3 & 6 & 1 \\ 1 & 2 & 1 & 1 \end{pmatrix}.$$

Because the sum has only nonzero entries, it shows that the graph is connected and the length of shortest walk between some vertices is 1 and between some other vertices is 2, and thus every vertex can be connected to any other vertex.

18.5 Euler Circuits and Hamilton Circuits

An **Euler circuit** for the connected graph G is a sequence of adjacent vertices and edges in G that has at least one edge, starts and ends at the same vertex, uses every vertex in G at least once, and uses every edge of G exactly once. A graph G has an Euler circuit if and only if G is connected and the degree of every vertex is even.

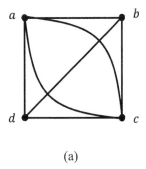

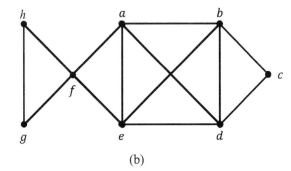

(a)   (b)

Fig. 18.11 Graphs for Example 18.9.

Example 18.9
Determine if each of the graphs shown in Fig. 18.11 has an Euler circuit.

Solution
(a) The graph in Fig. 18.11a does not have an Euler circuit, as not every vertex of the graph has an even degree.
(b) The graph in Fig. 18.11b has an Euler circuit, for every vertex has an even degree. However, not every circuit in the graph is an Euler circuit. For instance, the circuit *abcdefghfa* is not an Euler circuit, but *abcdefghfadbea* is an Euler circuit.

A *Hamilton circuit* for a graph G is a sequence of adjacent vertices and distinct edges in which every vertex of G appears exactly once, except for the first and the last, which are the same. There is no test to determine if a simple graph has a Hamilton circuit. However, there are sufficient conditions for the existence of Hamilton circuits that depend on the degrees of vertices of being sufficiently large.

Suppose G is a simple graph with $n \geq 3$ vertices. G has a Hamilton circuit, as stated by *Dirac's theorem*, if the degree of every vertex in G is at least $\frac{n}{2}$, or as stated by *Ore's theorem*, if $\deg(u) + \deg(v) \geq n$ for every pair of nonadjacent vertices u and v.

Example 18.10
Determine if each of the graphs shown in Fig. 18.12 has a Hamilton circuit.

Solution
(a) In Fig. 18.12a, we have $n = 5$. Dirac's theorem does not apply, as there is a vertex of degree $2 < \frac{5}{2}$, and Ore's theorem does not apply either, for there are two nonadjacent vertices of degree 2, so the sum of their degrees is less than 5. However, the circuit *acdbea* is a Hamilton circuit. This reflects the fact that neither of the sufficient conditions for the existence of a Hamilton circuit given in these theorems is necessary.

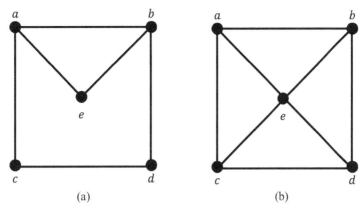

Fig. 18.12 Graphs for Example 18.10.

(b) Noting $n = 5$ in Fig. 18.12b, all the vertex degrees are either $3 > \frac{5}{2}$ or $4 > \frac{5}{2}$, and the sum of the degrees for every pair of nonadjacent vertices is greater than or equal to 5, both Dirac's theorem and Ore's theorem then guarantee the existence of a Hamilton circuit.

18.6 Shortest-Path Problem

Many problems can be modeled using graphs with weights associated to their edges. In this context, weights may mean distances, time periods, levels of risk, costs, or any other criterion of interest. Noting the weight of a path in a weighted graph is the sum of the weights of the edges along the path, the shortest-path problem is then about finding the path of minimum weight from one specified vertex to another vertex of interest. The minimum weight of interest in a graph may thus be the shortest distance, the quickest, the least risk, or the lowest cost. The applications of such problems are numerous, such as routing of phone calls and the Internet packets in telecommunication networks, transport and storage of raw materials and distribution of finished goods across a country, and air travel in the national/global aviation transportation system.

There are several algorithms that find a shortest path between any two vertices in a weighted graph. One such algorithm is known as a **brute-force approach**, by which every path between the two vertices of interest is examined. However, when there exists a large number of edges in the graph, such an approach is extremely impractical and seemingly impossible.

Our focus here is on **Dijkstra's algorithm**, which solves the shortest-path problem in undirected weighted graphs with positive weights along the edges. However, it is easy to adapt it to solve the shortest-path problem in directed weighted graphs. Note that the correctness of Dijkstra's algorithm can be shown by mathematical induction. This algorithm allows to find the length of the shortest path from the starting vertex to all other vertices, and not just to the ending vertex. Dijkstra's algorithm is a greedy algorithm: it seeks to find the global optimum (i.e., the shortest path) by making locally optimal choices (edges with minimum weights) at each stage of the algorithm. Dijkstra's algorithm uses $O(n^2)$ operations (additions and comparisons) to find the length of a shortest path between two vertices in a connected, simple, undirected, weighted graph with n vertices.

We assume the starting vertex is a and the ending vertex is z, and the goal is to find the shortest path between a and z in the graph G. The algorithm works outward from the vertex a, adding vertices and edges one by one to construct a tree T. The algorithm is based on an iterative procedure, and a labeling is carried out at each iteration. At each iteration, the only vertices in G that are candidates to join T are those that are adjacent to at least one vertex of T.

Each vertex u of G is given a label $L(u)$, which indicates the best estimate of the length of the shortest path from a to u. Initially, we label a with zero, denoted by $L(a) = 0$, and all other vertices, such as u, with ∞, denoted by $L(u) = \infty$, a number greater than the sum of the weights of all the edges of G. After each iteration, the values of $L(u)$ are changed and eventually become the actual lengths of the shortest paths from a to u. Therefore when the iteration procedure terminates, $L(z)$ is the length of the shortest path from a to z.

We now highlight the steps of the algorithm. We maintain two sets of vertices: one set, denoted by V, contains vertices already visited and included in the tree T, the other set, denoted by X, contains vertices either not yet visited or already visited. In either case, they are not included in the tree T. Initially, V includes the vertex a and X includes all other vertices in the graph. Note that the only vertices from X that are candidates to join V, the so-called candidate vertices, are those that are adjacent to at least one vertex in V. Of the candidate vertices, the one that is chosen to be added to V is the one for which the length of the shortest path to it from a is a minimum among all the candidate vertices. Each time a vertex is added to V, it is removed from the set of candidate vertices and the vertices adjacent to it are added to the set of candidate vertices if they are not already in the set of candidate vertices or in the set V.

Assuming the vertex v is the most recently vertex added to V, the only candidate vertices for which a shorter path from a might be found are those that are adjacent to v. The reason lies in the fact that the length of the path from a to v was a minimum among all the paths from a to vertices in what was then the candidate vertices. To this effect, after each addition of a vertex v to V, each candidate vertex u adjacent to v is examined. Noting that $w(u, v)$ is the weight of the edge connecting u and v, the current value of $L(u)$ and the value of $L(v) + w(u, v)$ are compared. If $L(v) + w(u, v) < L(u)$, then $L(u)$ is changed to $L(v) + w(u, v)$. After finding a vertex among the candidate vertices, which has the smallest label, the vertex is added to the set V. This procedure is iterated by successively adding vertices to the set V until z is added. When z is added, its label, $L(z)$, is then the length of the shortest path.

Example 18.11
Use Dijkstra's algorithm to find the length of the shortest path between the vertices a and h in the weighted graph shown in Fig. 18.13 (Step 0).

Solution
The steps used by Dijkstra's algorithm to find the shortest path between a and h are summarized in Table 18.4 and shown in Fig. 18.13, where the number next to a line connecting two adjacent vertices represents the length between them, and the length from a to each vertex is in a square next to the vertex. We find the shortest path from a to z is a, c, d, e, g, h, with length 32.

Exercises
(18.1)
Consider the undirected graph G in Fig. 18.14.
(a) Determine the degree of each vertex.
(b) List all its subgraphs and the total degree of each.

(18.2)
Determine which of the following walks in Fig. 18.15 are trails, paths, or circuits.
(a) $v_1 e_1 v_2 e_3 v_3 e_4 v_3 e_5 v_4$
(b) $e_1 e_3 e_5 e_5 e_6$
(c) $v_2 v_3 v_4 v_5 v_3 v_6 v_2$
(d) v_1

(18.3)
Determine the adjacency matrix for the following cases.
(a) The undirected graph shown in Fig. 18.16a
(b) The directed graph shown in Fig. 18.16b

(18.4)
(a) Show that the pair of graphs shown in Fig. 18.17a and 18.17b is not isomorphic.
(b) Show that the pair of graphs shown in Fig. 18.17c and 18.17d is isomorphic.

(18.5)
Consider an undirected graph shown in Fig. 18.18. Is it bipartite? Explain your answer.

(18.6)
Show the graph shown in Fig. 18.19 is connected.

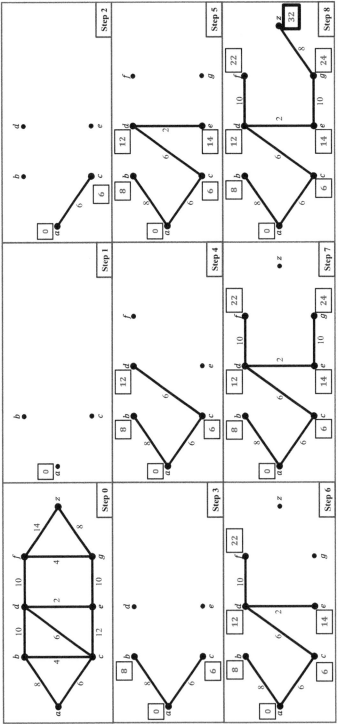

Fig. 18.13 Dijkstra's algorithm for Example 18.11.

Table 18.4 Steps for Example 18.11.

Step	Vertices of path T	Edges of path T	Candidate vertices	$L(a)$	$L(b)$	$L(c)$	$L(d)$	$L(e)$	$L(f)$	$L(g)$	$L(z)$
0	$\{a\}$	$\varnothing$		**0**	∞	∞	∞	∞	∞	∞	∞
1	$\{a, c\}$	$\{\{a,c\}\}$	$\{b, c\}$		8	**6**					
2	$\{a, c, b\}$	$\{\{a,c\}, \{a,b\}\}$	$\{b, d, e\}$		**8**		12	18			
3	$\{a, c, b, d\}$	$\{\{a,c\}, \{a,b\}, \{c,d\}\}$	$\{d, e\}$				**12**	18			
4	$\{a, c, b, d, e\}$	$\{\{a,c\}, \{a,b\}, \{c,d\}, \{d,e\}\}$	$\{e, f\}$					**14**	22		
5	$\{a, c, b, d, e, f\}$	$\{\{a,c\}, \{a,b\}, \{c,d\}, \{d,e\}, \{d,f\}\}$	$\{f, g\}$						**22**	24	
6	$\{a, c, b, d, e, f, g\}$	$\{\{a,c\}, \{a,b\}, \{c,d\}, \{d,e\}, \{d,f\}, \{e,g\}\}$	$\{g, z\}$							**24**	36
7		$\{\{a,c\}, \{a,b\}, \{c,d\}, \{d,e\}, \{d,f\}, \{e,g\}, \{g,z\}\}$	$\{z\}$								**32**
8	$\{a, c, b, d, e, f, g, z\}$										

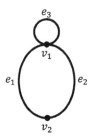

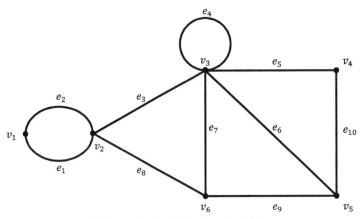

Fig. 18.14 Graphs for Exercise 18.1.

Fig. 18.15 Graph for Exercise 18.2.

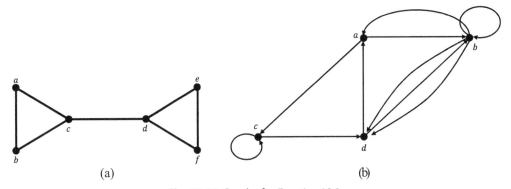

(a)   (b)

Fig. 18.16 Graphs for Exercise 18.3.

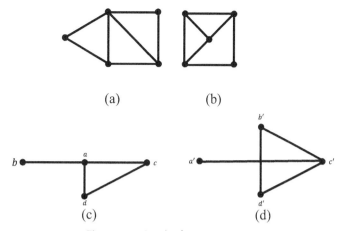

Fig. 18.17 Graphs for Exercise 18.4.

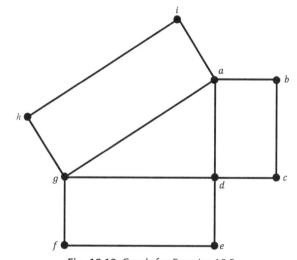

Fig. 18.18 Graph for Exercise 18.5.

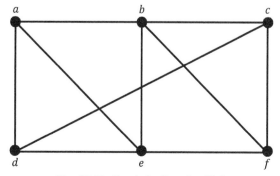

Fig. 18.19 Graph for Exercise 18.6.

(18.7)

(a) Does every symmetric zero-one matrix with zeros on its diagonal represent the adjacency of a simple graph? Explain your answer.

(b) Determine the sum of the entries in a column of the incidence matrix for an undirected graph.

(18.8)

Determine if any one of the following set of graphs has a Hamilton circuit.

(a) Complete graphs (mesh topology)

(b) Cycle graphs (ring topology)

(c) Wheel graphs (hybrid topology)

(18.9)

Determine if any one of the following graphs has an Euler circuit.

(a) Complete graphs (mesh topology)

(b) Cycle graphs (ring topology)

(c) Wheel graphs (hybrid topology)

(18.10)

Use Dijkstra's algorithm to find the length of the shortest path between the vertices a and f in the weighted graph shown in Fig. 18.20.

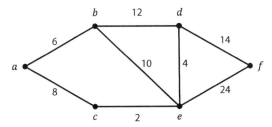

Fig. 18.20 Graphs for Exercise 18.10.

CHAPTER 19

Trees

Contents

Trees are the most important class of graphs. Trees in general can provide an overview of the road map through which a deeper and more intuitive understanding of the goal can be realized. Trees are modeling tools that are widely used to solve problems and construct efficient algorithms in a variety of areas, such as sorting and searching, computing numeric expressions, storing data, modeling hierarchical structures, and designing networks. Trees also have applications in the study of a wide variety of games, such as developing strategies and accommodating constraints in chess and counting games in elimination tournaments.

19.1 Basic Definitions and Terminology

It is of paramount importance to note that tree terminology is not standardized. A *tree* is defined as a connected undirected graph with no simple circuits. As a tree does not contain loops or multiple edges, any tree is a simple graph. A tree has a simple path between any two of its vertices. A *trivial tree* is a graph consisting of a single vertex. A graph is called a *forest* if and only if it is circuit free and not connected. All trees are assumed to be finite (i.e., they have a finite number of vertices). A tree with n vertices has exactly $n - 1$ edges and a connected graph with n vertices and $n - 1$ edges is a tree.

Example 19.1
Determine if the graphs shown in Fig. 19.1 are trees.

Solution
(a) Graph (a) is connected and has no simple circuits; therefore it is a tree.
(b) Graph (b) is circuit free but is not connected; therefore it is not a tree.
(c) Graph (c) is connected but has a simple circuit; therefore it is not a tree.
(d) Graph (d) is not connected and has a simple circuit; therefore it is not a tree.

Discrete Mathematics
ISBN 978-0-12-820656-0, https://doi.org/10.1016/B978-0-12-820656-0.00019-8

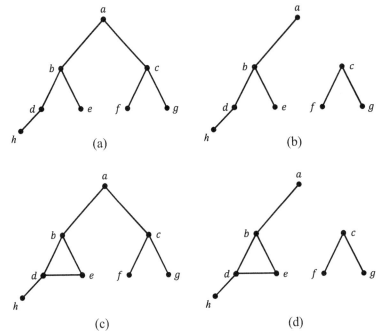

Fig. 19.1 Trees for Example 19.1.

A **rooted tree** is a tree in which one vertex has been designated as the **root**, and every edge is implicitly directed away from the root. The **level** or **depth** of a vertex in a rooted tree is the number of edges along the unique path from the root to the vertex. The **height** of a rooted tree is the maximum level of any vertex of the tree. A tree with exactly one vertex has height zero. In summary, the height is a feature of a tree and the level is a feature of an individual vertex.

A **child** of a vertex v in a rooted tree is a vertex that is the immediate successor of v on a path away from the root. The **parent** of a vertex v in a rooted tree is a vertex that is the immediate predecessor of v on the path to v away from the root. Note that a root is a vertex with no parent but has at least a child. Two distinct vertices that are both children of the same parent are called **siblings**. Given two distinct vertices v and w, if v lies on the unique path between w and the root, then v is an **ancestor** of w and w is a **descendant** of v. A vertex of a rooted tree is called a **leaf** if it has no children. Vertices that have children are called **internal vertices**.

A **subtree** is a subgraph of the tree consisting of a vertex and its descendants and all edges incident to these descendants (i.e., a subtree is a subgraph of a tree that is also a tree). An **ordered rooted tree** is a rooted tree where the children of each internal vertex are linearly ordered. The term *rooted tree* generally implies ordered rooted tree.

A rooted tree in which every node has at most m children is called an ***m-ary tree***. An m-ary tree of height h is ***balanced*** if its leaves lie at levels $h - 1$ or h (i.e., if its leaves lie on adjacent levels). There are at most m^h leaves in an m-ary tree of height h. A tree is called a ***full m-ary tree*** if every internal vertex has exactly m children. A full m-ary tree with i internal vertices contains $n = mi + 1$ vertices, $e = n - 1$ edges, and $l = ((m-1)n+1)/m$ leaves.

An m-ary tree with $m = 2$ is called a ***binary tree***. Binary trees are often used to store information on a computer. The height of the tree determines how quickly information can be retrieved. The height h of a binary tree with n vertices, which is a nonnegative integer, is $\log_2(n+1) - 1 \leq h \leq n - 1$.

In an ordered binary tree, the children are denoted as the left child and the right child. The convention is to regard the ***left child*** as the first child (the elder child) and the ***right child*** as the second child (the younger child). The subtree at a left child u is the ***left subtree*** rooted at u, and the subtree rooted at a right child v is the ***right subtree*** rooted at v.

Example 19.2
Identify the relevant tree terminology for the tree shown in Fig. 19.2.

Solution
The vertex a is the root of the tree and the height of the tree is 4. The vertices $\{b, c\}$ each have a level (or depth) of 1, the vertices $\{h, i, j\}$ each have a level of 3. The vertex d is a child to the vertex b, a sibling to the vertex e, and a parent to the left child h and the right child i. The vertex b is an ancestor of the vertex h, and the vertex h is a descendant of the vertex b. The vertices $\{a, b, d, h, c, g, j\}$ are called internal vertices. The vertices $\{e, i, k, l, f, m, n\}$ are called leaves. The tree is a binary tree but not a full one. The vertices $\{a, b, d, h, c, j\}$ each have two children, but the vertex g has only one child. The left subtree rooted at

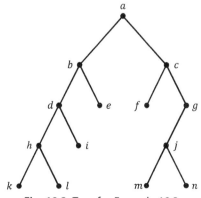

Fig. 19.2 Tree for Example 19.2.

the vertex b consists of the vertices $\{b, d, e, h, i, k, l\}$, and the right subtree rooted at the vertex c consists of the vertices $\{c, f, g, j, m, n\}$. The subtree at the vertex d is a balanced subtree as its leaves lie at levels 1 or 2, but the subtree at the vertex b is not a balanced subtree because its leaves lie at levels 1, 2, or 3.

19.2 Tree Traversal

Ordered rooted trees can be used to store data or algebraic expressions involving numbers, variables, and operations. The process of visiting every vertex of an ordered rooted tree in a systematic way is called ***tree traversal***. Trees are inherently recursive, as every node can be considered the root of its subtree. The simplest mechanism for systematically visiting every node is also recursive. As binary trees are the most important class of m-ary trees and have a wide range of applications, our focus here is on binary tree traversals. There are three elegant recursively defined methods for traversing a nonempty binary tree, which are as follows:

- In a ***preorder traversal***, the parent node is processed before the children. The nodes are visited in the order parent, left child, right child. In other words, a preorder traversal algorithm recursively visits the root, traverses the left subtree, and then traverses the right subtree. This process continues until the last subtree is traversed in preorder.
- In an ***inorder traversal***, the left child is processed first. The nodes are visited in the order left child, parent, right child. In other words, an inorder traversal algorithm recursively traverses the left subtree, visits the root, and then traverses the right subtree. This process continues until the last subtree is traversed in inorder.
- In a ***postorder traversal***, the parent node is processed after the children. The nodes are visited in the order left child, right child, parent. In other words, a postorder traversal algorithm recursively traverses the left subtree, traverses the right subtree, and then visit the root. This process continues until the last subtree is traversed in postorder.

Note that the names reflect the position that the parent node takes relative to the left and right children. In a transversal method, we assume the left child is given precedence over the right child.

Example 19.3

Give the output from traversing the binary tree shown in Fig. 19.3, using the following methods:

(a) Preorder traversal.

(b) Inorder traversal.

(c) Postorder traversal.

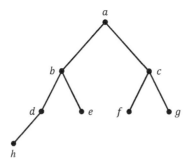

Fig. 19.3 Tree for Example 19.3.

Solution

(a) Root node $(a) \rightarrow a$'s left child $(b) \rightarrow b$'s left child $(d) \rightarrow d$'s left child $(h) \rightarrow$ b's right child $(e) \rightarrow a$'s right child $(c) \rightarrow c$'s left child $(f) \rightarrow c$'s right child (g). A preorder traversal thus visits the nodes in the order of $a\ b\ d\ h\ e\ c\ f\ g$.

(b) a's left child's left child's left child $(h) \rightarrow h$'s parent $(d) \rightarrow d$'s parent $(b) \rightarrow b$'s right child $(e) \rightarrow b$'s parent $(a) \rightarrow a$'s right child's left child $(f) \rightarrow f$'s parent $(c) \rightarrow c$'s right child (g). An inorder traversal thus visits the nodes in the order of $h\ d\ b\ e\ a\ f\ c\ g$.

(c) a's left child's left child's left child $(h) \rightarrow h$'s parent $(d) \rightarrow d$'s parent's right child $(e) \rightarrow e$'s parent $(b) \rightarrow a$'s right child's left child $(f) \rightarrow f$'s parent's right child $(g) \rightarrow g$'s parent $(c) \rightarrow c$'s parent (a). A postorder traversal thus visits the nodes in the order of $h\ d\ e\ b\ f\ g\ c\ a$.

An algebraic expression can be represented by a binary tree, where the internal vertices represent operations, and the leaves represent the variables or numbers. The operators in algebraic expressions have the following priorities from the highest to the lowest: (i) exponentiation ($\uparrow$), (ii) multiplication ($*$) and division ($/$), and (iii) addition ($+$) and subtraction ($-$). Parentheses representing arithmetic expressions can override the precedence rules (i.e., parenthesized subexpressions have the highest priority).

There are three types of notations to evaluate an algebraic expression—infix form, prefix form, and postfix form—where the prefixes pre-, in-, and post- indicate the location of the operator with respect to the operands. They are as follows:

- **Prefix notation** is the form of an arithmetic expression obtained from a preorder traversal of a binary tree representing the expression. For instance, to add a and b, we have "$+ab$" as a prefix expression: the operator as the parent precedes the two operands. We evaluate an expression in prefix form by working from right to left. Whenever an operation is performed, the result becomes a new operand.

- **Infix notation** is the form of an arithmetic expression obtained from an inorder traversal of a binary tree representing the expression. For instance, to add a and b, we have $a + b$ as an infix expression: the operator as the parent lies between the two operands.

- **Postfix notation** is the form of an arithmetic expression obtained from a postorder traversal of a binary tree representing the expression. For instance, to add a and b, we have "$ab+$" as a postfix expression: the operator as the parent follows the two operands. We evaluate an expression in postfix form by working from left to right. Whenever an operation is performed, the result becomes a new operand.

Note that an expression in prefix notation or postfix notation is unambiguous, so no parentheses are needed in such an expression.

Example 19.4

(a) Determine the value of the prefix expression $+ \ - \ * \ 4 \ 6 \ 10 \ / \ \uparrow \ 4 \ 6 \ 8$.

(b) Determine the value of postfix expression $14 \ 4 \ 6 \ * \ - \ 4 \ \uparrow \ 18 \ 6 \ / \ +$.

Solution

(a) $+ - \ * \ 4 \ 6 \ 10 \ / \ \uparrow \ 4 \ 6 \ 8 \ = \ + \ - \ * \ 4 \ 6 \ 10/ \ 4096 \ 8 \ =$
$+ \ - \ * \ 4 \ 6 \ 10 \ 512 \ = \ + \ - \ 24 \ 10 \ 512 \ = \ + \ 14 \ 512 \ = \ 526.$

(b) $14 \ 4 \ 6 \ * \ - \ 4 \ \uparrow \ 18 \ 6 \ / \ + \ = \ 14 \ 24 \ - \ 4 \ \uparrow \ 18 \ 6 \ / \ + \ =$
$10 \ 4 \ \uparrow \ 18 \ 6 \ / \ + \ = \ 10000 \ 18 \ 6 \ / \ + \ = \ 10000 \ 3 \ + \ = \ 10,003.$

19.3 Spanning Trees

A **spanning tree** of a graph G is a subgraph of G that is a tree and contains every vertex of G. Any two spanning trees for a graph have the same number of edges. Spanning trees are useful in visiting the vertices of a graph. Every connected simple graph has a spanning tree with a path between any two vertices. As an example, Fig. 19.4 shows all spanning trees for a graph with four vertices that are all connected to one another (a graph of a square with its two diagonals drawn). We now discuss two algorithms to find a spanning tree in a connected graph, namely, the depth-first search and the breadth-first search.

An outline of the recursive **depth-first search (DFS) algorithm**, also known as the **backtracking algorithm** consists of the following steps:

(i) Choose a vertex of the graph as the root.

(ii) Form a path starting at this vertex by successively adding vertices and edges, where each new edge is incident with the last vertex in the path and a vertex not already visited.

(iii) Continue the process by adding vertices and edges to the path.

(iv) Check if the path goes through all vertices:
- If it does, then the tree consisting of this path is a spanning tree.

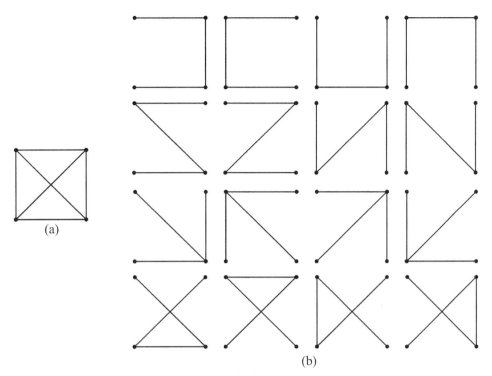

Fig. 19.4 (a) Graph, (b) Spanning trees.

- If it does not, then more vertices and edges must be added by moving back to the next to the last vertex in the path to form a new path starting at this vertex passing through vertices not already visited; if not, move back another vertex in the path, and try again. Repeat this procedure until all vertices have been visited.

In summary, each stage of the DFS traversal seeks to move to an unvisited neighbor of the most recently visited vertex, and backtracks only if there is none available.

An outline of the **breadth-first search (BFS) algorithm** consists of the following steps:

(i) Choose a vertex of the graph as the root, and mark it as visited.

(ii) Visit all unvisited vertices adjacent to this vertex, the new vertices added at this stage become the vertices at level 1.

(iii) Visit all unvisited vertices adjacent to each of them, this in turn produces the vertices at level 2.

(iv) Follow the same procedure until all the vertices in the tree have been visited.

In summary, after the BFS traversal visits a vertex, all of the previously unvisited neighbors of that vertex go to the queue, then the transversal removes from the queue whatever vertex is at the front of the queue and visits that vertex.

Example 19.5

Find a spanning tree for the graph shown in Fig. 19.5a, using the following algorithms:

(a) DFS algorithm.

(b) BFS algorithm.

Solution

(a) The steps of the DFS algorithm, as shown in Fig. 19.5b, are as follows:

   (1) Out of all vertices, we choose a to start the process.

   (2) Out of the vertices adjacent to a (i.e., b and d), neither has been visited yet; we choose b.

   (3) Out of the vertices adjacent to b (i.e., a, c, and f), a has been visited; we choose c.

   (4) Out of the vertices adjacent to c (i.e., b and f), b has been visited; we must then choose f.

   (5) Out of the vertices adjacent to f (i.e., b, c, e, and i), b and c have been visited; we choose e.

   (6) Out of the vertices adjacent to e (i.e., d, f, and h), f has been visited; we choose d.

   (7) Out of the vertices adjacent to d (i.e., a, e, g, and h), a and e have been visited; we choose g.

   (8) Out of the vertices adjacent to g (i.e., d and h), d has been visited; we must then choose h.

   (9) We then backtrack to f. Out of the vertices adjacent to f (i.e., b, c, e, and i), i has not been visited; we must then choose i.

(b) The steps of the BFS algorithm, as shown in Fig. 19.5c, are as follows:

   (1) Out of all vertices, we choose a to start the process.

   (2) Out of the vertices adjacent to a (i.e., b and d), we must visit them both; we first choose b.

   (3) Out of the vertices adjacent to a (i.e., b and d), b has been visited; we must then choose d.

   (4) Out of the vertices adjacent to b (i.e., a, c, and f), a has been visited; we must visit c and f; we choose c.

   (5) Out of the vertices adjacent to b (i.e., a, c, and f), a and c have been visited; we must then choose f.

   (6) Out of the vertices adjacent to d (i.e., a, e, g, and h), a has been visited; we must visit e, g, and h; we choose e.

   (7) Out of the vertices adjacent to d (i.e., a, e, g, and h), a and e have been visited; we must visit g and h; we choose g.

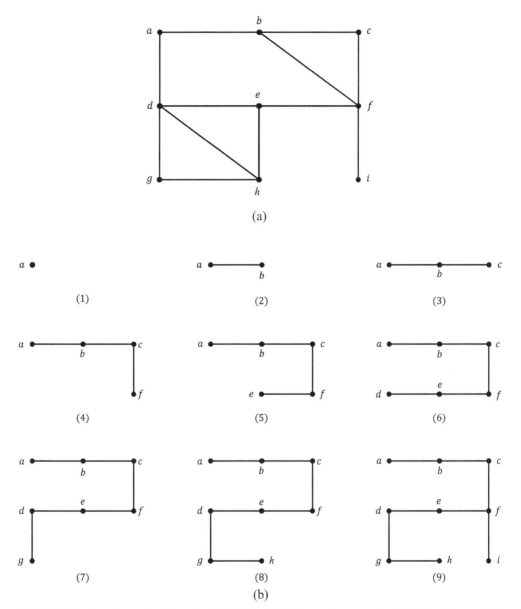

Fig. 19.5 Graph and trees for Example 19.5: (a) Graph; (b) Spanning tree using DFS algorithm; and (c) Spanning tree using BFS algorithm.

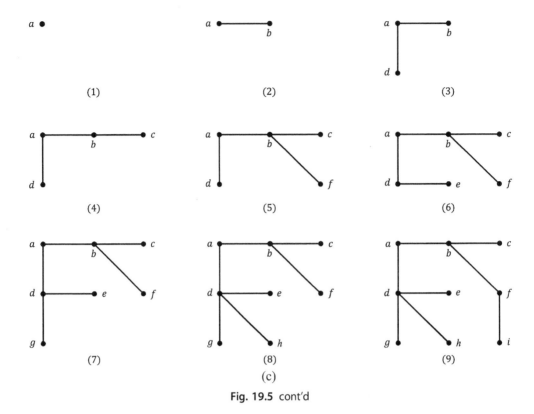

Fig. 19.5 cont'd

(8) Out of the vertices adjacent to d (i.e., a, e, g, and h), a, e, and g have been visited; we must then choose h.

(9) Out of the vertices adjacent to f (i.e., b, c, e, and i), i has not been visited; we must then choose i.

19.4 Minimum Spanning Trees

A *minimum spanning tree* in a connected weighted graph is a spanning tree for which the sum of weights of its edges is the least compared to all other spanning trees for the graph. This problem arises in a wide range of important applications, such as designing physical systems where the system components are geographically dispersed or distanced. Some specific applications may include (i) construction of a pipeline/road network connecting a significant number of places to reduce the total cost, (ii) construction of a digital computer system composing of high-frequency circuitry to reduce delay effects, and (iii) design of a backbone network of high-capacity links to support high-throughput, low-delay Internet traffic.

There are many ways to find a spanning tree for a graph. One such method is the brute force approach, in that all spanning trees for the graph are listed, the total weight of each is computed, and one for which this total is a minimum is chosen. Although by the well-ordering principle for the integers, the existence of such a minimum total is guaranteed, this solution is extremely inefficient and often becomes simply impossible to find. For instance, a complete graph with n vertices has n^{n-2} spanning trees.

We now discuss two algorithms: Prim's algorithm, which was first discovered by Jarnik and later made known by Prim, and Kruskal's algorithm, to construct a minimum spanning tree in a connected graph, where each adds an edge with the smallest weight to the current configuration based on only local information. They both guarantee to produce a minimum spanning tree. These two algorithms employ a greedy algorithm to make an optimal choice at each step, nevertheless, they produce optimal solutions. Note that a connected weighted graph may have more than one minimal spanning tree. Nonetheless, they all weigh the same.

An outline of **Prim's algorithm** consists of the following steps:

(i) Pick a vertex.

(ii) Choose an edge with the least weight connected to the selected vertex.

(iii) Add successively to the tree edges of minimum weight that are incident to a vertex already in the tree while avoiding to make a simple circuit with the edges already in the tree.

(iv) Repeat step (iii) until $n - 1$ edges have been added, where n is the number of vertices in the graph.

An outline of **Kruskal's algorithm** consists of the following steps:

(i) Arrange the edges in nondecreasing order of weights.

(ii) Choose an edge with the least weight.

(iii) Add an edge of the least weight while avoiding to make a simple circuit with the edges already in the tree.

(iv) Repeat step (iii) until $n - 1$ edges have been added, where n is the number of vertices in the graph.

Note that in Prim's algorithm, edges become eligible for inclusion in the tree gradually, whereas in Kruskal's algorithm, the eligible edges are known from the start.

Example 19.6

Find a minimum spanning tree for the graph shown in Fig. 19.6a, using the following algorithms:

(a) Prim's algorithm.

(b) Kruskal's algorithm.

Solution

There are two minimum spanning trees as shown in Figs. 19.6b and c.

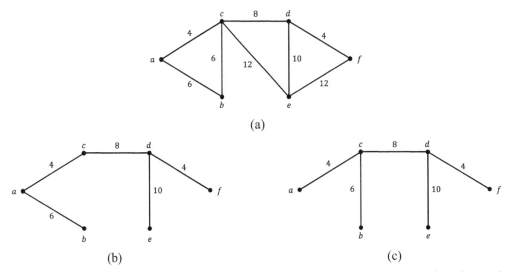

Fig. 19.6 Graph for Example 19.6: (a) Graph; (b) Minimum spanning tree using Prim's algorithm; and (c) Minimum spanning tree using Kruskal's algorithm.

(a) Starting at vertex a, edges are added in one of the following two orders:
   (i) $\{a, c\}, \{a, b\}, \{c, d\}, \{d, f\}, \{d, e\}$.
   (ii) $\{a, c\}, \{b, c\}, \{c, d\}, \{d, f\}, \{d, e\}$.

(b) Edges are added in one of the following two orders:
   (i) $\{d, f\}, \{a, c\}, \{a, b\}, \{c, d\}, \{d, e\}$.
   (ii) $\{d, f\}, \{a, c\}, \{b, c\}, \{c, d\}, \{d, e\}$.

19.5 Applications of Trees

Applications of trees are numerous. Here we just introduce three diverse problems that can be studied and solved using trees, namely the magic square of order 3, a best–of–seven game series, and the Huffman coding.

A *magic square* is an $n \times n$ array of distinct positive integers so that the sum of the numbers is the same in each row, column, and main diagonal. Note that if the array includes just the positive integers 1, 2, 3, ..., n^2, the magic square is said to be normal. The integer n (where n is the number of integers along one side) is the order of the normal magic square and the constant sum, called the magic sum, is as follows:

$$S = \frac{n(n^2 + 1)}{2}.$$

Any magic square can be rotated and reflected to produce eight different squares. In magic square theory, all of these are generally deemed equivalent, and the eight such squares are said to make up a single equivalent class. Equivalent squares are not considered as distinct. Note that the number of distinct magic squares (excluding those obtained by rotation and reflection) of order $n = 1, 2, 3, 4$, and 5 are 1, 0, 1, 880, and $275, 305, 224$, respectively, while noting that the number of magic squares of order $n \geq 6$ is not exactly known, though it can be approximated.

Example 19.7

Consider the smallest nontrivial case of a normal magic square (i.e., $n = 3$), where the sum must be 15. Identify the locations of the numbers 1, 2, 3, 4, 5, 6, 7, 8, and 9 in such a magic square.

Solution

Fig. 19.7 shows how a tree is built in order to find out which number goes where. We start with 9, as it is the largest integer among all and thus significantly restrict the locations of the other eight numbers. By considering symmetric positions equivalent, we only need to consider three possible positions for 9, namely, at the center, in a corner, and a position where it is not at the center or in a corner. We now examine each of these three options.

If 9 is at the center, then we cannot put 8 in any one of the remaining eight positions, and thus cannot go any further to place other numbers in the square. If 9 is in a corner, then there is only one position that 8 can go, and following that, there is only one position that 7 can go. With 9, 8, and 7 in these positions, we cannot put 6 in any one of the remaining six positions, and thus we cannot go any further to place other numbers in the square. To this effect, we must put 9 in a position where it is not at the center or in a corner.

With 9 in its right position, there are two distinct positions for 8, and for each, there is only one position for 7. We thus have two distinct options, where in each 9, 8, and 7 have already been placed. For each of these two options, there are two ways to place 6. We thus end up with four possible options, where in each option we have placed 9, 8, 7, and 6. In each of these four options, there is only one way to put the remaining five numbers 5, 4, 3, 2, and 1. We then check every option to find out if the sum of every row, every column, and every diagonal is 15. We notice that in three of these four options, there is one diagonal whose sum is not 15, thus being unacceptable. For the only remaining option, we see the sum requirement is fully met.

In order to reduce the storage requirements in digital systems as well as the transmission time requirements in digital networks, data must be encoded so fewer bits are used to represent a data symbol. For efficient coding (i.e., effective data compression), data symbols that occur more frequently should be encoded using shorter bit sequences, and longer bit sequences should be used to encode rarely occurring data symbols. Therefore

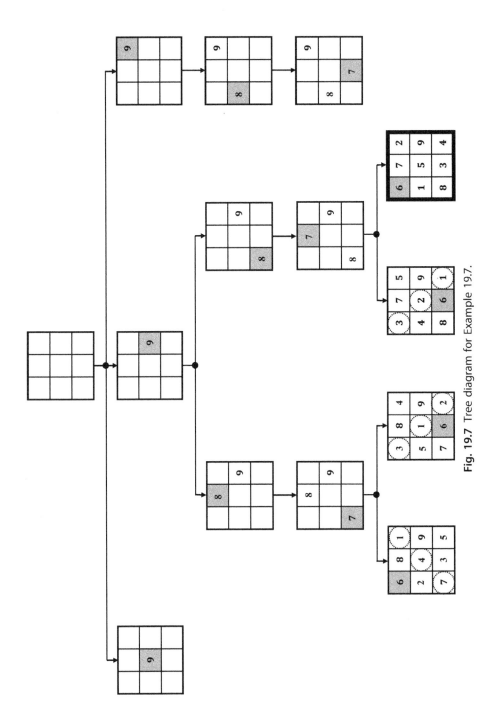

Fig. 19.7 Tree diagram for Example 19.7.

data symbols are encoded using varying numbers of bits. When the statistics regarding data symbols are available, the Huffman code is the optimum algorithm. The **Huffman code** is a prefix-free (instantaneous) code, where no codeword is a prefix of another codeword, as such a Huffman code can be represented using a rooted binary tree. It is important to note that the Huffman code is widely used in compression of image files.

The Huffman coding algorithm begins with a forest of trees, each consisting of a single vertex, where each vertex shows a data symbol and its probability of occurrence. It is essential to put the vertices in the order of increasing probabilities, that is, the first vertex indicates the least likely symbol, and the last vertex reflects the most likely symbol. At each step, we combine two trees with the least total probability into a single tree by introducing a new root and placing the tree with larger weight as one of its subtrees and the tree with smaller weight as the other subtree. Moreover, the sum of the two probabilities associated with the two subtrees is assigned as the total probability of the tree. If necessary, reorder the probabilities of trees, including the newly formed one, so they are still in increasing order. There are many ways to come up with a Huffman code for a given set of data symbols and their probabilities of occurrences. However, they will all have the same average number of bits per symbol for a given set of data symbols.

Example 19.8

Use Huffman coding to encode seven symbols (A, B, C, D, E, F, and G) whose probabilities of occurrences are as follows: A, 0.02; B, 0.03; C, 0.05; D, 0.10; E, 0.15; F, 0.25; and G, 0.40. Determine the average number of bits used to encode a symbol.

Solution

The steps, as shown in Fig. 19.8, are as follows:

(1) Put the vertices in increasing order of their probabilities.

(2) Combine the two least likely trees (vertices) in Step 1, B and A, where they are labeled by 1 and 0, respectively.

(3) Combine the two least likely trees, vertex C and the tree just formed in Step 2, where C is then labeled by a 1 and the subtree in Step 2 by a 0. Hence B and A are labeled as 01 and 00, respectively.

(4) Combine the two least likely trees, vertex D and the tree just formed in Step 3, and then reorder the probabilities to have them all in an increasing order, where D is then labeled by a 1 and the subtree in Step 3 by a 0. Hence C, B, and A are labeled as 01, 001, and 000, respectively.

(5) Combine the two least likely trees, vertex E and the tree just formed in Step 4, and then reorder the probabilities to have them all in an increasing order, where E is then labeled by a 1 and the subtree in Step 4 by a 0. Hence D, C, B, and A are labeled as 01, 001, 0001, and 0000, respectively.

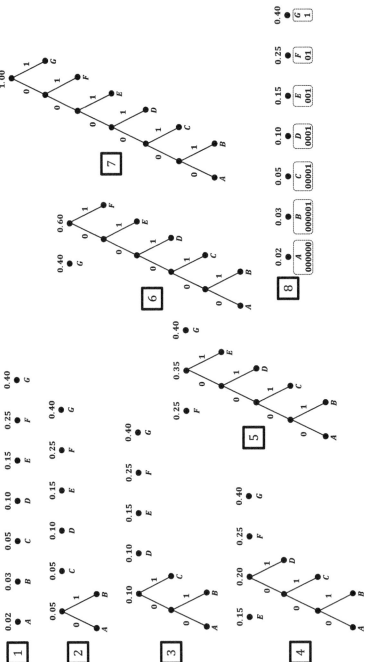

Fig. 19.8 Huffman algorithm for Example 19.8.

(6) Combine the two least likely trees, vertex F and the tree just formed in Step 5, and then reorder the probabilities to have them all in an increasing order, where F is then labeled by a 1 and the subtree in Step 5 by a 0. Hence E, D, C, B, and A are labeled as 01, 001, 0001, 00001, and 00000, respectively.

(7) Combine the two least likely trees, vertex G and the tree just formed in Step 6, where G is then labeled by a 1 and the subtree in Step 6 by a 0. Hence F, E, D, C, B, and A are labeled as 01, 001, 0001, 00001, 000001, and 000000, respectively.

(8) Present the Huffman code, whose average number of bits per symbol is $2.3 \, (= 0.02 \times 6 + 0.03 \times 6 + 0.05 \times 5 + 0.1 \times 4 + 0.15 \times 3 + 0.25 \times 2 + 0.40 \times 1)$.

Another important application of trees is to keep systematic track of all possibilities in scenarios in which events occur in order, but in a finite number of ways. There are some counting problems that cannot be directly solved using basic rules of counting. For instance, certain problems that require trees with asymmetric structures to solve cannot easily use the counting rules, as there are some conditions in these problems that must be met.

Example 19.9

In a sports championship, the eastern and western teams play against one another in a best-of-seven series, that is, the first team that wins four games wins the series and becomes the champion. Identify the tree highlighting all possibilities in which a champion can be determined and determine the number of ways that the series can end.

Solution

We use the tree diagram to find the number of ways that the championship can occur, noting that the series can end in four games, five games, six games, or seven games. The subtree if the first game is won by the eastern team and the subtree if the first game is won by the western team are identical in terms of structure. As shown in Fig. 19.9, a right branch indicates a win by the eastern team, and a left branch indicates a win by the western team. No further branching occurs when a team wins four games altogether. The total number of ways that a champion is determined is 70 $(= 2 \times 35)$.

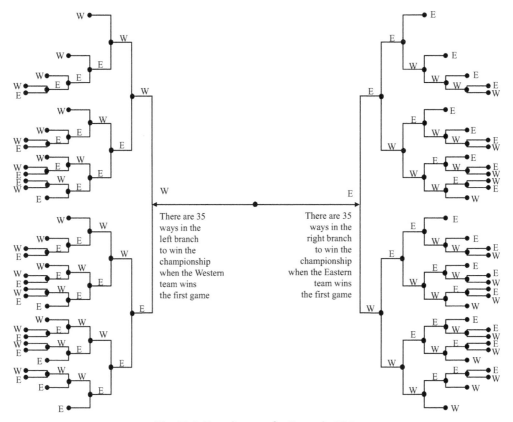

There are 35
ways in the
left branch
to win the
championship
when the Western
team wins
the first game

There are 35
ways in the
right branch
to win the
championship
when the Eastern
team wins
the first game

Fig. 19.9 Tree diagram for Example 19.9.

Example 19.10

Three high school teachers—a 9th-grade teacher, an 11th-grade teacher, and a 12th-grade teacher—are to be chosen from among four teachers A, B, C, and D. Due to a number of reasons, such as their educational backgrounds, personal interests, and teaching experiences, A cannot be a 12th-grade teacher, and either C or D must be a 9th-grade teacher. How many ways can the teachers be chosen?

Solution

The most practical way to determine all possible choices is to build a tree diagram. Note that at each level the two constraints imposed must be respected. Depending on the order of the selection, there are 6 ($= 3!$) different trees to help find all possible ways, Fig. 19.10 shows two of such trees. In any event, there are eight different ways (paths from the root to the leaves) to select the three teachers.

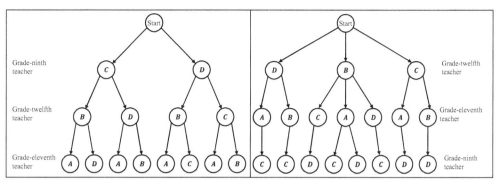

Fig. 19.10 Tree diagrams for Example 19.10.

Exercises

(19.1)

Which of the graphs shown in Fig. 19.11 are trees?

(19.2)

Answer the following questions about the rooted tree shown in Fig. 19.12.

(a) Which vertex is the root?
(b) Which vertices are internal?
(c) Which vertices are leaves?
(d) Which vertices are children of *g?*
(e) Which vertex is the parent of *g?*
(f) Which vertices are the siblings of *g?*
(g) Which vertices are ancestors of *g?*
(h) Which vertices are descendants of *g?*

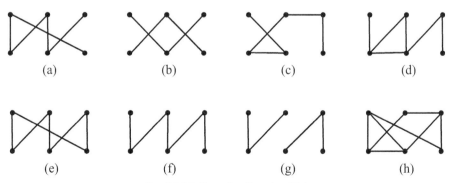

Fig. 19.11 Trees for Exercise 19.1.

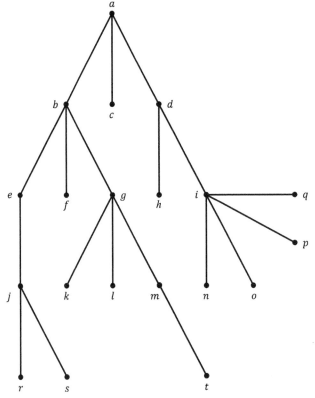

Fig. 19.12 Tree for Exercise 19.2.

(19.3)
Consider a full binary tree. Answer the following questions:
(a) Assuming there are 31 vertices, how many edges does the tree have?
(b) Assuming there are 16 leaves, how many vertices does the tree have?
(c) Assuming there are 15 internal vertices, how many leaves does the tree have?

(19.4)
(a) Determine the value of the prefix expression * + 3 + 3 ↑ 3 + 3 3 3.
(b) Determine the value of postfix expression 3 2 * 2↑5 3 − 8 4/ * −.

(19.5)
Give the output from traversing the binary tree shown in Fig. 19.13, using the following methods:
(a) Preorder traversal.
(b) Inorder traversal.
(c) Postorder traversal.

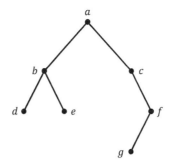

Fig. 19.13 Tree for Exercise19.5.

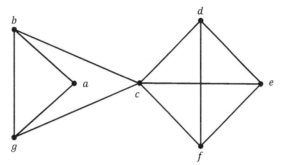

Fig. 19.14 Graph for Exercise 19.6.

(19.6)

Find a spanning tree for the graph shown in Fig. 19.14, using the following algorithms:

(a) DFS algorithm.

(b) BFS algorithm.

(19.7)

Find a minimum spanning tree for the graph shown in Fig. 19.15, using the following algorithms:

(a) Prim's algorithm.

(b) Kruskal's algorithm.

(19.8)

Find a minimum spanning tree for the graph shown in Fig. 19.16, using the following algorithms:

(a) Prim's algorithm.

(b) Kruskal's algorithm.

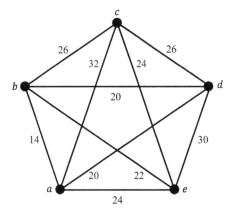

Fig. 19.15 Graph for Exercise 19.7.

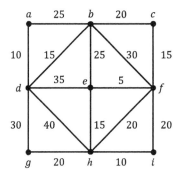

Fig. 19.16 Graph for Exercise 19.8.

(19.9)

Use Huffman coding to encode five symbols A, B, C, D, and E whose probabilities of occurrences are as follows: A, 0.7; B, 0.1; C, 0.1; D, 0.05; and E, 0.05. Determine the average number of bits used to encode a symbol.

(19.10)

Two teams are to play a soccer championship. The first team to win two matches in a row or wins a total of three matches wins the championship. Find the number of ways the tournament can occur.

CHAPTER 20

Finite-State Machines

Contents

The phrase *finite-state machine* refers to a finite-state model. It means the machine has to transition from one state to another so as to perform the required action. Finite-state machines can be used to model various problems in a number of disciplines, including mathematics, artificial intelligence, video games, and linguistics. For instance, they are the design plan of numerous electronic control devices from smart wristwatches to electric cars. This chapter briefly discusses the basics of finite-state machines.

20.1 Types of Finite-State Machines

A *finite-state machine* is a mathematical model of computation based on a hypothetical machine made of different states that can be used to simulate sequential logic in order to represent and control execution flow. Only one single state of a finite-state machine can be active at a given time. Finite-state machines accomplishing specific tasks are in essence computer programming, and there is no specific method for carrying them out, as there are many machines that can accomplish the same task.

In the context of finite-state machines, a **string** is a finite sequence of elements; an **alphabet** is a finite, nonempty set that contains elements used to form strings; the **length of a string** is the number of elements that make up the string; and a **language** is a subset of the set of all strings over an alphabet. As a simple example to illustrate some basic terms, Fig. 20.1 shows a finite-state machine with two final states, namely, "error" and "correct." This machine can parse the string "yes," whose length is 3 while noting that the alphabet consists of 26 letters in the English language.

A finite-state machine has a set of finite states, including a starting state, an input alphabet, and a transition function by which a next state is assigned to every pair of a state and an input. Due to its finite states, a finite-state machine has a limited memory. A finite-state machine is an abstract machine that can be in exactly one of a finite number of states at any given time, where a state changes to another in response to some input. Fig. 20.2 shows the finite-state machines discussed in this chapter.

Discrete Mathematics
ISBN 978-0-12-820656-0, https://doi.org/10.1016/B978-0-12-820656-0.00020-4

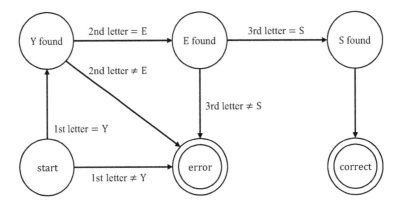

Figure 20.1 A finite-state machine.

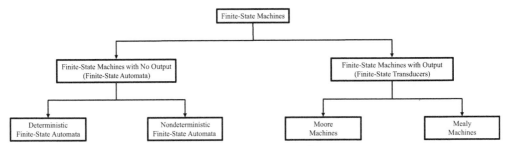

Figure 20.2 Finite-state machines.

A *finite–state machine with no output*, also known as a *finite–state automaton (FSA)*, models the changes of states within a system until it achieves one of a collection of desired states. The finite-state automata (the plural of automaton) do not produce output, but they have a set of final states. They recognize input strings that take the starting state to a final state. These machines can be used, for instance, to model ATMs, traffic lights, parity check bits, subway turnstiles, and DVD players. They are also known as language recognizers and thus play a central role in the design and construction of compilers for programming languages. Finite-state automata are categorized into deterministic FSA and nondeterministic FSA.

In a *finite–state machine with output*, also referred to as a *finite–state transducer (FST)*, each transition has an associated output that either provides some information about the state of the machine or outputs a stream of information as the machine is intended to produce. In a finite-state machine, there are therefore no final states. These machines can be used, for instance, to model vending machines, delay devices, binary adders, pattern finders, network protocols, language and speech recognizers, and spelling and grammar checking. FSTs are categorized into Moore machines and Mealy machines.

Finite-state machines are represented using either **state tables**, which are easier to present, or **state diagrams** (directed graphs with labeled edges), which are easier to understand. In a state diagram for a finite-state machine, the initial state is indicated by means of an arrow that terminates at the initial state but has no initial vertex. Note that every state has a transition for every input. A transition may result in a loop back to the same state. If from some state an input is impossible, then no transition corresponding to that input should be added to the state diagram. A state diagram contains transitions for all possible inputs at each state.

20.2 Finite-State Machines with No Output

A finite-state machine with no output, referred to as an FSA, is an abstract model of a machine that accepts input values but does not produce output values, yet it has a set of final states. In a state diagram for an FSA, the final states are shown by using double circles.

There are two types of finite-state automata: one is the **deterministic FSA**, where for each pair of state and input values there is a unique next state given by the transition function; the other is the **nondeterministic FSA**, in which there is a list of possible next states for each pair of input value and state. The adjective deterministic is often used to emphasize that an FSA is not nondeterministic.

A deterministic FSA $M = (S, I, f, s_0, F)$ is a model that consists of the following five characterizing parts:
- A finite set S of **states**.
- A finite set I of **input alphabet**.
- A **transition function** $f : S \times I \rightarrow S$, which maps state–input pairs to states.
- An **initial state** (or **start state**) s_0.
- A subset F of S consisting of **final states** (or **accepting states**).

The transition function specifies the actions of the system. For instance, if a system with half a dozen states is in state s_3, and the system receives the input i where $f(s_3, i) = s_4$, then the system will change to s_4. Note that state changes occur as a result of a sequence of input alphabets. When the input string consisting of single elements causes an FSA to land in a final state, the string is said to be **recognized** or **accepted**; otherwise, it is **rejected** by the automaton.

Two finite-state automata are called **equivalent** if they recognize the same language. However, two equivalent finite-state automata may have different numbers of states. As the memory space required to store an FSA with n states is approximately proportional to n^2, it is thus important to construct an FSA with the fewest possible states among all finite-state automata equivalent to a given FSA. In addition, simplifying an FSA to have fewer states make it easier to write a computer algorithm based on it. In general, simpli-fication of an FSA involves identifying equivalent states that can be combined while not affecting the action of the FSA on input strings. In summary, an equivalence relation on the set of states of the automaton is defined, and a new automaton whose states are the equivalence classes of the relation is formed.

Example 20.1

Design a deterministic FSA that accepts strings of 0s and 1s as input for which the number of 1s is divisible by 3.

Solution

This is a deterministic FSA with three states. The state s_0 is the initial state reflect-ing the number of 1s is zero (as 0 is a multiple of 3), and it is also the accepting state when the number of 1s is a multiple of 3, that is when it is equal to $3k$, where $k \geq 0$ is an integer. The state s_1 is the state that the number of 1s is $3k + 1$, and the state s_2 is the state that the number of 1s is $3k + 2$. As shown in Fig. 20.3, if the input is a 1, then the state changes (i.e., from s_0 to s_1 or from s_1 to s_2 or from s_2 to s_0), but if the input is a 0, then the state remains unchanged (i.e., there is a loop at each state).

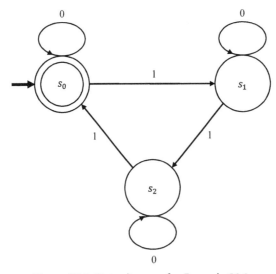

Figure 20.3 State diagram for Example 20.1.

Example 20.2

Consider the deterministic FSA defined by the state diagram shown in Fig. 20.4.
(a) Construct the corresponding state table.
(b) Determine the path that this automaton goes through if the input string is 0101100.
(c) Provide five input strings whose lengths are 2 bits, 3 bits, 4 bits, 5 bits, and 6 bits, such that from the initial state they all go to the accepting state.

Solution

(a) Table 20.1 presents the state table.
(b) The input string 0101100 takes the FSA to the following states: $s_0 \rightarrow s_1 \rightarrow s_2 \rightarrow s_1 \rightarrow s_2 \rightarrow s_0 \rightarrow s_1 \rightarrow s_1$.
(c) The candidate strings of various lengths are as follows: 01, 001, 1001, 01101, and 101101.

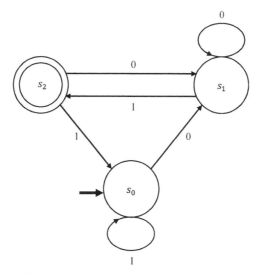

Figure 20.4 State diagram for Example 20.2.

Table 20.1 State table for Example 20.2.

S	f : Input 0	f : Input 1
s_0	s_1	s_0
s_1	s_1	s_2
s_2	s_1	s_0

Example 20.3

A deterministic FSA is modeled by the state diagram shown in Fig. 20.5. Identify the pattern that this automaton can recognize.

Solution

The initial state is s_0, and the state does not change as long as the input is 1. The final state is s_3, to get to s_3 there is only one way, and that is from s_2 when the input is 1. There is also one way to get to s_2, and that is from s_1 when the input is 1. There are four ways to get to s_1, which can be from s_0, s_1, s_2, or s_3 only when the input is 0. The only way state s_3 is reached when the input sequence is 011 while noting the other input sequences are ignored.

Example 20.4

Design a deterministic FSA that models a system, such as an ATM or a cell phone, requiring the 4-digit password $d_1 d_2 d_3 d_4$.

Solution

The state diagram for such an automaton is shown in Fig. 20.6, where d represents a digit. The states are as follows:

s_0: It is the initial state waiting for the first digit.

s_1: It is when the first digit is correct (i.e., d_1 has been pressed) and waiting for the second digit.

s_2: It is when the second digit is correct (i.e., d_2 has been pressed) and waiting for the third digit.

s_3: It is when the third digit is correct (i.e., d_3 has been pressed) and waiting for the fourth digit.

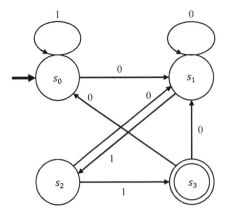

Figure 20.5 State diagram for Example 20.3.

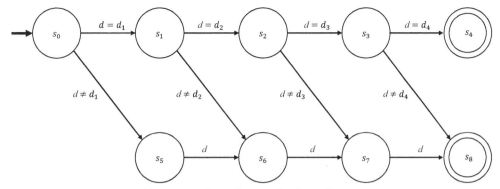

Figure 20.6 State diagram for Example 20.4.

s_4: It is when the fourth digit is correct (i.e., d_4 has been pressed) and represents the accepting state; a screen is thus displayed reflecting that the correct password was entered.

s_5: It is when the first digit is incorrect (i.e., d_1) was not pressed.

s_6: It is when at least one of the first two digits was incorrect.

s_7: It is when at least one of the first three digits was incorrect.

s_8: It is when at least one of the four digits was incorrect and represented the trap state; a screen is thus displayed reflecting an incorrect password was entered.

In a nondeterministic FSA, there is at least one state where there are multiple outgoing edges all having the same input x. A nondeterministic FSA $M = (S, I, f, s_0, F)$ is a model that consists of the following five characterizing parts:

- A finite set S of **states**.
- A finite set I of **input alphabet**.
- A **transition function** $f : S \times I \rightarrow P(S)$ that maps state-input pairs to states, where $P(S)$ denotes the power set of S (the set of all subsets of S).
- An **initial state** (or **start state**) s_0.
- A subset F of S consisting of **final states** (or **accepting states**).

In a nondeterministic FSA, each state-input pair is linked with a set of states, where a set of states can be the null set. A nondeterministic FSA can be represented by a state diagram, where an edge from each state to all possible next states are included, and a state table, where each pair of state and input values for a list of possible next states are included.

The language recognized by an automaton is the set of input strings that take the initial state to a final state of the automaton. In view of this, if a language is recognized by a nondeterministic FSA, then that language is also recognized by a deterministic

FSA; thus a deterministic FSA can be equivalent to a nondeterministic FSA in recognition of a language.

Example 20.5

(a) Determine the state diagram for the nondeterministic FSA, where Table 20.2 presents its state table, the initial state is s_0, and the final states are s_1 and s_2.

(b) Show that the input sequence 00100111 is accepted by this automaton through locating the path that ends at the final state s_1.

(c) Show that the input sequence 0110 is not accepted by the automaton.

Solution

(a) The state diagram for the nondeterministic FSA is shown in Fig. 20.7.

(b) The path is as follows: $s_0 \rightarrow s_0 \rightarrow s_0 \rightarrow s_2 \rightarrow s_2 \rightarrow s_1 \rightarrow s_1 \rightarrow s_1 \rightarrow s_1$.

Table 20.2 State table for Example 20.5.

S	f : Input 0	f : Input 1
s_0	s_0, s_1	s_2
s_1	$\varnothing$	s_1
s_2	s_1, s_2	$\varnothing$

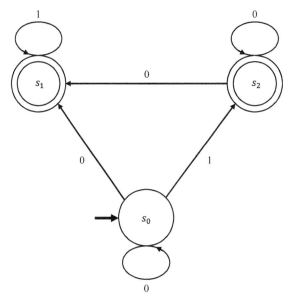

Figure 20.7 State diagram for Example 20.5.

(c) Starting at the initial state s_0, with input 0, there are two alternative paths, namely, either go to s_1 or remain at s_0. By going to s_1 and having two inputs 1 and 1, we remain at s_1. With the final 0, there is no edge along which to move. By remaining at s_0 and having input 1, we go to s_2. With the next 1, there is no edge along which to move. The input string 0110 is thus not accepted.

Example 20.6

The state diagram for a nondeterministic FSA is shown in Fig. 20.8. Show the path through which the input sequence 11011 is accepted by this automaton.

Solution

The path $s_0 \rightarrow s_0 \rightarrow s_0 \rightarrow s_1 \rightarrow s_1 \rightarrow s_2$, which ends at an accepting state, represents 11011. Note that the path $s_0 \rightarrow s_0 \rightarrow s_0 \rightarrow s_1 \rightarrow s_1 \rightarrow s_1$ also represents 11011 but does not end at an accepting state. An input string fails to be accepted if no path represents the string or a path representing the string ends at a nonaccepting state.

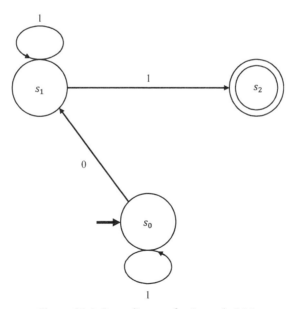

Figure 20.8 State diagram for Example 20.6.

Example 20.7

Show that the finite-state automata shown in Fig. 20.9 are equivalent, as they recognize the same language.

Solution

An ad hoc approach is usually a good way to determine the language recognized by a machine. For the nondeterministic FSA shown in Fig. 20.9a, there is only one accepting state s_2, and there are three possible ways to get there. For the deterministic FSA shown in Fig. 20.9b, there are three accepting states (s_1, s_2, and s_4), while noting that the state s_5 is a graveyard (i.e., there is no way out of it). In either of these two automata, the patterns 0, 01, and 11 are recognizable.

20.3 Finite-State Machines with Output

There are fundamentally two types of finite-state machines with output:

i) Moore machines, where the output is determined only by the state before transition.

ii) Mealy machines, where outputs correspond to transitions between states.

A **Moore machine** $M = (S, I, O, f, g, s_0)$ is a model that consists of the following six characterizing parts:

- A finite set S of **states**.
- A finite set I of **input alphabet**.
- A finite set O of **output alphabet**.
- A **transition function** $f : S \times I \rightarrow S$ that maps state-input pairs to states.
- An **output function** $g : S \times I \rightarrow O$ that maps an output to a state.
- An **initial state** (or **start state**) s_0.

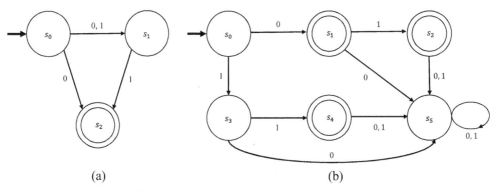

(a)   (b)

Figure 20.9 State diagrams for Example 20.7.

Example 20.8

Show the state diagram for the Moore machine where Table 20.3 presents its state table, and then determine the output string if the input string is 0111.

Solution

As shown in Fig. 20.10, the transitions are labeled with input values and states are labeled with the output values. Note that the output for a Moore machine is one bit longer than the input, as it always starts with the output for the state s_0, which is 0 for this machine. Therefore the states that are visited after s_0 are s_0, s_2, s_1, and s_0, the output string is thus 00110.

Table 20.3 State table for Example 20.8.

S	f : Input 0	f : Input 1	g
s_0	s_0	s_2	0
s_1	s_3	s_0	1
s_2	s_2	s_1	1
s_3	s_2	s_0	1

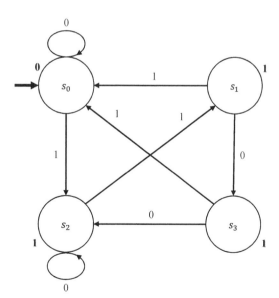

Figure 20.10 State diagram for Example 20.8.

A *Mealy machine* $M = (S, I, O, f, g, s_0)$ is a model that consists of the following six characterizing parts:

- A finite set S of *states*.
- A finite set I of *input alphabet*.
- A finite set O of *output alphabet*.
- A *transition function* $f : S \times I \to S$ that maps state-input pairs to states.
- An *output function* $g : S \times I \to O$ that maps state-input pairs to outputs.
- An *initial state* (or *start state*) s_0.

The input sequence takes the starting state through a sequence of states, as specified by the transition function. As each transition produces an output, an input sequence produces an output sequence. More specifically, the machine takes the input sequence $a_1 a_2 \dots a_n$ one by one and simultaneously changes through a sequence of states $s_0 s_1 s_2 \dots s_n$, starting with the initial state s_0, to produce the output sequence $o_1 o_2 \dots o_n$ one by one, while noting that $s_i = f(s_{i-1}, a_i)$, $o_i = g(s_{i-1}, a_i)$, and $i = 1, 2, \dots, n$.

Example 20.9

Show the state diagram for the Mealy machine where Table 20.4 presents its state table, and then determine the output string if the input string is 11111.

Solution

As shown in Fig. 20.11, the transitions are labeled with input and output values. Therefore the states that are visited after s_0 are s_2, s_3, s_4, s_1, and s_2, and the output string is thus 10011.

Example 20.10

Construct a Mealy machine that delays an input string by one bit, assuming 0 as the first bit of output.

Solution

The state diagram shown in Fig. 20.12 reflects that we need three states: the initial state s_0 to get started so as to account for the delay, and the states s_1 and s_2

Table 20.4 State table for Example 20.9.

S	f : Input 0	f : Input 1	g : Input 0	g : Input 1
s_0	s_1	s_2	1	1
s_1	s_3	s_2	0	1
s_2	s_4	s_3	1	0
s_3	s_0	s_4	0	0
s_4	s_0	s_1	1	1

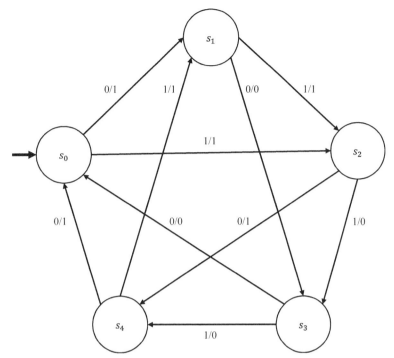

Figure 20.11 State diagram for Example 20.9.

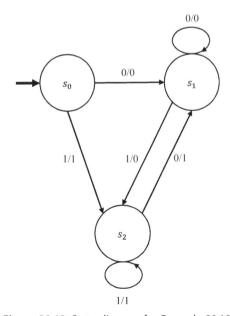

Figure 20.12 State diagram for Example 20.10.

correspond to the last bit having been 0 and 1, respectively. With the first output always 0, both edges leaving s_0 must yield 0. If the previous input was 0, the machine moves to state s_1 and outputs 0; if it was 1, it moves to state s_2 and outputs 1. For instance, the input 11011 gives rise to 01101, as it goes to states s_0, s_2, s_2, s_1, s_2, and s_2.

Exercises
(20.1)
Design a deterministic FSA that accepts a random bit sequence of 0s and 1s as input and recognizes any bit sequence containing at least two 1s in a row (i.e., two adjacent 1s).

(20.2)
Show the state diagram of a finite-state machine with accepting states when it recognizes a pattern of 101 in its input alphabet.

(20.3)
Determine the state diagram for the nondeterministic FSA, where Table 20.5 presents its state table, the initial state is s_0, and the final state is s_3. Provide four different input strings with lengths of 3 bits, 4 bits, 5 bits, and 6 bits such that from the initial state s_0, they all go to the accepting state s_3 after also visiting s_1 and s_2.

(20.4)
Design a deterministic FSA containing an even number of 0s when its input is a string of 1s and 0s.

(20.5)
Show that the finite-state automata shown in Fig. 20.13 are equivalent.

(20.6)
Analyze the operation of a token-operated turnstile in the context of deterministic finite-state automata.

Table 20.5 State table for Exercise 20.3.

S	f : Input 0	f : Input 1
s_0	s_0, s_1	$\varnothing$
s_1	s_1	s_2, s_3
s_2	s_3	s_2
s_3	$\varnothing$	$\varnothing$

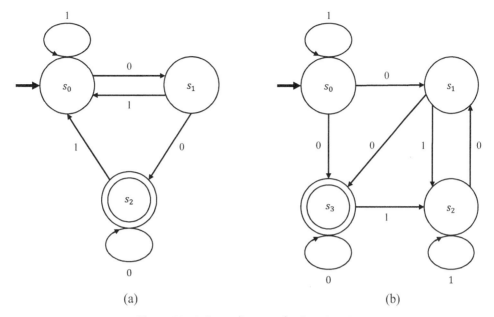

Figure 20.13 State diagrams for Exercise 20.5.

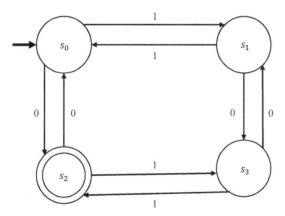

Figure 20.14 State diagram for Exercise 20.7.

(20.7)
Identify the bit patterns recognized by the deterministic FSA whose state diagram is shown in Fig. 20.14.

(20.8)
Show the state diagram of a finite-state machine whose output is a 1 when it recognizes a pattern of 101 in its input, and state how the machine works. Determine the output sequence when the input sequence is as follows: 00111010111101000.

Table 20.6 State table for Exercise 20.10.

S	f : Input 0	f : Input 1	g : Input 0	g : Input 1
s_0	s_1	s_0	1	0
s_1	s_3	s_0	1	1
s_2	s_1	s_2	0	1
s_3	s_2	s_1	0	0

(20.9)

Construct a Moore machine that gives an output of 1 whenever the number of bits in the input string read is divisible by 5 and an output of 0 otherwise.

(20.10)

Show the state diagram for the Mealy machine where Table 20.6 presents its state table, and then determine the output sequence if the input sequence is 0000.

List of Symbols

Chapter	Symbol	Meaning	Page
1	$p \equiv q$	logical equivalence of p and q	3
	$\bar{p}$	negation of p	3
	$p \wedge q$	conjunction of p and q	4
	$p \vee q$	disjunction of p and q	4
	$p \oplus q$	exclusive or of p and q	5
	$p \rightarrow q$	p implies q	7
	$p \leftrightarrow q$	p if and only if q	9
	$\mathbf{T}$	tautology	15
	$\mathbf{F}$	contradiction	15
2	$P(x)$	propositional function	21
	$\forall x P(x)$	universal quantification of $P(x)$	23
	$\exists x P(x)$	existential quantification of $P(x)$	23, 24
	$\exists ! x P(x)$ or $\exists_1 x P(x)$	uniqueness quantification of $P(x)$	24
3	$\therefore$	therefore	39
4	—	—	—
5	$x \in A$	x is an element of set A	67
	$x \notin A$	x does not belong to set A	67
	$\mid$	such that	68
	U	universal set	68
	$A = B$	equal sets	68
	$B \subseteq A$	B is a subset of A or A is a superset of B	68
	$B \subset A$	B is a proper subset of A	70
	$\mathbf{P}$	set of prime numbers	71
	$\mathbf{N}$	set of natural numbers	71
	$\mathbf{W}$	set of whole numbers	71
	$\mathbf{Z}$	set of integers	71
	$\mathbf{Q}$	set of rational numbers	71
	$\mathbf{R}$	set of real numbers	71
	$\mathbf{C}$	set of complex numbers	71
	$[a, b]$	closed interval	72
	$[a, b)$	closed-open interval	72
	$(a, b]$	open-closed interval	72
	(a, b)	open interval	72
	$\varnothing$	empty set or null set	70
	$A \cup B$	union of two sets A and B	73
	$A \cap B$	intersection of two sets A and B	73
	$A - B$	difference of sets A and B	73

Continued

Chapter	Symbol	Meaning	Page
6	A^c	complement of set A	73
	$A \oplus B$	symmetric difference of sets A and B	73
	$|A|$	cardinality of A	80
	$P(A)$	power set of A	82
	$A \times B$	Cartesian product of A and B	82
	$\mu_A(x)$	membership function of x in fuzzy set	87
	(a_{ij})	matrix with entries a_{ij}	93
	$\boldsymbol{I}_n$	identity matrix of size n	94
	$\boldsymbol{0}_{m \times n}$	zero matrix of size $m \times n$	95
	$\boldsymbol{A}^\mathrm{T}$	transpose of $\boldsymbol{A}$	95
	$\boldsymbol{A} + \boldsymbol{B}$	matrix sum of $\boldsymbol{A}$ and $\boldsymbol{B}$	96
	$k\boldsymbol{A}$	scalar multiplication of $\boldsymbol{A}$ by k	96
	$\boldsymbol{A} - \boldsymbol{B}$	matrix difference of $\boldsymbol{A}$ and $\boldsymbol{B}$	96
	$\boldsymbol{AB}$	matrix product of $\boldsymbol{A}$ and $\boldsymbol{B}$	98
	$\boldsymbol{A}^r$	rth power of square matrix $\boldsymbol{A}$	100
	$\boldsymbol{A}^{-1}$	inverse of $\boldsymbol{A}$	100
	$\boldsymbol{A} \vee \boldsymbol{B} = (a_{ij} \vee b_{ij})$	join of $\boldsymbol{A}$ and $\boldsymbol{B}$	104
	$\boldsymbol{A} \wedge \boldsymbol{B} = (a_{ij} \wedge b_{ij})$	meet of $\boldsymbol{A}$ and $\boldsymbol{B}$	104
7	$f : \boldsymbol{X} \to \boldsymbol{Y}$	function from $\boldsymbol{X}$ to $\boldsymbol{Y}$	113
	$f(x)$	value of function f at x	113
	$(f + g)(x)$	sum of functions f and g	116
	$(fg)(x)$ or $fg(x)$	product of functions f and g	116
	$|x|$	absolute value of x	118
	$\lfloor x \rfloor$	floor function of x	118
	$\lceil x \rceil$	ceiling function of x	118
	$m!$	m factorial	119
	$\binom{n}{r} = \frac{n!}{r!(n-r)!}$	n choose r	120
	a^x	exponential function of x to the base a	120
	$\log_a x$	logarithmic function of x to the base a	120
	$\approx$	almost equal to	120
	$e \approx 2.718281828459$	Euler's number (Napier's constant)	120
	$(g \circ f)(x) = g(f(x))$	composition of functions f and g	123
	f^{-1}	inverse function of f	126
8	$x + y$	Boolean sum of x and y	133, 134
	$x \cdot y$	Boolean product of x and y	133, 134
	0	zero element	131
	1	unit element	131
	$\bar{x}$	complement of Boolean variable x	131
	$\{B, +, \cdot, ', 0, 1\}$	Boolean algebra	131
	B^n	Boolean function of degree n	133
	$x \uparrow y \triangleq \overline{xy}$	x NAND (not AND) y	138
	$x \downarrow y \triangleq \overline{x + y}$	x NOR (not OR) y	138

Chapter	Symbol	Meaning	Page
9	(a, b)	ordered pair of elements a and b	155
	$(a, b) \in R$	a is related to b	155
	R^{-1}	inverse relation	156
	$\overline{R}$	complementary relation	156
	$R \cup S$	union of two relations R and S	165
	$R \cap S$	intersection of two relations R and S	165
	$R - S$	difference of two relations R and S	165
	$S \circ R$	composition of two relations R and S	166
	R^n	nth power of relation R	166
	$M_{S \circ R}$	zero-one matrix for relation $S \circ R$	167
	$M_{S \circ R} = M_R \odot M_S$	Boolean product of matrices M_R and M_S	167
	Δ_A	diagonal relation on A	168
	$[a]$	equivalence class of a	169, 170
	(A, R)	poset with set A and partial ordering R	171
	$a \preccurlyeq b$	a is less than or equal to b	171
10	$b = ac$	a is a factor of b or b is a multiple of a	178
	$a \mid b$	a divides b	178
	$a \nmid b$	a does not divide b	178
	$q = a \operatorname{div} d$	quotient when a is divided by b	178, 179
	$r = a \bmod d$	remainder when a is divided by b	178, 179
	$\gcd(a, b)$	greatest common divisor of a and b	181
	$\operatorname{lcm}(a, b)$	least common multiple of a and b	184
	$\min(x, y)$	minimum of x and y	182
	$\max(x, y)$	maximum of x and y	184
	$a \equiv b \pmod{m}$	a is congruent to b modulo m	186
	$ax \equiv b \pmod{m}$	linear congruence in one variable	188
	$\overline{a} a \equiv 1 \pmod{m}$	$\overline{a}$, an inverse of a modulo m	188
	$(a_k a_{(k-1)} \cdots \ a_1 \ a_0)_b$	base$-b$ expansion	193
11	$\{n, e\}$	public key in RSA cryptosystem	206
	$\{n, d\}$	private key in RSA cryptosystem	206
	$c = m^e \bmod n$	encryption of m in RSA cryptosystem	206
	$m = c^d \bmod n$	decryption of c in RSA cryptosystem	206
12	$f(x) = O(g(x))$	$f(x)$ is big-Oh of $g(x)$	217
	$f(x) = \Omega(g(x))$	$f(x)$ is big-Omega of $g(x)$	221
	$f(x) = \Theta(g(x))$	$f(x)$ is big-Theta of $g(x)$	221
13	—	—	
14	a_n	nth term of sequence	249
	$\sum_{i=k}^{m} a_i$	sum of $a_k, a_{k+1}, \ldots, a_m$	250
	$\prod_{j=k}^{m} a_j$	product of $a_k, a_{k+1}, \ldots, a_m$	250
15	$\cong$	approximately equal to	281

Continued

Chapter	Symbol	Meaning	Page
16	$\varnothing$	empty (null) event	286
	$P(A)$	probability of A	287
	$P(A,\, B)$	joint probability of A and B	289
	$P(B|A)$	conditional probability of A given B	289
17	$F_X(x)$	cumulative distribution function of X	308
	$p_X(x)$	probability mass function of X	308
	$E[X]$ or μ_X	expected value of X	310
	σ_X^2	variance of X	313
	σ_X	standard deviation of X	313
	$F_X(x|B)$	conditional cdf of X given B	315
	$p_X(x|B)$	conditional pmf of X given B	316
18	$G = (V, E)$	graph consists of vertex set V and edge set E	327
	$|V|$	order of graph (number of its vertices)	327
	$|E|$	size of graph (number of its edges)	327
	$\{u, v\}$	undirected edge	327
	(u, v)	directed edge	328
	$\deg(u)$	number of edges incident with u	328
	$\deg^-(v)$	number of edges with v as their terminal vertex	328
	$\deg^+(u)$	number of edges with u as their initial vertex	328
	$G_1 \cup G_2$	graph union of two simple graphs G_1 and G_2	330
	$G_1 \cap G_2$	graph intersection of two simple graphs G_1 and G_2	330
	K_n	complete graph (or mesh topology)	331
	C_n	cycle graph (or ring topology)	332
	S_{n+1}	star graph	332
	W_{n+1}	wheel graph (or hybrid topology)	333
	L_n	linear graph (or bus topology)	333
	$G_{m,n}$	m by n grid graph	333
	Q_n	n-dimensional hypercube or n-cube	333
19	$+\, ab$	addition of a and b in prefix notation	355
	$a + b$	addition of a and b in infix notation	355, 356
	$ab\, +$	addition of a and b in postfix notation	356
20	$M = (S, I, f, s_0, F)$	finite-state machine with no output	374
	$M = (S, I, O, f, g, s_0)$	finite-state machine with output	374

Glossary of Terms

Chapter 1—Propositional Logic

Biconditional statement $p \leftrightarrow q$: It is true when p and q have the same truth values and is false otherwise.

Compound proposition: A proposition that is composed of more than one simple proposition.

Conditional statement $p \rightarrow q$: It is false when the premise p is true but the conclusion q is false and is true otherwise.

Conjunction of two propositions: It is true when both propositions are true and is false otherwise.

Contingency: A compound proposition that is neither a tautology nor a contradiction.

Contradiction: A compound proposition that is always false.

Contrapositive statement of $p \rightarrow q$: It is the conditional statement $\bar{q} \rightarrow \bar{p}$.

Converse statement of $p \rightarrow q$: It is the conditional statement $q \rightarrow p$.

Disjunction of two propositions: It is false when both propositions are false and is true otherwise.

Dual of a compound proposition: It is obtained by replacing each $\vee$ by $\wedge$, each $\wedge$ by $\vee$, each T by F, and each F by T.

Exclusive or of two propositions: It is true when exactly one of them is true.

Inverse statement of $p \rightarrow q$: It is the conditional statement $\bar{p} \rightarrow \bar{q}$.

Logic: Formal principles of reasoning, strict criteria of validity, and necessary rules of thought.

Logic puzzle: A problem that can be solved through deductive reasoning.

Logical operators: Connectives used to combine simple propositions.

Logically equivalent: When two compound propositions have identical truth tables.

Negation of proposition: It is not the truth value of the proposition.

Proposition: A declarative statement, which is either true or false, but not both.

Propositional form: An expression consisting of propositional variables and logical operators.

Propositional logic: Logic that deals with propositions.

Propositional variable: A variable that represents a proposition.

Satisfiable compound proposition: When there is at least one assignment of truth values to its variables for which the compound proposition is then true.

Simple proposition: A proposition that cannot be broken down into simpler propositions.

Tautology: A compound proposition that is always true.

Truth table: A table presenting all possible truth values of propositions.

Chapter 2—Predicate Logic

Binary predicate: It is a predicate with two variables.

Bound variable: A variable that is quantified.

Domain (or universe) of discourse: It is the set of all values of a variable in a propositional function that can replace it.

Existential quantification: It indicates a predicate is true for at least one value of a variable in a given domain.

Free variable: A variable that is not bound by a quantifier, nor is it equal to a particular value.

Nested quantifier: When a quantifier is within the scope of another.

Predicate: A property that the subject of the statement can have or gives information about the subject.

Predicate logic: When logic uses predicates and quantified variables.

Propositional function: A statement with one or more variables that becomes a proposition when each variable is assigned a value.

Quantifier: A word that reveals for how many elements a given predicate is true.

Scope of a quantifier: A portion of a statement where the quantifier binds its variable.

Universal quantification: It indicates that a predicate is true for all values of a variable in a given domain.

Chapter 3—Rules of Inference

Abduction: A reasoning in which the major premise is certain, but the minor premise and therefore the conclusion is only probable.

Argument: It is a sequence of propositional statements.

Argument form: It is a sequence of compound propositions involving propositional variables

Converse error: An argument that when an implication and its conclusion are both true, then its hypothesis is true.

Fallacy: An invalid argument where it resembles a rule of inference, but it is based on a contingency rather than a tautology.

Inverse error: An argument that when an implication and the negation of its hypothesis are both true, then the negation of its conclusion is true.

Rule of inference: A valid argument form that can be used in the demonstration that arguments are valid.

Sound argument: When an argument is valid and all of its premises are true.

Valid argument: It is a sequence of propositions where the truth of all the premises implies the truth of the conclusion.

Valid argument form: The conclusion is true if all the premises are true.

Chapter 4—Proof Methods

Axiom: A self-evident true statement that is accepted on its intrinsic merit without proof.

Conjecture: A statement that is being proposed to be a true statement but not proven yet.

Constructive existence proof: A proof that an element with a specified property exists that explicitly finds such an element.

Corollary: A proposition that can be proven as an immediate consequence of some other theorems.

Definition: A statement expressing the essential nature of a concept.

Direct proof: An implication is constructed with the assumption that the premise is true and a series of intermediate implications leads to the conclusion be true.

Lemma: A less important theorem that can help prove a more important theorem.

Mathematical proof: An inferential argument for a mathematical statement showing that the stated assumptions methodically and logically lead to guarantee the conclusion.

Nonconstructive existence proof: A proof that an element with a specified property exists that does not explicitly find such an element.

Proof: A sequence of logically valid statements to demonstrate the validity of some precise statement.

Proof by cases: $(p \rightarrow q) \leftrightarrow ((p_1 \vee \ldots \vee p_n) \rightarrow q) \leftrightarrow ((p_1 \rightarrow q) \wedge \ldots \wedge (p_n \rightarrow q))$

Proof by contradiction: A conditional statement is true by showing if the premise is false, then it leads to a contradiction.

Proof by contraposition: A conditional statement is true by showing the premise must be false when the conclusion is false.

Proof by counterexample: An example in the domain of discourse for which the hypothesis is true and the conclusion is false.

Proof by exhaustion: A proof that establishes a result by checking a list of all possible cases.

Proof of a disjunction: Proving $p \rightarrow (q \vee r)$ by proving either $(p \wedge \overline{q}) \rightarrow r$ or $(p \wedge \overline{r}) \rightarrow q$ is true.

Proofs of equivalence: $(p_1 \leftrightarrow p_2 \leftrightarrow \cdots \leftrightarrow p_n) \leftrightarrow ((p_1 \rightarrow p_2) \wedge \ldots \wedge (p_n \rightarrow p_n))$

Theorem: A mathematical statement that can be shown (proven) to be true.

Trivial proof: If the conclusion can be shown to be true, the implication is true by default.

Uniqueness proof: A proof that there is exactly one element satisfying a specified property.

Vacuous proof: If the hypothesis can be shown to be false, the implication is true by default.

Chapter 5—Sets

Axiomatic set theory: A theory based on the rules of inference provided by formal logic.

Cardinality of a set: The number of elements in the set.

Cartesian product of A and B: The set of all ordered pairs (a, b), where $a \in A$ and $b \in B$.

Complement of A: The set of elements in the universal set that are not in A.

Difference of A and B: The set containing those elements that are in A but not in B.

Difference of two multisets: A multiset in which the multiplicity of an element is the difference between the multiplicities of the element in the two multisets unless the difference is negative in which case the multiplicity is 0.

Disjoint sets: The sets that have no common elements.

Empty (null) set: The set with no elements.

Finite set: A set whose number of elements is a nonnegative integer.

Fundamental product of X_1, X_2, ..., X_n: A set defined as $Y_1 \cap Y_2 \cap ... \cap X_n$, where Y_i is either the set X_i or its complement, for $i = 1, 2, ..., n$.

Fuzzy set: A set where each member of the set is defined by the degree of fuzziness (membership).

Infinite set: A set that is not finite.

Intersection of two multisets: A multiset in which the multiplicity of an element is the minimum of its multiplicities in those two multisets.

Intersection of two sets: The set containing those elements that are in both sets.

Membership function in a fuzzy set: A mapping from the universal set U to the unit interval $[0, 1]$ to show the degree of membership for each member.

Membership table: A table displaying the membership of elements in sets.

Multiplicity: The number of occurrences for each element in a multiset.

Multiset (bag): An unordered collection of objects where an object can occur as a member of a set more than once.

Naïve set theory: A theory based on intuitive notion of an object and a set as defined informally in natural language.

Paradox: A logical inconsistency.

Power set of a set: The set of all subsets of the set.

Set: An unordered collection of distinct objects that are called elements or members of the set.

Set B is a proper subset of set A: Every member of B is also a member of A, but there is at least one element of A that is not an element of B.

Set B is a subset of set A: Every member of B is also a member of A.

Set builder notation: A method through which some property held only by all members of the set is clearly and completely described.

Set identity: An equality between two set expressions that is true for all elements of the sets involved in the identity.

Set roster method: A method by which all the elements of the set are listed.

Singleton (unit) set: A set with one element.

Sum of two multisets: A multiset in which the multiplicity of an element is the sum of multiplicities in those two multisets.

Symmetric difference of A and B: The set of elements that belong to A or B, but not to both.

Union of two multisets: A multiset in which the multiplicity of an element is the maximum of its multiplicities in those two multisets.

Union of two sets: The set containing those elements that are in at least one of the two sets.

Universal set: A set that includes all elements in a given setting as well as every set under consideration.

Venn diagram: A group of simple closed curves arranged in the plane to visually illustrate collections of sets and their logical relationships.

Chapter 6—Matrices

Boolean product of A and B: It is denoted by $A \odot B$, where we have $A \odot B = C \to (c_{ik}) = ((a_{i1} \wedge b_{1k}) \vee \ldots \vee (a_{in} \wedge b_{nk}))\ i = 1, \ldots m\ \ \&\ \ k = 1, \ldots, r$.

Column vector: A matrix with only one column.

Columns of a matrix: Vertical sets of numbers in a matrix.

Diagonal matrix: A matrix whose all entries off its main diagonal are zero.

Equal matrices: Two matrices of the same size and the corresponding entries in every position in the two matrices are equal.

Identity matrix: A square matrix with 1s on the main diagonal and 0s off the main diagonal.

Inverse of a matrix A: It is a matrix A^{-1}, where the product of A and A^{-1} is the identity matrix.

Join of A and B: It is denoted by $A \vee B = (a_{ij} \vee b_{ij})$.

Matrix: A rectangular array of numbers.

Matrix addition: The sum of any two matrices of the same size is obtained by adding entries in the corresponding positions.

Matrix multiplication: If the matrix A is an $m \times n$ matrix and the matrix B is an $n \times r$ matrix, then the product of A and B is the $m \times r$ matrix whose entry in the ith row and the kth column is the sum of the product of the corresponding entries from the ith row of A and the kth column of B.

Matrix subtraction: The difference between any two matrices of the same size is obtained by subtracting entries in the corresponding positions.

Meet of A and B: It is denoted by $A \wedge B = (a_{ij} \wedge b_{ij})$.

Null (or zero) matrix: A matrix whose entries are all 0s.

Row vector: A matrix with only one row.

Rows of matrix: Horizontal sets of numbers in a matrix.

Scalar: A quantity described by a real number.

Size of a matrix: It is represented by its number of rows and number of columns.

Square matrix: A matrix with the same number of rows as columns.

Trace: The sum of entries on the main diagonal of a square matrix.

Transpose of a matrix: When the rows and columns of the matrix are interchanged.

Zero-one (Boolean or logical) matrix: A matrix whose entries are either 0 or 1 and subject to the Boolean operations.

$m \times n$ matrix: A matrix with m rows and n columns, which has a total of $m \times n$ entries.

Chapter 7—Functions

Absolute-value function: $f(x) = x$, if $x \geq 0$, and $f(x) = -x$, if $x < 0$.

Ceiling (the least integer) function: It assigns to the real number x the smallest integer that is greater than or equal to x.

Codomain of the function from X to Y: It is the set Y.

Composition of the functions f and g: A function that assigns $g(f(x))$ to x.

Decreasing function $f(x)$: If $f(x_1) > f(x_2)$, whenever $x_1 < x_2$.

Domain of the function from X to Y: It is the set X.

Euler's number (Napier's constant): $e = \lim\limits_{n \to \infty} \left(1 + \frac{1}{n}\right)^n \approx 2.718281828459$.

Exponential function f: $R \to R^+$: It is defined by $f(x) = a^x$, where $a \in R^+$ and $a \neq 1$.

Floor (the greatest integer) function: It assigns to the real number x the largest integer that is less than or equal to x.

Function from X to Y: A relation from X to Y, where every element in X is related to some element in Y, and no element in X is related to more than one element in Y.

Increasing function $f(x)$: If $f(x_1) < f(x_2)$, whenever $x_1 < x_2$.

Integer-valued function: A function whose codomain is the set of integers.

Inverse function: A function that assigns to an element belonging to Y the unique element in X.

Logarithmic function f: $R^+ \to R$: It is defined by $f(x) = \log_a x$, where $a \in R^+$ and $a \neq 1$.

Nondecreasing function $f(x)$: If $f(x_1) \leq f(x_2)$ whenever $x_1 < x_2$.

Nonincreasing function $f(x)$: If $f(x_1) \geq f(x_2)$, whenever $x_1 < x_2$.

One-to-one correspondence, bijection: A function that is both one-to-one and onto.

One-to-one function, injection: $f(x_1) = f(x_2)$ implies that $x_1 = x_2$ for all elements in X.

Onto function, surjection: The range and codomain of the function are the same.

Piecewise-defined function: A function defined by more than one formula.

Range of the function from X to Y: It is the set of all possible values of the function.

Real-valued function: A function whose codomain is the set of real numbers.

Unit step function: $f(x) = 0$, if $x < 0$, and $f(x) = 1$, if $x \geq 0$.

Chapter 8—Boolean Algebra

AND gate: A device that accepts the values of two or more Boolean variables as input and produces their Boolean product as output.

Binary expression simplification rule: Using $ef + \bar{e}f = f$, where e and f are binary expressions, iteratively to reduce an expression into a simpler, but equivalent, expression.

Bits: The set of binary digits $\{0, 1\}$.

Boolean algebra: A set B with two binary operations $\vee$ and $\wedge$, elements 0 and 1, and a complementation operator that satisfies the identity, complement, commutative, and distributive laws.

Boolean expression: It consists of Boolean variables and Boolean operators.

Boolean function of degree n: A function from B^n to B where $B = \{0, 1\}$.

Boolean product of x and y: It has the value 1 when both x and y have the value 1 and the value 0 otherwise.

Boolean sum x and y: It has the value 1 when either x or y has, or both have, the value 1, and 0 otherwise.

Boolean variable: A variable that assumes only the values 0 and 1.

Combinational circuit: It is made up of different types of logic gates, where its present output is a function of only the present input; thus it has no memory capabilities.

Complement of x: It is 1 when x is 0, and 0 when x is 1.

Don't care condition: A combination of input values for a circuit that is not possible or never occurs.

Dual of a Boolean expression: The expression obtained by interchanging sums and products and interchanging 0s and 1s.

Functionally complete: A set of Boolean operators if every Boolean function can be represented using these operators.

Gates: Basic elements of logic circuits.

Inverter: A device that accepts the value of a Boolean variable as input and produces the complement of the input.

K-map for n variables: A rectangle divided into 2^n cells where each cell represents a minterm of the variables.

Literal of the Boolean variable x: It is either x or $\bar{x}$.

Minimization of a Boolean function: Representing a Boolean function with the fewest products of literals such that these products contain the fewest literals possible among all sums of products.

Minterm of the Boolean variables $x_1, x_2, ..., x_n$: It is the Boolean product $y_1, y_2, ..., y_n$, where $y_i = x_i$ or $y_i = \bar{x}_i$, and $i = 1, 2, ..., n$.

OR gate: A device that accepts the values of two or more Boolean variables as input and produces their Boolean sum as output.

Principle of duality: The duals of both sides of an identity are another identity.

Sum-of-products expansion (disjunctive normal form): The representation of a Boolean function as a disjunction of minterms.

Unit element: The symbol 1.

Zero element: The symbol 0.

x NAND y: The expression that has the value 0 when both x and y have the value 1 and the value 1 otherwise.

x NOR y: The expression that has the value 0 when either x or y has or both have the value 1 and the value 0 otherwise.

Chapter 9—Relations

Antisymmetric: A relation R on A if $a = b$ whenever $(a, b) \in R$ and $(b, a) \in R$.

Binary relation from A to B: The relation R is a subset of $A \times B$, where the set A is called the domain of the relation and the set B is called the range of the relation.

Comparable: Elements a and b in the poset (A, R) and if either $a \preccurlyeq b$ or $b \preccurlyeq a$.

Complementary relation: $\bar{R} = \{(a, b) | (a, b) \notin R\}$.

Composite key: The Cartesian product of domains of an n-ary relation such that an n-tuple is uniquely determined by its values in these domains.

Difference of two relations R and S: $R - S = \{(a, b) | (a, b) \in R \text{ and } (a, b) \notin S\}$.

Equivalence relation: A reflexive, symmetric, and transitive relation.

Hasse diagram: A graphical representation of a poset where loops and all edges resulting from the transitive property are not shown, and the direction of the edges is indicated by the position of the vertices.

Incomparable: Elements in a poset that are not comparable.

Intersection of two relations R and S: $R \cap S = \{(a, b) | (a, b) \in R \text{ and } (a, b) \in S\}$.

Inverse relation: $R^{-1} = \{(b, a) | (a, b) \in R\}$.

One-to-one binary relation from A and B: If no element of B appears as a second coordinate in more than one ordered pair in R.

Onto relation binary relation from A and B: If every element of B appears as a second coordinate in at least one ordered pair in R.

Ordered pair of elements a and b: (a, b).

Partial ordering: A relation that is reflexive, antisymmetric, and transitive.

Poset (S, R): A set S and a partial ordering R on this set.

Primary key: A domain of an n-ary relation such that an n-tuple is uniquely determined by its value for this domain.

Reflexive: A relation R on A if $(a, a) \in R$ for all $a \in A$.

Reflexive closure of relation R: The smallest relation R_r, such that $R \subset R_r$ and R_r is reflexive on the set A.

Relation on A: A binary relation from A to itself, that is a subset of $A \times A$.

Relational data model: A model for representing databases using n-ary relations.

Symmetric: A relation R on A if $(b, a) \in R$ whenever $(a, b) \in R$.

Symmetric closure of relation R: The smallest relation R_s, such that $R \subset R_s$ and R_s is symmetric on the set A.

Total (linear) order relation: If R is a partial order relation on a set A, and every pair of elements in A is comparable.

Transitive: A relation R on A is transitive if $(a, b) \in R$ and $(b, c) \in R$ implies that $(a, c) \in R$.

Transitive closure of relation R: The smallest relation R_t, such that $R \subset R_t$ and R_t is transitive on the set A with n elements.

Union of two relations R and S: $R \cup S = \{(a, b) | (a, b) \in R$ and/or $(a, b) \in S\}$.

Zero-one matrix of relation R: Each entry that belongs to the set of the ordered pairs in the relation is set to 1; otherwise it is set to 0.

n-ary relation on $A_1, A_2, ..., A_n$: A subset of $A_1 \times A_2 \times ... \times A_n$.

Chapter 10—Number Theory

Bezout coefficients of integers a and b: Integers s and t such that the **Bezout's identity** $\gcd(a, b) = sa + tb$ holds.

Binary representation: The base 2 representation of an integer.

Composite: An integer that is not prime.

Decimal representation: The base 10 representation of an integer.

Divisibility test: A quick way to determine whether an integer is divisible by a smaller integer without performing the division.

Division (quotient–remainder) theorem: $a = dq + r$, where a (**dividend**), d (**divisor**), q (**quotient**), and r (**remainder**) are integers with $0 \le r < d$.

Euclid's theorem: There are infinitely many primes.

Euclidean algorithm: An algorithm determining the great common divisor of two integers through successive application of the division algorithm.

Euler's theorem: For every a and n that are relatively prime, we have $a^{\varphi(n)} \equiv 1 \pmod{n}$, where $\varphi(n)$ is the Euler's totient function.

Euler's totient function: The number of positive integers less than a number and relatively prime to that number.

Fermat's last theorem: The equation $x^n + y^n = z^n$, where x, y, and z are integers and $xyz \ne 0$, has no solutions for an integer $n > 2$.

Fermat's little theorem: If p is prime and a is an integer not divisible by p, then $a^{p-1} \equiv 1 \pmod{p}$.

Fundamental theorem of arithmetic: Every integer greater than 1 is either prime or the product of two or more primes.

Goldbach's conjecture: Every even integer greater than two is the sum of two primes.

Greatest common divisor of two integers: The largest integer that divides both of them.

Hashing function: It assigns memory location $h(k)$ to the record that has k as its key.

Hexadecimal representation: The base 16 representation of an integer.

Hindu-Arabic numeral system: Numerals are represented by ten distinct symbols.

Inverse of a modulo m: A unique integer $\bar{a}$, such that $\bar{a}a \equiv 1 \pmod{m}$.

Least common multiple of two integers: The smallest positive integer that is divisible by both integers.

Numeral: Any symbol used to represent a number.

Octal representation: The base 8 representation of an integer.

One's complement of a binary number: Invert each bit from 1 to 0 and from 0 to 1.

Prime: An integer greater than 1 which is divisible only by 1 and itself.

Relatively prime integers: Integers whose greatest common divisor is 1.

Roman numeral system: Numerals are represented by the seven distinct letters $I = 1, V = 5, X = 10, L = 50, C = 100, D = 500$, and $M = 1000$.

Twin prime conjecture: There are infinitely many twin primes (pairs of primes that differ by 2).

Two's complement of a binary number: Obtain the one's complement of the binary number and then add 1 to the least significant bit.

a divides b: a is a **factor** of b, a is a **divisor** of b, b is **divisible** by a, or b is a **multiple** of a.

a is congruent to b modulo m: $a - b$ is divisible by m.

Chapter 11—Cryptography

Affine cipher: The numerical equivalent of each letter is shifted by the integer b to encrypt the plaintext letter p using $f(p) = (mp + b) \bmod n$, where $1 \leq p \leq n$, b, m, and n are all integers, and $\gcd(m, n) = 1$.

Block cipher: A cipher that encrypts blocks of characters of a fixed size.

Cipher: The algorithm used for encryption and decryption.

Classical cryptography: Symbols, characters, letters, and digits are directly manipulated with the sole goal to provide secrecy through obscurity.

Decryption: The process of returning a secret message to its original form.

Encryption: The process of making a message secret.

Encryption key: A secret key (a number) that the cipher operates on.

Modern cryptography: It operates on binary bit sequences and relies on publicly known algorithms for encoding the message, and secrecy is obtained through a secrete key which is used as the seed for the algorithms.

Private key cryptography: Encryption where both encryption keys and decryption keys must be kept secret.

Public key cryptography: Encryption where encryption keys are public knowledge, but decryption keys are kept secret.

RSA: The public key $\{n, e\}$ is used to encrypt the plaintext m by using $c = m^e \bmod n$, and the private key $\{n, d\}$ is used to decrypt the cyphertext c by using $m = c^d \bmod n$.

Shift cipher: The numerical equivalent of each letter is shifted by the integer b to encrypt the plaintext letter p using $f(p) = (p + b) \bmod n$, where $1 \leq p \leq n$, b, and n are all integers.

Transposition cipher: A cipher in which the order of letters in a block of letters is rearranged (reordered) according to a fixed permutation.

Chapter 12—Algorithms

Algorithm: A finite unambiguous sequence of steps that involves the repetition of an operation for performing a task in a finite amount of time.

Algorithmic efficiency: A property of an algorithm that relates to the amount of computational resources used by the algorithm.

Average-case time complexity: The average amount of time required for an algorithm to solve a problem of a given size.

Backtracking algorithm: An algorithm that incrementally builds candidates to the solutions and abandons a candidate as soon as it determines that the candidate cannot possibly be a part of a valid solution.

Brute-force algorithm: An algorithm that iterates all possible solutions to search for one or more than one solution that may solve a problem without any regard to the heavy computational requirements.

Divide-and-conquer algorithm: An algorithm that works by recursively breaking down a problem into subproblems of the same or related type, until these become simple enough to be solved easily.

Dynamic programming: An algorithm that can be effectively used for solving a complex problem by recursively breaking down the problem.

Greedy algorithm: An algorithm that makes the optimal choice at each step as it attempts to find the minimum or maximum value of some parameter.

Halting problem: A procedure that takes as input a computer program and the input to the program and determines whether the program will ultimately stop or continue to run forever.

Probabilistic algorithm: An algorithm that makes some random choices at some steps, which may lead to different output in much fewer steps, but with a tiny probability that the final answer may not be correct.

Search algorithm: An algorithm that locates an element in a list.

Sorting algorithm: An algorithm that puts elements of a list in a certain order.

Space complexity: The maximum amount of computer memory needed in the execution of an algorithm.

Time complexity: The number of key operations using the size of the input as its argument.

Worst-case time complexity: The greatest amount of time required for an algorithm to solve a problem of a given size.

$f(x)$ **is** $O(g(x))$: There are real constants C and k such that $|f(x)| \leq C|g(x)|$ whenever $x > k$.

$f(x)$ **is** $\Theta(g(x))$: There are real constants C_1, C_2, and k such that $C_1|g(x)| \leq |f(x)| \leq C_2|g(x)|$ whenever $x > k$.

$f(x)$ **is** $\Omega(g(x))$: There are real constants C and k such that $|f(x)| \geq C|g(x)|$ whenever $x > k$.

Chapter 13—Induction

Deductive reasoning: The process of concluding that something must be true because it is a specific case of a general principle that is already known to be true.

Inductive reasoning: The process of reasoning that a general principle is true because the special cases are true.

Principle of mathematical induction: The statement $\forall n P(n)$ is true if $P(1)$ is true and $\forall k(P(k) \rightarrow P(k+1))$ is true.

Strong induction: The statement $\forall n P(n)$ is true if $P(1)$ is true and $\forall k((P(1) \wedge P(2) \wedge \ldots \wedge P(k-1) \wedge P(k)) \rightarrow P(k+1))$ is true.

Well-ordering principle: Every nonempty set of positive integers has a least element.

Chapter 14—Recursion

Arithmetic progression: A sequence with the general term $a_n = a + nd$, where the initial term a and the common difference d are real numbers, and $n > 0$ is an integer.

Characteristic roots of a linear homogeneous recurrence relation with constant coefficients: The roots of the polynomial associated with a linear homogeneous recurrence relation with constant coefficients.

Generating function: It is the infinite series $G(z) \triangleq a_0 + a_1 z + a_2 z^2 + \ldots = \sum_{n=0}^{\infty} a_n z^n$, where the sequence $a_0, a_1, a_2 \ldots$ are real numbers.

Geometric progression: A sequence with the general term $a_n = ar^n$, where the initial term a and the common ratio r are real numbers, and $n > 0$ is an integer.

Initial conditions for a recurrence relation: The values of the terms of a sequence satisfying the recurrence relation before the relation takes effect.

Iteration: A procedure based on the repeated use of operation in a loop.

Iterative algorithm: It evaluates the value of a function at the base cases and successively applies the recursive definition to find values of the function at larger integers.

Linear homogeneous recurrence relation with constant coefficients: A recurrence relation that expresses the terms of a sequence, except for initial terms, as a linear combination of previous terms.

Linear nonhomogeneous recurrence relation with constant coefficients: A recurrence relation that expresses the terms of a sequence, except for initial terms, as a linear combination of previous terms plus a function.

Recurrence relation: A formula expressing a term of a sequence as a function of prior terms in the sequence.

Recursion: The process of defining a problem or the solution to a problem in terms of a simpler version of itself.

Recursive algorithm: It evaluates the value of a function at a positive integer in terms of the values of the function at smaller integers.

Recursively defined function: It refers to itself and its domain is a subset of the set of positive or nonnegative integers.

Sequence: A function whose domain is either all the integers between two given integers or all the integers greater than or equal to a given integer.

Chapter 15—Counting Methods

Binomial coefficient (m choose r): $\binom{m}{r} \triangleq \frac{m!}{r!(m-r)!}$.

Combination: An unordered selection of distinguishable objects to make a group.

Counting method: A way to determine the total number of equally likely outcomes in a random experiment, without actually listing the outcomes.

Fundamental principle of counting (product rule for counting): When a task can be divided into a sequence of k independent subtasks, there is a total of $n_1 \times n_2 \times \ldots \times n_k$ distinct ways to carry out the task.

Generalized pigeonhole principle: If k and n are positive integers and k pigeonholes are occupied by $m = kn + 1$ or more pigeons, then at least one pigeonhole is occupied by $n + 1$ or more pigeons.

Permutation: An ordered arrangement of distinguishable objects to make a list.

Pigeonhole principle: If $k > 1$ is an integer and k pigeonholes are occupied by $m = k + 1$ or more pigeons, then at least one pigeonhole is occupied by more than one pigeon.

Selection with replacement (repetition, substitution): An object, once selected, is returned and thus available for future selections.

Selection without replacement (repetition, substitution): An object, once selected, is not available for future selections.

Subtraction rule for counting (principle of inclusion-exclusion): When a task can be accomplished in k sets of ways, the number of distinct ways to accomplish the task is $n_1 + n_2 + \ldots + n_k$ minus the number of common ways that have been overcounted.

Sum (or addition) rule for counting: When a task can be done in k mutually exclusive sets of ways, there is a total of $n_1 + n_2 + \ldots + n_k$ distinct ways to carry out the task.

Tree diagram: A tree structure to keep systematic track of all possibilities in cases in which events occur in sequence but in a finite number of ways.

Chapter 16—Discrete Probability

A posteriori probability: The probability of an event after the experiment has been performed.

A priori probability: The probability of an event before the experiment is performed.

Axiom I of probability: The probability of an event is nonnegative.

Axiom II of probability: The probability of all possible outcomes is one.

Axiom III of probability: The total probability of a number of nonoverlapping events is the sum of the individual probabilities.

Bayes' rule: When one conditional probability is given, but the reversed conditional probability is required
$$\left(P(B_1|A) = \frac{P(A,B_1)}{P(A)} = \frac{P(A|B_1)\ P(B_1)}{P(A|B_1)\ P(B_1) + P(A|B_2)\ P(B_2) + \cdots + P(A|B_n)\ P(B_n)} \right).$$

Complement of an event: All outcomes that are not included in the event.

Conditional probability: The probability of an event when it is known that another event has occurred.

Discrete sample space: A sample space that is countable.

Disjoint (mutually exclusive) events: If the occurrence of one event excludes the occurrence of the other.

Event: A collection of one or more than one outcome.

Independent trial: The outcome of a trial is independent of the outcomes of the past and future trials.

Intersection of two events (joint events): The set of all outcomes that are in both events.

Joint probability of events: The probability that the events simultaneously occur.

Law of total probability: The probability of an event expressed as a combination of the probabilities of the mutually exclusive events that form the partition of the sample space ($P(A) = P(A \cap B_1) + \cdots + P(A \cap B_n) = P(A|B_1)\ P(B_1) + \cdots + P(A|B_n)\ P(B_n)$).

Marginal probability: The probability of the occurrence of a single event, irrespective of the probabilities of other events.

Mutually exclusive events: When the joint probability of events is zero.

Null event: An event that never occurs.

Outcome: The end result of an experiment.

Partitioning a sample space: Dividing the sample space into mutually exclusive events.

Probability: A numerical measure of how likely an event is to occur or be the case.

Random experiment: The outcome is always unpredictable and the conditions under which it is performed cannot be known in advance.

Sample points: The outcomes of a random experiment that cannot occur simultaneously.

Sample space: The set of all possible outcomes of an experiment.

Statistically independent events: When the joint probability of two events is equal to the product of individual probabilities.

Sure event: An event that always occurs.

Trial: A repetition of an experiment.

Union of two events: The set of all outcomes that are in either one of them or in both of them.

Chapter 17—Discrete Random Variables

Bernoulli distribution: It takes the value of 1 with probability p and the value of 0 with probability $1 - p$.

Binomial distribution: The number of times 1 occurs in n independent Bernoulli trials, where each occurrence of 1 is assumed to have probability p.

Chebyshev's inequality: The probability of a large deviation from the expected value is inversely proportional to the square of the deviation.

Cumulative distribution function of a random variable: The probability that the random variable is no larger than x.

Discrete uniform distribution: It occurs when outcomes are equally likely.

Domain of a random variable: The sample space of the random variable.

Expected value of a discrete random variable: It is obtained by multiplying each possible value by its respective probability and then summing these products over all the values that have nonzero probabilities.

Geometric distribution: In a sequence of independent Bernoulli trials with a success probability p, the random variable that denotes the number of trials performed until the first success occurs.

Hypergeometric distribution: With a finite population of N items of which K possess a certain attribute, it describes the probability that a sample of n items, without replacement, is selected of which x possess the attribute.

Markov's inequality: An upper bound on the probability that a value of a nonnegative random variable is greater than or equal to some positive constant.

Median of a random variable: The particular value for which the sum of the probabilities of all values greater than the median and the sum of the probabilities of all values less than the median are equal.

Mode of a random variable: The value of the random variable that occurs most often.

Pascal distribution: It represents the number of Bernoulli trials that take place until one of the two outcomes is observed a certain number of times.

Poisson distribution: It represents the number of occurrences of events independently occurring within certain specified boundaries.

Probability mass function of a discrete random variable: The set of the probabilities only at the values belonging to the range of the random variable.

Random variable: A deterministic function that assigns a real number to each outcome in the sample space.

Range of a random variable: The set of all values taken on by the random variable.

Standard deviation of a random variable: The square root of its variance.

Variance of a random variable: The mean square of the difference between a random variable and its mean (expected value).

Chapter 18—Graphs

Adjacency matrix: A matrix representing a graph using the adjacency of vertices.

Adjacent: Two vertices are adjacent if there is an edge between them.

Bipartite graph: A graph with a vertex set that can be partitioned into two subsets so that each edge connects a vertex in one subset and a vertex in the other subset.

Circuit: A closed trail that begins and ends on the same vertex, does not contain a repeated edge, but may have repeated vertices.

Closed walk: A walk when the starting vertex is the same as the ending vertex.

Complement (inverse) of a simple graph: A graph that has the same vertices as the simple graph and has edges joining every pair of vertices that are not joined in the simple graph.

Complete bipartite graph: The graph with vertex set partitioned into two subsets with two vertices connected by an edge if and only if one vertex is in one subset and the other vertex is in the second subset.

Complete graph: An undirected graph where each pair of vertices is connected by an edge.

Connected graph: An undirected graph with the property that there is a path between every pair of vertices.

Cycle: A path that begins and ends on the same vertex and no other vertices are repeated.

Degree of the vertex v in an undirected graph: The number of edges incident with v with loops counted twice.

Dijkstra's algorithm: A procedure for finding the shortest path between two vertices in a weighted graph.

Directed edge: An edge associated to an ordered pair (u, v), where u and v are vertices.

Directed graph: A set of vertices together with a set of directed edges, each of which is associated with an ordered pair of vertices.

Directed multigraph: A graph with directed edges that may contain multiple directed edges.

Euler circuit: A circuit that contains every edge of a graph exactly once.

Graph intersection of two simple graphs: A graph whose vertices are in both graphs and edges are in both graphs.

Graph union of two simple graphs: A graph whose vertices are in either or both graphs and edges are in either or both graphs.

Hamilton circuit: A circuit in a graph that passes through each vertex exactly once.

Handshaking theorem: The sum of the degrees of the vertices is twice the number of edges.

Incidence matrix: A matrix representing a graph using the incidence of edges and vertices.

Incident: An edge is incident with a vertex if the vertex is an endpoint of that edge.

In-degree of the vertex v in a graph with directed edges: The number of edges with v as their terminal vertex.

Invariant for graph isomorphism: A property that isomorphic graphs either both have or both do not have.

Isolated vertex: A vertex of degree zero.

Isomorphic graphs: Two simple graphs with the same structure and hence the same properties, where there is a one-to-one correspondence between vertices of two isomorphic graphs preserving the adjacency relationship.

Loop: An edge connecting a vertex with itself.

Multigraph: An undirected graph that may contain multiple edges but no loops.

Multiple directed edges: Distinct directed edges associated with the same ordered pair (u, v), where u and v are vertices.

Multiple edges: Distinct edges connecting the same vertices.

Open walk: A walk when the starting vertex is not the same as the ending vertex.

Order: The number of vertices of a graph.

Out-degree of the vertex v in a graph with directed edges: The number of edges with v as their initial vertex.

Path: A trail that does not include any vertex twice.

Pendant vertex: A vertex of degree one.

Pseudograph: An undirected graph that may contain multiple edges and loops.

Regular graph: A graph where all vertices have the same degree.

Shortest-path problem: The problem of determining the path in a weighted graph such that the sum of the weights of the edges in the path is a minimum over all paths between specified vertices.

Simple directed graph: A directed graph without loops or multiple directed edges.

Simple graph: An undirected graph with no multiple edges or loops.

Size: The number of edges of a graph.

Subgraph: All vertices and edges in a subgraph are in the graph, and every edge in the subgraph has the same endpoints as it has in the graph.

Trail: A walk that does not pass over the same edge twice.

Undirected edge: An edge associated to a set $\{u, v\}$, where u and v are vertices.

Undirected graph: A set of vertices and a set of undirected edges, each of which is associated with a set of one or two of these vertices.

Walk: Any route through a graph from vertex to vertex along edges.

Weighted graph: A graph with numbers assigned to its edges.

Chapter 19—Trees

Ancestor of a vertex v in a rooted tree: Any vertex on the path from the root to v.

Balanced tree: A tree in which every leaf is at level h or $h - 1$, where h is the height of the tree.

Binary tree: An m-ary tree with $m = 2$.

Child of a vertex v in a rooted tree: Any vertex with the vertex v as its parent.

Descendant of a vertex v in a rooted tree: Any vertex that has v as an ancestor.

Forest: An unconnected graph with no simple circuits.

Full m-ary tree: A tree with the property that every internal vertex has exactly m children.

Height of a tree: The largest level of the vertices of a tree.

Huffman code: A prefix-free (instantaneous) code, where no codeword is a prefix of another codeword and it can be represented using a rooted binary tree.

Infix notation: The form of an expression obtained from an inorder traversal of the binary tree representing this expression.

Inorder traversal: A listing of the vertices of an ordered rooted tree defined recursively—the first subtree is listed, followed by the root, followed by the other subtrees in the order they occur from left to right.

Internal vertex: A vertex that has children.

Leaf: A vertex with no children.

Level of a vertex: The length of the path from the root to the vertex.

Magic square: An $n \times n$ array of distinct positive integers so that the sum of the numbers is the same in each row, column, and main diagonal.

Minimum spanning tree: A spanning tree with the smallest possible sum of weights of its edges.

Parent of a vertex v in a rooted tree: Any vertex that is the immediate predecessor of v on the path to v away from the root.

Parent of v in a rooted tree: The vertex u such that (u, v) is an edge of the rooted tree.

Postfix notation: The form of an expression obtained from a postorder traversal of the tree representing the expression.

Postorder traversal: A listing of the vertices of an ordered rooted tree defined recursively—the subtrees are listed in the order they occur from left to right, followed by the root.

Prefix notation: The form of an expression obtained from a preorder traversal of the tree representing this expression.

Preorder traversal: A listing of the vertices of an ordered rooted tree defined recursively—the root is listed, followed by the first subtree, followed by the other subtrees in the order they occur from left to right.

Rooted tree: A directed graph with a specified vertex, called the root, such that there is a unique path to every other vertex from the root.

Sibling of a vertex v in a rooted tree: A vertex with the same parent as v.

Spanning tree: A tree containing all vertices of a graph.

Subtree: A subgraph of a tree that is also a tree.

Tree: A connected undirected graph with no simple circuits.

Tree traversal: A listing of the vertices of a tree.

Trivial tree: A graph consisting of a single vertex.

m-ary tree: A tree with the property that every internal vertex has no more than m children.

Chapter 20—Finite-State Machines

Deterministic finite-state automaton $\mathbf{M} = (S, I, f, s_0, F)$: A model that consists of a finite set S of states, a finite set I of input alphabet, a transition function $f: S \times I \to S$ that maps state-input pairs to states, an initial state (or start state) s_0, and a subset F of S consisting of final states (or accepting states).

Finite-state machine: A mathematical model of computation based on a hypothetical machine made of different states that can be used to simulate sequential logic in order to represent and control execution flow.

Finite-state machine with no output (finite-state automaton): It models the changes of states within a system until it achieves one of a collection of desired states. The finite-state automata (the plural of automaton) do not produce output, but they have a set of final states.

Finite-state machine with output (finite-state transducer): Each transition has an associated output that either provides some information about the state of the machine or outputs a stream of information as the machine is intended to produce.

Mealy machine $M = (S, I, O, f, g, s_0)$: A model that consists of a finite set S of states, a finite set I of input alphabet, a finite set O of output alphabet, a transition function $f: S \times I \to S$ that maps state-input pairs to states, an output function $g: S \times I \to O$ that maps state-input pairs to outputs, and an initial state (or start state) s_0.

Moore machine $M = (S, I, O, f, g, s_0)$: A model that consists of a finite set S of states, a finite set I of input alphabet, a finite set O of output alphabet, a transition function $f: S \times I \to S$ that maps state-input pairs to states, an output function $g: S \times I \to O$ that maps an output to a state, and an initial state (or start state) s_0.

Nondeterministic finite-state automaton $\mathbf{M} = (S, I, f, s_0, F)$: A model that consists of a finite set S of states a finite set I of input alphabet, a transition function $f: S \times I \to P(S)$ that maps state-input pairs to states, where $P(S)$ denotes the power set of S (the set of all subsets of S), an initial state (or start state) s_0, and a subset F of S consisting of final states (or accepting states).

State diagram: A directed graph with labeled edges that contains transitions for all possible inputs at each state.

Bibliography

Books

Anton H, Rorres C: *Elementary linear algebra applications version*, ed 10, 2010, John Wiley & Sons. ISBN: 978-0-470-43205-1.

Biggs NL: *Discrete mathematics*, ed 2, 2002, Oxford University Press. ISBN: 978-0-19-850717-8.

Conradie W, Goranko V: *Logic and discrete mathematics: a concise introduction*, 2015, Wiley & Sons. ISBN: 978-1-118-75127-5.

Das MK: *Discrete mathematical structures for computer scientists and engineers*, 2007, Alpha Science. ISBN: 978-1-84265-298-5.

Elaydi S: *An introduction to difference equations*, ed 3, 2005, Springer. ISBN: 0-387-23059-9.

Ensley DE, Crawley JW: *Discrete mathematics: mathematical reasoning and proof with puzzles, patterns, and games*, 2006, Wiley & Sons. ISBN: 978-0-471-47602-3.

Epp SS: *Discrete mathematics with applications*, ed 5, 2019, Cengage. ISBN: 978-1337694193.

Goodaire EG, Parmenter MM: *Discrete mathematics with graph theory*, ed 3, 2006, Prentice Hall. ISBN: 0-13-167995-3.

Gossett E: *Discrete mathematics with proof*, ed 2, 2009, John Wiley & Sons. ISBN: 978-0-470-45793-1.

Grami A: *Probability, random variables, statistics, and random processes*, 2020, John Wiley & Sons. ISBN: 9781119300816.

Hein JL: *Discrete structures, logic, and computability*, ed 4, 2017, Jones and Bartlett Learning. ISBN: 978-1-284-09986-7.

Hunter DJ: *Essentials of discrete mathematics*, ed 3, 2017, Jones and Bartlett Learning. ISBN: 978-1-284-05624-2.

Johnsonbaugh R: *Discrete mathematics*, ed 8, 2018, Pearson Education. ISBN: 978-0-321-96468-7.

Khoussainov B, Khoussainova N: *Lectures on discrete mathematics for computer science*, 2012, World Scientific. ISBN: 978-981-4340-50-2.

Kolman B, Busby RC, Ross SC: *Discrete mathematical structures*, ed 6, 2009, Pearson Prentice Hall. ISBN: 978-0-13-229751-6.

Koshy T: *Discrete mathematics with application*, 2004, Elsevier. ISBN: 978-0-124-21180-3.

Lewis H, Zax R: *Essential discrete mathematics for computer science*, 2019, Princeton University Press. ISBN: 978-0691179292.

Liben-Nowell D: *Discrete mathematics for computer science*, 2017, John Wiley & Sons. ISBN: 978-1119441854.

Lipschutz S: *Set theory and related topics*, ed 2, 1998, McGraw-Hill. ISBN: 0-07-038159-3.

Lipschutz S, Lipson M: *Discrete mathematics*, ed 3, 2007, McGraw-Hill. ISBN: 978-0-07-161586-0.

Lovasz L, Pelikan J, Vesztergombi K: *Discrete mathematics: elementary and beyond*, 2003, Springer. ISBN: 0-387-955585-2.

Rosen KH: *Discrete mathematics and its applications*, ed 8, 2019, McGraw Hill. ISBN: 978-1-260-09199-1.

Rosen KH: *Handbook of discrete and combinatorial mathematics*, ed 2, 2018, CRC Press. ISBN: 978-1-5848-8780-6.

Scheinerman ER: *Mathematics: a discrete introduction*, ed 3, 2013, Brooks/Cole, Cengage Learning. ISBN: 978-0-8400-4942-1.

Stein C, Drysdale RL, Bogart K: *Discrete mathematics for computer scientists*, 2011, Addison-Wesley. ISBN: 978-0-13-212271-9.

Vaughn L, MacDonald C: *The power of critical thinking*, 2010, Oxford. ISBN: 978-0-19-543122-3.

Websites

https://www.britannica.com.
https://www.khanacademy.org.
https://ocw.mit.edu.
https://en.wikipedia.org.
https://www.wolframalpha.com.

Answers/Hints to Exercises

(1.1)

(a) It is a true proposition.
(b) It is not a proposition.
(c) It is not a proposition.
(d) It is a false proposition.

(1.2)

(a) $p \wedge q$.
(b) $p \wedge \bar{q}$.
(c) $\bar{p} \wedge \bar{q}$.
(d) $p \vee q$.
(e) $p \rightarrow q$.
(f) $(p \vee q) \wedge (p \rightarrow \bar{q})$.
(g) $p \leftrightarrow q$.

(1.3)

(a) He studied hard for the final exam and he got an A^+ in the course.
(b) He did not study hard for the final exam or he got an A^+ in the course.
(c) If he studied hard for the final exam, then he did not get an A^+ in the course.
(d) If he did not get an A^+ in the course, then he studied hard for the final exam.
(e) He did not study hard for the final exam, then he did not get an A^+ in the course.
(f) He studied hard for the final exam if and only if he did not get an A^+ in the course.
(g) He did not study hard for the final exam and either he studied hard for the final exam or he did not get an A^+ in the course. Note that the parentheses were incorporated by using the word *either*.

(1.4)

(a) $\bar{p} \wedge q$.
(b) $\bar{r} \wedge \bar{s}$.
(c) $t \wedge (u \vee w)$.

(1.5)

(a)

$\overline{(p \wedge q)}$	$\bar{p} \vee \bar{q}$
F	F
T	T
T	T
T	T

 Equivalent

(b)

$\overline{(p \vee q)}$	$\bar{p} \wedge \bar{q}$
F	F
F	F
F	F
T	T

 Equivalent

(1.6)

(a)

$(p \vee q) \rightarrow (p \oplus q)$
F
T
T
T

(b)

$(p \leftrightarrow q) \oplus (\bar{p} \leftrightarrow q)$
T
T
T
T

(1.7)

Converse: If you fail in life, then you do not work hard in life.
Inverse: If you work hard in life, then you do not fail in life.
Contrapositive: If you do not fail in life, then you work hard in life.

(1.8)

$(p{\wedge}q{\wedge}\bar{r}) \vee (p{\wedge}\bar{q}{\wedge}r) \vee (\bar{p}{\wedge}q{\wedge}r).$

(1.9)

(a) It is a contingency.
(b) It is a tautology.

(1.10)

(a) $(p \vee q){\wedge}r.$
(b) $p \leftrightarrow (q \to r).$
(c) $(\bar{p} \vee \bar{q}) \to (r{\wedge}s).$

(2.1)

(a) It is false.
(b) It is false.

(2.2)

(a) This is true.
(b) This is false.

(2.3)

(a) It is true.
(b) It is false.
(c) It is true.
(d) It is false.
(e) It is true.

(2.4)

(a) $\forall x \exists y (x + y = 100).$
(b) $\exists x \forall y (x + y = y).$
(c) $\forall x \forall y (x + y = y + x).$
(d) $\exists x \exists y (x + y = 100).$

(2.5)

(a) $\forall y \forall x \exists z \overline{P(x,\ y,\ z)}$.

(b) No one is 90 years old or older or equivalently, all people are under 90.

(c) $\exists y \big((\exists x \exists z \overline{T(x,\ y,\ z)}) \vee (\forall x \forall z \overline{U(x,\ y,\ z)}) \big)$.

(2.6)

(a) $\forall x \exists y P(x,\ y)$.

(b) $\exists y \forall x P(x,\ y)$.

(c) $\overline{\exists x \forall y P(x,\ y)} \equiv \forall x \exists y \overline{P(x,\ y)}$.

(d) $\overline{\forall x \exists y P(x,\ y)} \equiv \exists x \forall y\ \overline{P(x,\ y)}$.

(e) $\forall x P(x, x)$.

(f) $\exists x \forall y (P(x,\ y) \leftrightarrow x = y)$.

(2.7)

(a) There exists a real number x such that for every real number y, $xy = y$. This is true.

(b) The product of two negative real numbers is always a positive real number. This is true.

(c) There exist real numbers x and y such that x^2 exceeds y but x is less than y. This is true.

(d) For every pair of real numbers x and y, there exists a real number z that is their sum. This is true.

(2.8)

Let $P(p,\ m)$ be "p has eaten m" and $Q(m,\ r)$ be "m is a meal in r."

(a) $\exists p \forall r \exists m (P(p, m) \wedge Q(m, r))$.

(b) $\forall p \exists r \forall m \big(\overline{P(p, m)} \vee Q(m,\ r) \big)$.

(2.9)

(a) $\forall x \exists y \exists z \exists w \exists u ((x > 0) \rightarrow x = y^2 + z^2 + w^2 + u^2)$, where the domain consists of all integers.

(b) $\forall x \big((x < 0) \rightarrow \overline{\exists y (x = y^2)} \big)$, where the domain consists of all real numbers.

(2.10)

(a) There is a pair of real numbers whose sum is zero. It is true.

(b) Every real number has an additive inverse. It is true.

(c) There is a universal additive inverse; that is, there is a real number that is an additive inverse for every real number. It is false.

(d) The sum of every pair of real numbers is zero. It is false.

(3.1)

To show expressions are rules of inference, show that they are tautologies using truth tables.

(3.2)

(a) "I did not drink and I did not have pain."

(b) "I am working out" and "I feel good about myself."

(3.3)

(a) Consider the two false premises: "If Cyrus is a good person, then Cyrus lives to be 1000 years old," and "Cyrus is a good person." We can thus conclude that "Cyrus lives to be 1000 years old," which is a false conclusion. However, the argument is valid by modus ponens.

(b) Consider the two true premises: "If it is sunny, then it is a hot day" and "It is a hot day." We can thus conclude that "It is sunny," which is a true conclusion. However, the argument is invalid by the converse error.

(3.4)

(a) There exists some x that makes $P(x)$ true, but it is not correct to conclude Neda is one such x. Then $P(\text{Neda})$ is false.

(b) There exists some w that makes $Q(w, \text{Mina})$ true, but it is not correct to conclude Mina is one such w. Therefore the argument is invalid.

(3.5)

(a) Employ universal generalization and then universal modus ponens.

(b) Use universal modus tollens.

(3.6)

(a) It is the fallacy of red herring.

(b) It is the fallacy of slippery slope.

(c) It is the fallacy of appeal to the person.

(3.7)

Use universal modus ponens.

(3.8)

Use universal modus ponens.

(3.9)

The conclusion is invalid as it shows the converse error.

(3.10)

The conclusion is invalid as it shows the inverse error.

(4.1)

(a) Using a trivial proof, the statement is true, as the conclusion is true.
(b) Using a vacuous proof, the statement is true, as the hypothesis is false.

(4.2)

(a) Square both sides of $\dfrac{x+y}{2} > \sqrt{xy}$.
(b) Square both sides of $\left(x^1 - x^{-1}\right)$.

(4.3)

(a) Suppose $\sqrt{3} = \dfrac{a}{b}$ for some integers a and $b \neq 0$ and square it.
(b) Suppose $\sqrt{5} = \dfrac{a}{b}$ for some integers a and $b \neq 0$ and square it.

(4.4)

(a) $x = \dfrac{a+b}{2}$.
(b) $p = 11$.

(4.5)

(a) Suppose the conclusion is false and we have $\overline{\left(x \geq \frac{m}{2}\right) \vee \left(y \geq \frac{m}{2}\right)}$.
(b) Suppose the conclusion of the conditional statement is false and assume n is even.

(4.6)

(a) As this theorem has the form "p if and only if q," where p is "n is even," and "q is "n^2 is even," show $p \rightarrow q$ and $q \rightarrow p$ are both true.

(b) Employ proof by contraposition and assume one of the integers is even.

(4.7)

(a) Having $|a - c| = |b - c|$ implies we have either $a - c = b - c \rightarrow a = b$ or $a - c = -b + c \rightarrow c = \dfrac{a + b}{2}$.

(b) Because $4^3 = 64$, for there to be positive solutions to this equation both x and y must be less than 4.

(4.8)

(a) Using a proof by exhaustion, start with small one-digit numbers.

(b) Using a proof by contraposition, assume it is not the case that $a < \sqrt[3]{n}$ or $b < \sqrt[3]{n}$ or $c < \sqrt[3]{n}$.

(4.9)

(a) Verify the inequality for only $n = 1, 2, 3,$ and 4.

(b) Suppose this were not the case; that is, suppose there are only finitely many primes, then there must be a last largest prime.

(4.10)

(a) There are two ways to measure out 4 liters:
   (i) $(8, 0, 0) \rightarrow (3, 5, 0) \rightarrow (3, 2, 3) \rightarrow (6, 2, 0) \rightarrow (6, 0, 2) \rightarrow (1, 5, 2) \rightarrow (1, 4, 3)$.
   (ii) $(8, 0, 0) \rightarrow (5, 0, 3) \rightarrow (5, 3, 0) \rightarrow (2, 3, 3) \rightarrow (2, 5, 1) \rightarrow (7, 0, 1) \rightarrow (7, 1, 0) \rightarrow (4, 1, 3)$.

(b) Use a proof by contradiction.

(5.1)

(a) Form the truth tables for $(A \cup B)$, $(A \cap B)$, and $A \oplus B$ first to prove the set identity.

(b) Form the truth tables for $(A - B)$, $(B - A)$, and $A \oplus B$ first to prove the set identity.

(c) Form the truth tables for (B^c), $(A \cap B^c)$, and $(A - B)$ first to prove the set identity.

(5.2)

(a) Form the truth tables for $(A \cup B)$, $(C - A)$, and $(B - C)$ first to prove the set identity.

(b) Use the duality rules; that is, replace U by $\varnothing$, $\cap$ by $\cup$, and vice versa.

(5.3)

(a) Replace the difference operation by its equivalent intersection operation.

(b) $A^c \cup B^c$.

(5.4)

(a) $A \times B \times C = \{(x, y, 1), (x, y, 2), (x, y, 3), (x, z, 1), (x, z, 2), (x, z, 3)\}$.

(b) $Z_0 = \{0, 4, 8, \ldots\}$, $Z_1 = \{1, 5, 9, \ldots\}$, $Z_2 = \{2, 6, 10. \ldots\}$, $Z_3 = \{3, 7, 11, \ldots\}$.

(5.5)

(a) $A_n = \{\ldots, -4, -3, -2, -1, 0, 1, 2, 3, 4, \ldots, n\}$.

(b) $A_1 = \{\ldots, -4, -3, -2, -1, 0, 1\}$.

(5.6)

(a) 1000.

(b) 560.

(5.7)

(a) $P(A) = \{\{a, b, c, d\}, \{a, b, c\}, \{a, b, d\}, \{a, c, d\}, \{b, c, d\}, \{a, b\}, \{a, c\},$
$\{a, d\}, \{b, c\}, \{b, d\}, \{c, d\}, \{a\}, \{b\}, \{c\}, \{d\}, \{\varnothing\}\}$.

(b) $\{\{a, b, c, d\}\}$.
$\{\{\{a\}, \{b, c, d\}\}, \{\{b\}, \{a, c, d\}\}, \{\{c\}, \{a, b, d\}\}, \{\{d\}, \{a, b, c\}\}, \{\{a, b\},$
$\{c, d\}\}, \{\{a, c\}, \{b, d\}\}, \{\{a, d\}, \{b, c\}\}\}$.
$\{\{\{a\}, \{b\}, \{c, d\}\}, \{\{a\}, \{c\}, \{b, d\}\}, \{\{a\}, \{d\}, \{b, c\}\}, \{\{b\}, \{c\}, \{a, d\}\},$
$\{\{b\}, \{d\}, \{a, c\}\}, \{\{c\}, \{d\}, \{a, b\}\}\}$.
$\{\{a\}, \{b\}, \{c\}, \{d\}\}$.

(5.8)

(a) Build a truth table, and its sixth column and eleventh column are identical.
(b) Use the first De Morgan law, the second De Morgan law, the commutative law for intersections, and the commutative law for unions.

(5.9)

(a) $A \cap B = \{b, e, g\}$.
(b) $A \cup B = \{a, b, c, e, g, h\}$.
(c) $A \oplus B = \{a, c, h\}$.
(d) $A^c = \{a, c, d, f\}$.
(e) $B - A = \{a, c\}$.

(5.10)

$R \cap C = \{a|\mu_{(R \cap C)}(a) = 0.9, b|\mu_{(R \cap C)}(b) = 0.7, c \mid \mu_{(R \cap C)}(c) = 0.5, d|\mu_{(R \cap C)}(d) = 0.1, e|\mu_{(R \cap C)}(e) = 0.01\}$.

(6.1)

There are no values of x and y satisfying all equations. Therefore the matrices A and B cannot be equal.

(6.2)

$$C = \begin{bmatrix} 1 & 1 & 15 \\ -22 & -2 & 24 \end{bmatrix}.$$

(6.3)

$$\begin{cases} x = -2 \\ y = 3 \end{cases}$$

(6.4)

$$A^{-1} = \begin{bmatrix} -0.5 & 1 \\ 1.5 & -2 \end{bmatrix}.$$

(6.5)

$$B = \begin{bmatrix} 0 & 0 \\ 0 & 0 \end{bmatrix}.$$

(6.6)
$$A^{-1} = \begin{bmatrix} -11 & 2 & 2 \\ -4 & 0 & 1 \\ 6 & -1 & -1 \end{bmatrix}.$$

(6.7)
$$u = \begin{bmatrix} 3a \\ 2a \end{bmatrix}, \text{ where } a \text{ can be any number.}$$

(6.8)
$$AB = \begin{bmatrix} 1 & 0 & 0 & 1 \\ 0 & 1 & 0 & 1 \\ 1 & 1 & 1 & 1 \end{bmatrix} \begin{bmatrix} 1 & 0 \\ 0 & 1 \\ 1 & 1 \\ 1 & 0 \end{bmatrix} = \begin{bmatrix} 1 & 0 \\ 1 & 1 \\ 1 & 1 \end{bmatrix}.$$

(6.9)
$$A \vee B = \begin{bmatrix} 1 & 1 \\ 1 & 1 \end{bmatrix}, \quad A \wedge B = \begin{bmatrix} 0 & 1 \\ 0 & 0 \end{bmatrix}, \quad \& \quad A \odot B = \begin{bmatrix} 1 & 1 \\ 1 & 0 \end{bmatrix}.$$

(6.10)
$$x = \begin{bmatrix} 1 \\ -1 \\ -2 \end{bmatrix}$$

(7.1)

(a) Domain of $f = \{a, b, c\}$. Codomain of $f = \{1, 2, 3, 4\}$. Range of $f = \{2, 4\}$.
(b) $f(a) = 2$. Inverse image of $1 = \emptyset$. Inverse image of $2 = \{a, c\}$. Inverse image of $3 = \emptyset$. Inverse image of $4 = \{b\}$.

(7.2)

(a) The set of odd integers.
(b) The set of positive odd integers.
(c) The set of real numbers.

(7.3)

Prove or give a counterexample for each of the following statements:
(a) As a hint, use the contrapositive proof.
(b) A counterexample can be $f(x) = c$, where c is a real number.
(c) As a hint, use the contrapositive proof.
(d) A counterexample can be $f(x) = c$, where c is a real number.

(7.4)

(a) 2.71.
(b) 3.142.

(7.5)

(a) $n = \lfloor b \rfloor - \lceil a \rceil + 1$.
(b) $n = \lceil b \rceil - \lfloor a \rfloor - 1$.
(c) $n = \lfloor b \rfloor - \lceil a \rceil + 1 = 9 - 5 + 1 = 5$ (they are 5, 6, 7, 8, 9) &
    $n = \lceil b \rceil - \lfloor a \rfloor - 1 = 9 - 4 - 1 = 4$ (they are 5, 6, 7, 8).

(7.6)

(a) It is one to one.
(b) It is not onto.
(c) It is not one-to-one correspondence.

(7.7)

(a) It is one to one.
(b) It is onto.
(c) It is one-to-one correspondence.

(7.8)

(a) $g \circ f = \{(a, t), (b, s), (c, t)\}$.
(b) $\text{Im}(f) = \{x, y\}$. $\text{Im}(g) = \{r, s, t\}$. $\text{Im}((g \circ f)) = \{s, t\}$.

(7.9)

$bc + d = ad + b$.

(7.10)

This function is not one-to-one correspondence, because its range is the set of positive real numbers rather than the set of all real numbers, which is its codomain. By restricting the codomain to be the set of positive real numbers, the function is then an invertible function.

(8.1)

(a) $1.0 + (0 + 1).1 = 0 + (1).1 = 0 + 1 = 1.$
(b) $\overline{(0+1)} + 0.1 = \overline{1} + 0 = 0 + 0 = 0.$

(8.2)

(a) Build a Boolean table with 14 columns step by step to prove the functions are equivalent.
(b) Build a Boolean table with eight columns step by step to prove the functions are equivalent.

(8.3)

(a) $E = xyz + x\bar{y}z + x\bar{y}\bar{z} + \bar{x}yz + \bar{x}\,\bar{y}z + \bar{x}\,\bar{y}\,\bar{z}.$
(b) $E = y\bar{x}\,\bar{y} + \bar{x}y\bar{z}.$
(c) $E = x\bar{y}z + x\bar{y}\,\bar{z} + \bar{x}yz + \bar{x}y\bar{z}.$

(8.4)

(a) All terms in $x\bar{z}$ are already part of E. Therefore we have $x\bar{z} + E = E.$
(b) There are terms in x that are not part of E. Therefore we have $x + E \neq E.$

(8.5)

$xyz + xyw + xyt + xzw + xzt + xwt + yzw + yzt + ywt + zwt.$

(8.6)

(a) $f(x, y, z) = x\bar{z}y + x\bar{z}\,\bar{y} + x\bar{y}z + \bar{x}yz + \bar{x}\,\bar{y}z + \bar{x}y\bar{z} + \bar{x}\,\bar{y}\,\bar{z}.$
(b) $f(x, y) = (\bar{x} + \bar{y})(x + y).$

(8.7)

(a) $c_i = xy$ and $s_i = (x + y)\bar{c}_i$
(b) $c_{i+1} = xy + c_i s_i$ and $s_{i+1} = c_i \bar{s}_i + \bar{c}_i s_i$

(8.8)

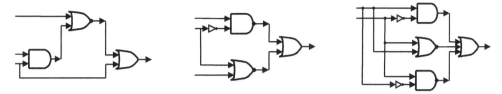

(8.9)

(a) $F = \bar{x}y + yw + x\bar{y}\,\bar{w} + \bar{x}\,\bar{z}\,\bar{w}.$
(b) $G = \bar{y}z + xy\bar{z} + y\bar{w}\,\bar{z}.$

(8.10)

$F(x, y, z, w) = w + \bar{x}y + \bar{x}\,\bar{z} + xz.$

(9.1)

It is reflexive, symmetric, and transitive. Therefore it is an equivalence relation.

(9.2)

(a) The domain of the relation R is the set $\{1,\ 3,\ 4\}$, and the range of R is the set $\{x,\ y,\ z\}$.
(b) $R^{-1} = \{(y,\ 1),\ (z,\ 1),\ (y,\ 3),\ (x,\ 4),\ (z,\ 4)\}.$

(9.3)

(a) $\{(2,\ 2),\ (2,\ 4),\ (2,\ 6),\ (2,\ 12),\ (3,\ 3),\ (3,\ 6),\ (3,\ 12),\ (4,\ 4),\ (4,\ 12),\ (6,\ 6),\ (6,\ 12),\ (12,\ 12)\}.$

(b) $\begin{pmatrix} 1 & 0 & 1 & 1 & 1 \\ 0 & 1 & 0 & 1 & 1 \\ 0 & 0 & 1 & 0 & 1 \\ 0 & 0 & 0 & 1 & 1 \\ 0 & 0 & 0 & 0 & 1 \end{pmatrix}.$

(c) The relation R is reflexive. The relation R is not symmetric. The relation R is transitive.

(9.4)

The relation $R = \{(2, 2), (2, 4), (2, 6), (2, 8), (2, 12), (3, 3), (3, 6), (3, 12),$ $(4, 4), (4, 8), (4, 12), (6, 6), (6, 12), (8, 8), (12, 12)\}$ is reflexive, is not symmetric, and is transitive.

(9.5)

(a) It is reflexive. It is symmetric. It is not transitive.
(b) It is not reflexive. It is not symmetric. It is transitive.

(9.6)

(a) The relations R, T, and U are transitive, but S is not transitive.
(b) R_1 is not an equivalence relation. R_2 is an equivalence relation. R_3 is an equivalence relation.

(9.7)

(a) $R_t = \{(a, b), (b, a), (a, a), (b, c), (a, c), (b, b)\}$.
(b) $S_t = R$, by default.
(c) $T_t = \emptyset$.

(9.8)
$R_t = \{(a, a), (a, b), (a, c), (a, d), (b, a), (b, b), (b, c), (b, d),$
$(c, a), (c, b), (c, c), (c, d), (d, a), (d, b), (d, c), (d, d)\}$.

(9.9)

(a) R is an equivalence relation, by showing it is reflexive, symmetric, and transitive.
(b) R is an equivalence relation, by showing it is reflexive, symmetric, and transitive.

(9.10)

(a) This relation is not symmetric. Therefore R is not an equivalence relation.
(b) This relation is not transitive. Therefore R is not an equivalence relation.

(10.1)

(a) $a = 30$

(b) $\begin{cases} m = 1 \rightarrow n = 12 \rightarrow a = 6 \text{ and } b = 72 \\ \\ m = 3 \rightarrow n = 4 \rightarrow a = 18 \text{ and } b = 24 \end{cases}$

(10.2)

$\begin{cases} \gcd(82320,\ 950796) = 4116 \\ \\ \text{lcm}(82320,\ 950796) = 19015920 \end{cases}$

(10.3)

(a) $(10000010)_2$.
(b) $(4FA1)_{16}$.
(c) 181.

(10.4)

(a) $q = 278\ \&\ r = 13$.
(b) $q = -88\ \&\ r = 2$.

(10.5)

(a) $\gcd(2310, 2431) = 11$.
(b) $\gcd(221, 209) = 1 \quad \rightarrow$ They are relatively prime.

(10.6)

Addition: $m + n = (10110)_2 + (1011)_2 = (100001)_2$.
Multiplication: $m \times n = (10110)_2 \times (1011)_2 = (11110010)_2$.
Subtraction: $m - n = (10110)_2 - (1011)_2 = (1011)_2$.
Division: $m \div n = (10110)_2 \div (1011)_2 = (10)_2$.

(10.7)

(a) 3.
(b) 9.
(c) 1.
(d) -2.

(10.8)

(a) $3^{302} \equiv 9 \pmod{11}$.

(b) $5^{2003} \equiv 8 \pmod{13}$.

(10.9)

$x_1 = 8, x_2 = 7, x_3 = 0, x_4 = 5, x_5 = 4, x_6 = 6, x_7 = 2, x_8 = 1, x_9 = 3,$
$x_{10} = 8$.

(10.10)

27,720.

(11.1)

WTAAD.

(11.2)

BEWARE OF MARTIANS $\rightarrow$ BEWA REOF MART IANS $\rightarrow$
EABW EFRO ATMR ASIN

(11.3)

As a hint, a brute-force approach using a computer can be the method of choice. For each
permutation, the resulting plaintext is examined.

(11.4)

$$\begin{cases} b = \dfrac{n}{2}, & \text{if } n \text{ is even} \\ \text{There is no key } b, \text{ if } n \text{ is odd} \end{cases}$$

(11.5)

307 is an inverse of 43 modulo 660.

(11.6)

As a hint, use $C^d \bmod n = (M^e \bmod n)^d \bmod n = M^{ed} \pmod{n} = M^{ed} \pmod{pq}$,
where $\gcd(e, (p-1)(q-1)) = 1 \rightarrow ed \equiv 1 (\bmod (p-1)(q-1)) \rightarrow ed = 1 + k(p-1)$
$(q-1)$ to show the results of interest.

(11.7)

$C = 11.$

(11.8)

$C = 88^7 \bmod 187 \;=\; 11 \quad \rightarrow \quad M = 11^{23} \bmod 187 \;=\; 88.$

(11.9)

$$M = \begin{cases} 0667^{937} \bmod 2537 \;=\; 1808 \\ 1947^{937} \bmod 2537 \;=\; 1121 \\ 0671^{937} \bmod 2537 \;=\; 0417 \end{cases}$$

(11.10)

Solve $x^2 + (\varphi(n) - n - 1)x + n = 0$ to obtain p and q.

(12.1)

The form of the comparison is subtraction or division, depending on the function:
(a) The increase is approximately equal to zero.
(b) The increase is the constant 1.
(c) The increase is a logarithmic function of n.
(d) The increase is an exponential function of n.
(e) The increase is twice as much.
(f) The increase is $(n + 1)$ times as much.

(12.2)

Split the list into two 8-element sublists. Compare 72 to the largest item in the first list. Because $72 > 36$, split the second 8-element sublist into two 4-element sublists, and continue the process.

(12.3)

$\dfrac{x^7 + x^4 + x^3 + x^2 + x + 1}{x^3 + 1}$ is $\Theta(x^4)$, where $C_1 = 3$ and $C_2 = 1$, for $k = 1$.

(12.4)

$x > 1 \quad \rightarrow \quad 8x^5 > 8x^4 \rightarrow k = 1 \;\&\; C = 8.$

(12.5)

We have $O(n^{2m})$, with witnesses $C = 2^m$ and $k = 0$.

(12.6)

(a) The number of comparisons increases from n to mn.
(b) The number of comparisons increases from $\log n$ to $\log_2 mn$, that is by $\log_2 m$.

(12.7)

$f(x, y)$ is less than $\left(4x^2 y^2\right)^4 = 256x^8 y^8$, where the witnesses are $C = 256$ and $k_1 = k_2 = 1$.

(12.8)

(a) The number of comparisons increases by m^2.
(b) The same as part (a).

(12.9)

$(\log n)^4$, $\sqrt[3]{n}\log n$, n^{100}, 2^n, 10^n, $(n!)^2$.

(12.10)

$g(n) = n^{k+1}$.

(13.1)

(a) As a hint, add $(a + kd) = \dfrac{(k+1)(2a + kd)}{2}$ to both sides of $P(k)$.
(b) As a hint, add ar^k to both sides of $P(k)$.

(13.2)

(a) As a hint, add $(k + 1)^2$ to both sides of $P(k)$.
(b) As a hint, add $(k + 1)^3$ to both sides of $P(k)$.

(13.3)

As a hint, factor out $P(k + 1)$ to obtain a multiple of $P(k)$.

(13.4)

As a hint, break $P(k+1)$ into $P(k)$ and $5(k-2)(k-1)k(k+1)$, then show each is a multiple of 120.

(13.5)

As a hint, break $P(k+1)$ into $P(k)$ and another term, and show each is a multiple of 6.

(13.6)

As a hint, an integer greater than 8 is the sum of at least three 3s, the sum of two 5s, or the sum of a combination of 3s and 5s.

(13.7)

As a hint, factor out $P(k+1)$ to obtain a multiple of $P(k)$.

(13.8)

As a hint, factor out $P(k+1)$ to obtain a multiple of $P(k)$.

(13.9)

As a hint, break $P(k+1)$ into $P(k)$ and another term, and show each is a multiple of 10.

(13.10)

(a) As a hint, assume the first player involves removing j matches from one pile, where $0 \leq j \leq k+1$.
(b) As a hint, use the proof by contradiction.

(14.1)

(a) $S = \frac{a(r^m - 1)}{(r-1)}$.
(b) $S = 0.5m(2a + (m-1)d)$.

(14.2)

Suppose a_n is the amount of money in the account at the end of the nth period, where a period is $\frac{1}{m}$ years and the interest during a period is thus $\frac{i}{m}$.

(a) $a_{mk} = p\left(1 + \frac{i}{m}\right)^{mk}$.

(b) $a_{mk} = pe^{ik}$.

(14.3)

(a) $a_n = 2^n - 1, \, n \geq 1$.

(b) $a_n = \frac{4}{3}(4^n - 1), \, n \geq 1$.

(14.4)

(a) $a_n = a_{n-1} + a_{n-2} + a_{n-3} + 2^{n-3}, \, n \geq 3, \, a_0 = a_1 = a_{n2} = 0$.

(b) $a_n = a_{n-1} + a_{n-2} + a_{n-3}, \, n \geq 3, \, a_0 = 1, \, a_1 = 2, \, a_2 = 4$.

(14.5)

(a) $a_n = 5 \times 2^n + 2 \times 3^n$.

(b) $a_n = 2^n - n2^{n-1}$.

(14.6)

$$a_n = \frac{1}{\sqrt{5}}\left(\frac{1+\sqrt{5}}{2}\right)^n - \frac{1}{\sqrt{5}}\left(\frac{1-\sqrt{5}}{2}\right)^n.$$

(14.7)

$$a_n = k\,m^n + \sum_{i=1}^{n} m^{n-i} g(i).$$

(14.8)

$$a_n = \left(b_{1,0} + b_{1,1}n + b_{1,2}n^2 + b_{1,3}n^3\right)(1)^n + \left(b_{2,0} + b_{2,1}n + b_{2,2}n^2\right)(-2)^n + \left(b_{3,0} + b_{3,1}n\right)(3)^n + \left(b_{4,0}\right)(4)^n.$$

(14.9)

$$a_n = (2 - n)3^n, \, n \geq 0.$$

(14.10)

$$a_n = \frac{n(n+1)(2n+1)}{6}.$$

(15.1)

The number of words is $210 \times 10 \times 24 = 50{,}400$.

(15.2)

To choose two balls of different colors, three mutually exclusive sets are considered. By the sum rule, there are therefore $12 + 15 + 20 = 47$ ways.

(15.3)

The total number of even four-digit numbers is $156 \ (= 96 + 60)$.

(15.4)

Number of passwords $= 36^3 = 46{,}656$.

(15.5)

$$\frac{26!}{(26-4)!} = \frac{26!}{22!}.$$

(15.6)

$$\frac{52!}{5!(52-5)!} = \frac{52!}{5!47!}.$$

(15.7)

$3 \times 4 + 3 \times 5 + 5 \times 4 = 47$ ways.

(15.8)

$k = 3 \ \& \ n = 7$.

(15.9)

(a) $\binom{12}{2} = 66$ ways.

(b) $15 + 6 + 1 = 22$ ways.

(15.10)

There are 10 possible ways to win the tournament.

(16.1)

0.2.

(16.2)

$\dfrac{29}{45}$.

(16.3)

The unfair coin is more likely to have been picked.

(16.4)

91%.

(16.5)

$x \leq 27$.

(16.6)

0.86.

(16.7)

(a) 0.56.
(b) 0.94.

(16.8)

21%.

(16.9)

$k \geq 138{,}148$.

(16.10)

$p \cong 0.0001$.

(17.1)

$\dfrac{1}{5}$.

(17.2)

$\dfrac{9}{25} = 0.36$.

(17.3)

$$\sum_{x=998}^{1000} \binom{1000}{x} (0.001)^x (1 - 0.001)^{1000-x}.$$

(17.4)

0.20324.

(17.5)

$j + 4$, where j is an integer.

(17.6)

$P(X = 4) \cong 0.075$.

$P(X = 0 | X \le 2) \cong 0.028$.

(17.7)

$E[X] = 3.7$.

$E[X^2] = 15.5$.

$\sigma^2 = 1.81$.

(17.8)

70%.

(17.9)

k	p
0	43.5965%
1	41.3019%
2	13.2378%
3	1.7650%
4	0.0969%
5	0.0018%
6	0.000007%

(17.10)

$$S_Y = \left\{ -1, -\frac{1}{2}, \frac{1}{2}, 1 \right\}.$$

$$p_Y(-1) = \frac{1}{6}, \ p_Y(1) = \frac{1}{6}, \ p_Y\left(-\frac{1}{2}\right) = p_Y\left(\frac{1}{2}\right) = \frac{1}{3}.$$

$$E[Y] = 0.$$

(18.1)

(a) The total degree of the graph G is thus 6.

(b) There are 11 subgraphs of G. Three of them each has a total degree of 0, four of them each has a total degree of 2, three of them each has a total degree of 4, and one has a total degree of 6.

(18.2)

(a) It is a walk with repeated vertex but does not have a repeated edge, so it is a trail from v_1 to v_4 but not a path.

(b) It is a walk from v_1 to v_5, but it is not a trail, as it has a repeated edge.

(c) It is a walk starting and ending at v_2, has at least one edge, and does not have a repeated edge, so it is a circuit.

(d) It is a closed walk from v_1 to v_1.

(18.3)

(a)
$$A_a = \begin{pmatrix} 0 & 1 & 1 & 0 & 0 & 0 \\ 1 & 0 & 1 & 0 & 0 & 0 \\ 1 & 1 & 0 & 1 & 0 & 0 \\ 0 & 0 & 1 & 0 & 1 & 1 \\ 0 & 0 & 0 & 1 & 0 & 1 \\ 0 & 0 & 0 & 1 & 1 & 0 \end{pmatrix}$$

(b)
$$A_b = \begin{pmatrix} 0 & 1 & 1 & 0 \\ 1 & 1 & 0 & 2 \\ 0 & 0 & 1 & 1 \\ 1 & 1 & 0 & 0 \end{pmatrix}$$

(18.4)

(a) Graph (a) has a vertex of degree 4, whereas the graph (b) does not.
(b) There is a function g, which is one to one and onto, which takes a to c', b to a', c to b', and d to d'.

(18.5)

It is not bipartite.

(18.6)

The graph is connected.

(18.7)

(a) Because the matrix is symmetric, it is square and thus represents a graph. With a zero-one matrix, it is a simple graph with no parallel edges and no loops.
(b) Noting each column represents an edge, the sum of the entries in the column is either 2, if the edge has two incident vertices (i.e., there is no loop), or 1 if it has 1 incident vertex (i.e., there is a loop).

(18.8)

(a) A complete graph has a Hamilton circuit for all $n \geq 3$.
(b) A cycle graph has a Hamilton circuit for all $n \geq 3$.
(c) A wheel graph has a Hamilton circuit for all $n \geq 3$.

(18.9)

(a) A complete graph has an Euler circuit if the number of vertices is odd, and it does not have an Euler circuit if the number of vertices is even.
(b) A cycle graph, regardless of the number of its vertices, has an Euler circuit (namely, itself) for $n \geq 3$.
(c) A wheel graph, regardless of the number of its vertices, does not have an Euler circuit.

(18.10)

28.

(19.1)

(a) It is a tree, as it is connected with no simple circuit.
(b) It is not a tree, as it is not connected.
(c) It is a tree, as it is connected with no simple circuit.
(d) It is not a tree, as it has a simple circuit.
(e) It is not a tree, as it has a simple circuit.
(f) It is a tree, as it is connected with no simple circuit.
(g) It is not a tree, as it is not connected.
(h) It is not a tree, as it has simple circuits.

(19.2)

(a) The vertex a, as it is at the top with no parent.
(b) The vertices with children, namely, a, b, d, e, g, i, j, m.
(c) The vertices without children, namely, c, f, h, k, l, n, o, p, q, r, s, t.
(d) The children of g are k, l, m.
(e) The parent of g is b.
(f) The siblings of g are e, f.
(g) The ancestors of g are b, a.
(h) The descendants of g are k, l, m, t.

(19.3)

(a) There are 30 edges.
(b) There are 31 vertices.
(c) There are 16 leaves.

(19.4)

(a) 2205.
(b) 32.

(19.5)

(a) A preorder traversal thus visits the nodes in the following order: $a\,b\,d\,e\,c\,f\,g$.
(b) An inorder traversal thus visits the nodes in the following order: $d\,b\,e\,a\,c\,g\,f$.
(c) A postorder traversal thus visits the nodes in the following order: $d\,e\,b\,g\,f\,c\,a$.

(19.6)

(a) The steps are as follows:
  – Begin at the vertex a.
  – Choose b (of the vertices adjacent to a).
  – Select c (of the vertices adjacent to b not yet visited).
  – Choose d, and then e, and then f.
  – Backtrack to e, and then to d, and then to c.
  – Visit g.
(b) As reflected in Figure 19.14(c), the steps are as follows:
  – Begin at the vertex a.
  – Visit both b and g.
  – Visit c, the only unvisited vertex adjacent to b.
  – Visit the unvisited vertices adjacent to c, namely, d, e, and f.

(19.7)

(a) Starting at vertex a, edges are added in the following order: $\{a,\,b\}$, $\{a,\,d\}$, $\{b,\,e\}$, $\{c,\,e\}$.
(b) Edges are added in the following order: $\{a,\,b\}$, $\{b,\,d\}$, $\{b,\,e\}$, $\{c,\,e\}$.

(19.8)

(a) Edges are added in the following order, where the total weight of the minimum spanning tree is 110: $\{e, f\}, \{c, f\}, \{e, h\}, \{h, i\}, \{b, c\}, \{b, d\}, \{a, d\}, \{g, h\}$.

(b) Edges are added in the following order, where the total weight of the minimum spanning tree is 110: $\{e, f\}, \{a, d\}, \{h, i\}, \{b, d\}, \{c, f\}, \{e, h\}, \{b, c\}, \{g, h\}$.

(19.9)

$A: 1, B: 00, C: 011, D: 0101, E: 0100$. The average number of bits per symbol is as follows:

$1 \times 0.7 + 2 \times 0.1 + 3 \times 0.1 + 4 \times 0.05 + 4 \times 0.05 = 1.6$ bits.

(19.10)

There are 10 ways for the championship to occur.

(20.1)

As shown in the figure, the state zero is the start state, and the state two is the final state.

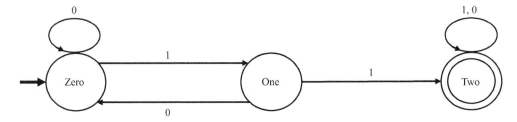

(20.2)

The figure shows the state diagram, which consists of four states: s_0, s_1, s_2, and s_3.

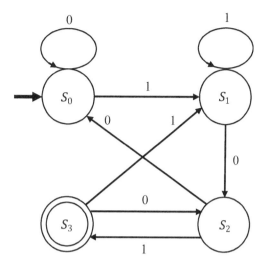

(20.3)

The state diagram for the nondeterministic FSA is shown in the figure. The strings are as follows: 010, 0110, 00010, and 001110.

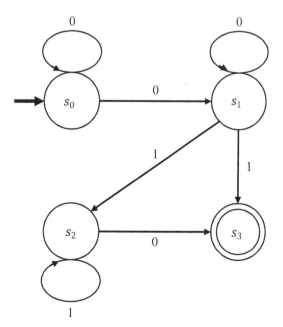

(20.4)

This automaton, also called a parity-check machine, contains either an even number of 0s or an odd number of 0s; it thus has only two states, E and O, as shown in the figure.

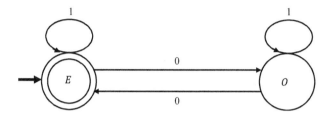

(20.5)

In either of the state diagrams, for a string of 1s and 0s, the accepting state is reached if at least two 0s occur in a row.

(20.6)

The figure presents the corresponding state diagram.

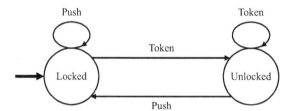

(20.7)

The deterministic FSA recognizes the set of bit strings containing an even number of 1s and odd number of 0s.

(20.8)

The figure shows the state diagram, which consists of three states (s_0, s_1, and s_2), where a transition from one state to another is accompanied by the input/output label. The output sequence is thus as follows: 00000010100001000.

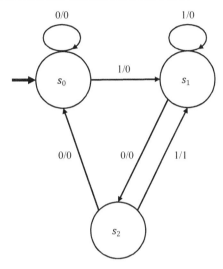

(20.9)

As shown in the figure, the Moore machine has five states.

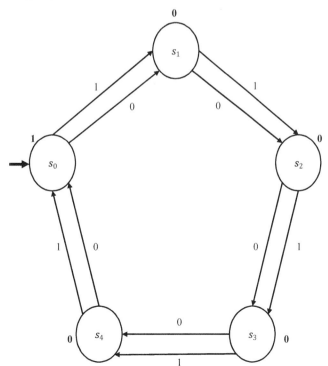

(20.10)

As shown in the figure, the transitions are labeled with input and output values. Therefore the output sequence is 1100.

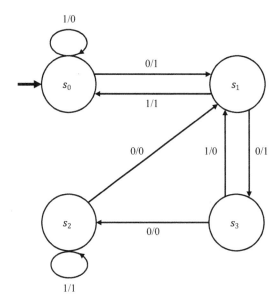

Index

Note: Page numbers followed by *f* indicate figures, *t* indicate tables and *b* indicates box.